NOVITAS MUNDI

Perception of The History of Being

NOVITAS MUNDI

Perception of The History of Being

D.G. Leahy

State University of New York Press

Published by
State University of New York Press, Albany

Printed in the United States of America

For information, address the State University of New York Press,
State University Plaza, Albany, NY 12246

Production by Christine Lynch
Marketing by Theresa Abad Swierzowski

Library of Congress Cataloging-in-Publication Data

Leahy, David G., 1937-
Novitas mundi: perception of the history of being/David G. Leahy.
p. cm.
Includes bibliographical references and index.
ISBN 0-7914-2137-6 (alk. paper), —ISBN 0-7914-2138-4 (pbk. alk. paper)
1. Philosophy. 2. History—Philosophy. 3. Time—Religious aspects—Christianity. 4. Science—Philosophy. 5. Consciousness. 6. Absolute, The 7. Philosophical theology. I. Title.
B61.L43 1994
111—dc20 93-45673
CIP

10 9 8 7 6 5 4 3 2 1

Dedicated to my wife

GENEVIEVE

beautiful
as the sun rising over the mountains
truly a blessing

CONTENTS

PREFACE

Two decades ago I was invited to give three series of lectures at the Catholic Center at New York University, in the years 1973-1975. In 1976 I was invited to address the Columbia University Faculty Seminar on Religious Studies. Those lectures and that talk are substantially, and respectively, Section B and Appendix ß of this book. Between the lectures and the talk, I composed the "Prolegomena in Comprehension of the History of Being," Section A, and the "Epilogue: The Essential Anticipation of the Finality of the Fact," Section C. It was at that point that I became powerfully, and for the first time, explicitly aware of the full import of the analysis of the essential history of thought undertaken in the lectures, namely, that there is an essentially new form of thought now actually existing for the first time: there is the beginning of a thinking the essence of which is the logic of faith itself. I prefaced the first printing of this book as follows: "You, reader, and I, writer, stand today in our common humanity face to face with existence itself. The substance of this book, the conception itself of the fact of history, is then, essentially no more my thought than it is your thought; but along with an indefinite number of others we may together witness what is, precisely, the intelligible power of life itself or its integral existence.

In this book is conceived the substance of history itself. Herein the absolute thought of existence itself is essentially differentiated into substance and thought. If the materialist conception of history originating with Karl Marx understood the fact that thought changes with changing material conditions, it nevertheless, neither did nor could understand what only now substantially exists and only so is able to be understood, namely, the fact that thought itself has changed in essence in the course of time. If before now in the materialist conception of history matter itself was essentially material (historical matter), still it was not yet itself thought itself (historical substance): now what is newly thought itself is material existence. This work is an explication of that perception of existence itself that is now unconditionally thought itself; directly, in terms of that thought itself now occurring (this especially in the "Prolegomena" and the appendices of the present text), and indirectly, in terms of the history of thought itself (this especially in the "Reflection on the History of Being"). The latter is comprehended in the former as perception is comprised both of itself and reflection. The perception of the history of being does not begin with critique of the materialist conception of history which will not be found within the compass of this work, although such critique is essentially implicit in what now occurs and is properly reserved for a subsequent presentation. (For this critique the reader is now referred to *Foundation: Matter the Body Itself,* in conjunction with whose publication the present work is being reprinted.)

"The engine of this perception is the substantial power of faith itself. At this point in history, when its truth is clearly of the greatest practical import, it is not possible in examining the foundation of scientific philosophy to set aside, as Descartes did formally, the content of faith itself. At the same time and for the analogous reason it is not possible for the substantial existence of faith to set aside, as did Kierkegaard, the formality of scientific philosophy. Insofar, then, as the material and formal conditions of perceiving the history of being necessitate it, this perception begins with an essentially historical critique of the essence of modern thought itself. Insofar as the *essential* condition of this perception is freedom itself, life itself terminates in the synthesis of life itself. Thought itself exists. There is thought without being beyond it. The history of being exists without necessity."

Since those words were first written fifteen years ago, the beginning of the new historical order, of the *novus ordo seclorum,* at once itself the beginning of the new thinking, has made manifest the necessity of thinking the essence of existence itself, of conceiving absolutely that "the history of being exists without necessity" and that existence itself exists for the first time. In and through and beyond the analysis of the essential history of thought this book sets out for the first time the transcendental limits of the essentially new form of thought which is what the new world's new thinking really is if it be really new. Further explication and development of the new thinking now beginning is contained in *Foundation: Matter the Body Itself* first published this year.

I remember with gratitude all those, some now deceased, whose encouragement or help was acknowledged in the original publication of this volume. And now, as then, I must express special thanks to Professor Thomas J.J. Altizer.

D.G. Leahy

NOVITAS MUNDI

Perception of The History of Being

Section A
PROLEGOMENA IN COMPREHENSION OF THE HISTORY OF BEING

καὶ οἱ χρώμενοι τὸν κόσμον ὡς μὴ καταχρώμενοι· παράγει γὰρ τὸ σχῆμα τοῦ κόσμου τούτου.

and they who occupy themselves with this world, let them be as those who deal with it not on *its* terms; for this world's understanding is Being diverted to its own end.

1 Corinthians 7:31

Now a critical occurrence to mankind in furtherance of the end of his history. In this new time, and in comprehension of the history of being as actually understood for the first time in essence, it is our first task to set forth the essential history of thought. This history is, in fact, what is this moment's thoroughgoing departure from thinking in the past, a radical critique of modern thought's essence. Comprehensively stated that essence is to be fashioned by Being for its own purposes. This being fashioned is evidence of Being at the source of history as Being's essence. It is of the essence of modern thought that the history of being is taken to be Being's very own history. That is, history is taken to be not essentially an occurrence to thought, but to Being. With respect to this Being, it is not, therefore, an occurrence in essence, but is no different than thought. That is, history, essentially like thought, is taken to be fashioned by Being for its own purposes. But it is the essence of what now occurs that, faced as it is with that absolute pretension of Being, it be without resource. Without Being beyond itself, without either a cause of its own or a reason to begin with, but also not beginning with self-denial, it displays the essential history of thought. In comprehension of the fact that the history of being is an occurrence in essence to thought (only now able to be thought to be what in fact it is), in absolute clarity about itself, it relates the history of thought (with which it does not begin) to the transcendental essence of existence itself with which in thought it is taken up. To be taken up in thought with the transcendental essence of existence itself is a new occupation for thinking; it is so occupied through that essence whose *appearance* is the essence of history. Therefore, that appearance in which the history of being is essentially prior to thought exists for it in absolute fact-evidence. In fact, the essence of history now occurs to thought as its departure from modernity's essence or thinking-in-the-past. That it occurs to thought in its essence is its not-being-its-own-effect. It is the transcendental essence of existence itself in its effect, thereby constituting this essentially historical thinking an absolute objectivity devoid of the encumbrance of being fashioned by Being. This unencumbered being now occurring in thought is that being which itself has a history, to which something has occurred during its being in the world. The transcendental essence of existence itself before now appeared in the world; but in the past it was not in its effect in essence as it now exists in thought; it was an occurrence to the essence of thought itself so as to be identically what had occurred. That being which itself has a history, being unencumbered, has that history not essentially as its own, but for another. Therefore, the conditions of appearing as being unencumbered for another taken upon itself by the transcendental essence of existence preclude its being in its effect in essence as it *now* exists for thought, until those conditions of its appearance should come to be known as *in themselves* not obstacles to the comprehension of the history of being in thought. That the necessary state of affairs with respect to these conditions has, as a matter of fact, come into being as a result of that thinking which begins with this very appearance is a manifest conclusion of the essential history of thought. The knowledge of the essential history of thought is, in itself, a *potential* knowing, which is actualized only in the event of that thought taken up with the transcendental essence of existence itself *now occurring.* Which is not to say that this occurrence is in doubt. But neither the emergence of the necessary state of affairs, nor its definitive arrangement in the essential history of thought is to be confused with that critical occurrence itself in its effect. Though they belong to it as matter and form to essence, *only the latter is the absolutely evident comprehension of the history of being.*

The essential history of thought demonstrates modern thought's existence. It brings it into existence in its essence as being encumbered, fashioned by Being for its own purposes, but inconceivably so except it be a *result* of the transcendental form of reason that *begins with* the *appearance* of the essence of existence itself, which transcendental form of reason was in its inception nothing other than its dependence upon the *fact of existence* first encountered in the transcendental form of conversion. Modern thought is the result of the taking up of the transcendental form of reason as the form of thought's very own existence. This subjective transcendentalism takes itself to be something other than its dependence upon the *fact* of existence; it takes itself to begin with in its own right to be the *thought* of existence. If the form of conversion first encounters the *fact* of transcendental creation in the *appearance* of the transcendental essence of existence itself, that is, if the fact itself of creation, otherwise visible to the mind's reflection upon the nature of the world, is first encountered in the *event of history itself* as the form in essence of the world's conversion, and if that essential form of the world should be known *in faith* as the *novitas mundi,* as the indemonstrable fact with which the transcendental form of reason itself begins within the world (understood to be *as a matter of fact* created), if the event of history is an objective occurrence to the form of reason (the transcendental form of faith), then it will have become possible, through the *otherwise* careful appropriation of the *appearance* of the transcendental essence of existence, and in the event of the identification of human subjectivity with the transcendental form of faith, to understand that appearance as a transformation of humanity's thinking into something of its own, or the essence of history as being fashioned by Being for its own purposes.

Indeed, the ultimate truth of this appropriation of the essence of history by thought is only now seen to have come to pass in the creation of that state of affairs that provides the material basis for that thinking now existing for the first time in history. But it therefore exists in essential discontinuity with modern thought, the thought which appropriated history as its own, displaying thereby its essence as existence-thinking-in-the-past. In light of this thinking now occurring it is clear that an uncreated world does not exist. Insofar as modern thought appropriated the essence of history as its own, the essence of the world was not its occurrence in the *fact* of existence (a fact previously known only in the event of conversion or in the habit of faith, but which, if it was to be then known, was only so in personal commitment), but the essence of this world is the open secret of unfulfilled promise. Its history, in essence, is *about to be.* Its present is the future of its past thinking. In the pure form of its self-assertion, wherein it takes its appearance as its actuality, it is a mockery of existence. In the sheer perpetuity of its self-denial, wherein it takes leave of its universality, it is a sterile exercise facing Nothing. If in the midst of an uncreated world, faced with that state of affairs provident matter for the essential history of thought in comprehension of the history of being, man foolish enough in essential impatience anticipated that *now occurring,* it would be necessary for him to will his existence together with that of all things. He would himself, in effect, be keeping the promise by fulfilling everything's secret intention as if it were his own. This happy generosity of will would exist above the nonexistent essence whose thinking is past, whose history is no longer about to be. But it is the essence of that now occurring in thought taken up with the transcendental essence of existence itself that it cannot be anticipated so. The purity of man's will, arising as it does out of an experience of the nothingness of the idea of history in modern thought, just so, does not stand merely as the contrary to that thinking now existing for the first time, as if it shared with it a common idea of the nothingness of history in modern thought. It is the contradictory position in its purest expression, which, because it is rooted in the nothingness of the idea, opposes nothing in the way of the thought now occurring in history, although this critical occurrence to mankind overcomes its purity at its root. Indeed, this thinking, taken up with the transcendental essence of existence itself, anticipates in its very essence that purity of will which, just so, wills unwittingly this very existence now occurring for the first time in thought. In this form it is unable to recognize it for what it is. The recognition of the essential priority of existence itself to will would be its downfall. But the thinking taken up with the transcendental essence of existence itself comprehends this purity of man's will within its anticipation of this world's nonexistence as that willing existence to everything in perpetual recollection of its opposition to the coming to pass of Nothing. The thinking for which the essence of history exists in absolute evidence is not so simply opposed to the event of Nothing, but rather comprehends that eventuality within the essential priority of the history of being to the history of thought. In the *essential* history of thought this thinking, now existing for the first time, clearly, radically separates that final event of Being fashioned for its own purposes, the Nothing, from this thought of history's absolutely creative essence. In this separation, Nothing's claim to exist in essence is acknowledged to be what it is: the ultimate recollection of Being upon itself; which includes within itself that perpetual self-opposition of the will to exist, as something finally overcome through the essential assimilation of its form to time. Just here it may be comprehended how infinitely transcendent must appear this thinking's transcendental essence: existence itself in its effect in thought, a pure transparency to essence, *in* essence not assimilable *to* essence; sheer *fact* of creation comprehended in the historical essence.

The historical essence is the appearance of the transcendental essence of existence itself now occurring to mankind in thought. Before now the essence of history is the *appearance* itself in its essential priority to thought. In the thinking which *begins with* this appearance *before now* creation is *known* as the *essential form* of the world *above* time, or it is *believed* to be the *conception* of the world *prior* to time, or it is *experienced* as the *appearance* of the world *in* time. But the appearance of the transcendental essence of existence itself is *essentially indifferent to time* (which is precisely not to say that this appearance is an eternal event or a conversion of temporality). This appearance or the essence of history is *in* time, *to* time its *essential difference.* The appearance of the transcendental essence of existence itself is, in its being identically what has occurred to it during the course of its worldly being or being in time, that which makes that time *to be* what it is, identifying it through its transcendental essence with existence itself. Thereby it brings to time such a difference in itself as to make possible at a later time its appearance in thought as the historical essence. This 'possibility at a later time' is *in the very essence of history impossible of perception prior to its now occurring* to mankind in thought. What we now see is truly seen for the first time. What we now see occurs through no necessity whatsoever. It is the manifest freedom of the fact of creation in history at this time. This critical occurrence of the transcendental essence of existence itself in thought is the essence of its history; manifested in the historical essence it is the *templative authority* of the history in essence of thought. In the essential history of thought, this world's *existence* is comtemplated for the first time (a fact made possible by history's *essential indifference* to time); for the first time, what has occurred presents itself in absolute evidence. This absolute objectivity is not an attribute claimed for itself by this thinking or assumed by it with universal consent, but this contemplative judgment beholds its authority through its own essence absolutely attentive to existence itself. Contemplating this essence of history at this time, it comprehends the latter's essential indifference to time to be *for it* in its potentiality an *essential indifference to point of view. It sees clearly that its departure at this time from thinking-in-the-past is its separation from modern thought's very essence* (*as it operates in the form of thought as*)*: the point of view.* It is of the nature of the point of view that it passes away, that its existence is merely temporary. Nor is the point of view left behind in Being's ultimate recollection upon itself. Indeed, it is to be comprehended in essence in this world's resource, together with that point of view in the form of will rightly assimilated to time as arising out of the nothingness of the idea. There is in modernity only a choice between two equally abstract ultimate alternatives: either *the apotheosis of the point of view in universality* (the hypothetical nature of this enterprise becoming clear in the event of Nothing), or *the incarnation of the point of view in perpetuity* (the hyper-liability of this reaction to the end of this world now apparent in its separation, as belonging to Being fashioned for its own purposes, from that thinking now occurring for the first time in history). These alternatives are *in essence* ultimate, that is, *qua* points of view (which formality *is* of the very essence of modernity), irreconcilable. In the final analysis now occurring *modernity's synthesis is seen to be impossible.* The final analysis now occurring is the synthesis, in absolutely radical discontinuity with every point of view in essence, of the fact itself of existence in history. This identity of the final analysis with the synthesis of the fact of existence now occurring, *should it account for itself,* were impossible; but it does not account for itself: It is thinking taken up with the transcendental essence of existence itself: It accounts *for this world's existence;* the *form* of its contemplation is ὁ λόγος καθολικός, its *essence* or *templative authority* is τὸ εἶναι καθολικόν. In the novelty of its thinking it is essentially liberated not only from the encumbrance of Being diverted to its own end, but now for the first time from the limitation of that thinking which *began with* the *appearance* of the transcendental essence of existence: the embarrassment of not being in its essence indifferent to time, that is, *qua* thinking, essentially historical. In its faithful attention to the essence of what occurred to being in time, this thinking now existing brings each object into existence on its own terms without making those terms in themselves its object but only the appearance in them of the transcendental essence of existence itself. It is in the historical essence, through which each object in perpetuity is at once made wholly itself within that existence accounting for this world's existence, that everything comes into existence on its own terms, identically terminated in *to einai katholikon's* absolute intention: the fact itself. The continuity of this intention of existence itself extends to historical existence. This is the absolute evidence of purely *factual* contingency. This absolute evidence is in the thinking now occurring radically discontinuous in essence with every point of view that encumbers existence with its own perspective, imposing upon it a *logos* of its own, that is, a purely *logical* contingency, essentially unhistorical, by which the past is bound to its thinking, the essence of which is termination in itself, or *world-determination.* In its antilogical form, it is will to exist or pure determination of existence. But the termination of the transcendental essence of existence itself is the termination of essence in existence; its *appearance,* or *the essence of history,* terminates in existence itself, not in its *de*termination. The transcendental essence of existence itself is not the determination of existence, except from the *point of view* which encumbers it with a logical contingency, understanding the object essentially on its own terms, thereby further creating the situation wherein in the event of Nothing the transcendental essence itself, so construed as the hypostasis of the point of view (when, as a matter of fact, it originally merely reflected the time-embarrassment of that thinking which began with the event of history), is necessarily understood by the thinking that occurs before now *to have itself succumbed in the passage of time;* this by a pure determination of existence which identifies the transcendental essence with itself. But in the essentially new thinking now occurring it is absolutely clear that the transcendental essence of existence itself is indifferent to the event of Nothing: *It does not contend with Nothing.* For this thinking for which the essence of history is absolutely evident, it is the very essence of its liberation from thinking-in-the-past that it *neither affirms nor denies the termination* of the essence in existence. Indeed, it *is* that termination in its effect in thought: It has nothing to say in its own behalf, except that it *is,* that it now exists. In the absence of knowledge of the historical essence with which, in comprehension of the history of being, we actively contemplate the history of thought, we would still be in that position occupied by many thinkers before us of having to contend with Nothing by *negating* modernity *in one way* or *another way.* But we would be seriously embarrassed by the fact that there did not yet exist a thinking *essentially* independent, that is, existing, *qua* thinking, in radical separation from the logos of the past. This is assuming that we were not so foolish as to attempt to understand the history of being in terms of that thought encumbered by Being diverted to its own end. But in fact we are not in the position of negating modernity (and most certainly not with its own modes of thought); we occupy no position whatsoever. We are fully occupied in contemplation of the world's existence taken up in thought with the transcendental essence of existence itself (sheer perception of the existence itself of *to einai katholikon*). We neither affirm not deny the existence of the Nothing *in its event:* It is untouched *in its essence* by this radical critique which perceives its existence in history. Modern thought itself is not to be disturbed in its essence by this final analysis of its position in the essential history of thought, by this synthesis of the fact of existence in history by which it is brought into existence by the absolute evidence of this thought now occurring. This, by its essence, absolutely *sheer* affirmative *lacks a contrary.* It is the *essentially new essence* which occurs now to mankind in thought, the *essence of history.* We understand the formal definition of the essence of history to be the identity of the storyteller with the story of what occurred to him during the course of his worldly being. With this formal definition before us, we easily notice that the essence of history is *truth occurring;* it brooks no opposition. *The thinking in which this truth of history is now occurring is not, by virtue of its essence, the fruit of opposition to modernity. It occurs in consequence of its essence.* True enough, *modernity* will perceive this new thinking as an *impossible* form of opposition. It will see, in the absolute evidence of this truth now occurring, itself as an object in its otherness. It will perceive this essentially new thinking as ultimate self-alienation. In its essential passivity, that is, its final inability to maintain itself face to face with its object in its otherness, its self as reflected to it in absolute evidence, seeing itself in its essential otherness it will deny that it sees anything. In this event, it will have been confirmed in its being encumbered more perfectly than if it had never encountered this truth of history now occurring. Those who love truth will be renewed in thinking. They will see in thinking its radical historicity. They will perceive that thinking is now in time essentially different than it was before now. They will see that it is now *not merely formally-materially* different from previous thinking as thinking might take it upon itself to be, but *essentially* different. Thinking is now radically conscious of its historical essence, therefore, that, in now occurring, thinking *does not take it upon itself to be,* but that it is in its effect the essence of history in thought, essentially comprehending the priority of the history of being to the history of thought. This new thinking leaves behind modern thought, together with the forms of its extenuation (including the form of its self-opposition in its purity), as not being *essentially* new, but bound in its essence to the past. This fact-evident thinking is the transcendental thought of existence itself appearing, entering into time as its historical essence. Time taken up as never before in this transcendental thought of existence is displayed in its essential otherness to history. Time's perpetuation is in fact historical as it now appears, in its pure objectivity, to this absolutely *active* contemplation of this world's existence. The historical perpetuation of time is that now-evident fact through the essential perception of which this thinking now occurring is related to previous thought as through the essence of potentiality. Therefore, it is related to modern thought as to its absolute essence or pure formality, that is, to the essence of possibility, as its contradiction in existence in fact. The absolute essence of modern thought is, in this now-occurring transcendental thought of existence, a pure abstraction, in the form of self, of history from time. The transcendental thought of existence actively contemplates the existence of this abstraction as the indifference of history to time existing in the form of self appearing to itself to be in opposition to an other's appearance in time. However, for this now-occurring historical thought for which *the existence of the appearance* is absolutely evident, the self appears to be another *not in form but in fact.* It is therefore evident that the form of previous thought, including the pure form of modern thought, is not in fact the form of what now occurs, but its *being-prepared* essentially in the matter of the form of this world (through its, at first, material, then, formal, and, finally, essential inadvertence to the historical perpetuation of time) for a cognition that neither time nor space, nor matter nor mind, but Nothing in existence lies between thinking in existence and its recognition of the fact of history in its own essence. This *cognition* being prepared is perpetuated in the essential history of thought as that purely potential form of knowledge which is actualized only in that thought which *recognizes* the appearance of the transcendental essence of existence in being absolutely taken up with it in essence, *recognizes it as that appearance with which previous thought began.* Between the cognition and the recognition stands Nothing, but the cognition is a potential *existence* only *in* the recognition, which recognition is *essentially not self-perpetuating.* Therefore the recognition requires, but does not effect, that the appearance be the existence of another in *its* effect for the first time in thought. To put it directly: This new thinking now existing is in no sense essentially a *return* to an earlier form of thinking, nor a *rethinking* of the appearance itself, *returning*, as it were, to that original appearance. It understands that any turning whatsoever on its part is essentially inadvertence to the historical perpetuation of time. It turns not; it essentially appears as the transcendental thought of existence itself at this time. Only by abstracting a moment from the form of this new thinking would it even be possible to think that the *appearance* of the transcendental essence of existence itself *returns.* However, such an abstraction of a moment from existence is essentially impossible to this thinking whose essence is the essence of history. For *this* thinking therefore the appearance *exists* in absolute fact-evidence. For *this* thinking all Being diverted lies essentially beyond its ken, posing or reposing in its self, in its essential passivity unable to recognize the fact of existence, which fact is the object of the absolutely active contemplation now occurring. What time is it at which this critical occurrence to mankind exists? When is this now? At what time precisely does this thinking conscious of its essential historicity exist? Essentially the answer has been given before the question has been asked. This thinking now occurring brings with it the question that has been asked as its essential misunderstanding. It is essentially the answer to the question that has not been asked, because, before the occurrence of this thinking, it could not have been thought, while, now that thinking is for the first time conscious of its radical historicity, it remains unasked as that question which belongs essentially to Being diverted to its own end. Meanwhile, the question that has been asked may be answered by saying: at no time prior to the active contemplation of the essential history of thought in comprehension of the essential priority of history to thought, or, at no time able to be determined by thinking in the past. In the thinking now occurring the entire history of thought is essentially submitted to the judgment of the transcendental essence of existence in its effect. In this thinking the truth of history occurs to thought in its essence. This thinking begins neither with *self-doubt* nor with *existence-doubt.* Rather, it is thoroughly prepared by modernity's absolute clarification of the essence of doubt as fashioned by Being for its own purposes (a clarification effected within the self-opposition of the absolute essence of existence, finally taking the form of the *doubt of the truth of appearances*). As a result, this new thinking begins with the essential recognition that doubt absolutely belongs to that world the existence of which is the object of its untiring contemplation. Taken up as it is with the transcendental essence of existence in thought, it is absolutely liberated from every form of doubt, including, *nota bene,* certainty itself. This new thinking is radically serious in the factuality of its existence, namely, that it is not *determined* to exist. The sheer *termination* of the fact in existence itself through the essence of existence frees transcendental historical thinking to be absolutely objective. Unlike the objectivity of the absolute essence of modern thought, which is grounded upon the essentially hypothetical self-evidence of reason itself in absolute self-relation, and unlike the objectivity of the pure ego that exists in abstraction from absolute self-relation infinitely interested in the pure formality of its own essence, the objectivity of this thinking now occurring is essentially in its being taken up with the transcendental essence of existence itself whose appearance exists for this thinking absolutely. That is, essentially *prior* to *determination,* it *exists,* this appearance, as the fact of history. *A priori: the essential lack of determination in the synthesis of the fact in history; in the absolute activity of the contempla-*

tive judgment the object appears in fact in existence through its historical essence. The historical essence (*ho logos katholikos*) is logically posterior to the essence of history (*to einai katholikon*), but the latter, in its essential indifference to the point of view in time or thinking in the past, terminates the former in its existence in thought, thereby displaying also its essential indifference to the *a priori* of past thinking. This thinking now occurring essentially departs from the contrariety of doubt and certainty in previous thinking. It so departs not to an eternal cognition of an essentially unhistorical being, nor to the temporal cognition of an eternal history of being. Being thoroughly prepared for through the final paralysis effected by the identification of thought with history in the event of Nothing, this thinking now occurring is free in essence from the horror of authority that flows from self-certainty, but free also from the fascination with authority that wells up from self-doubt. This new thinking judges every object with that authority manifest in its historical essence, essentially the authority of the fact itself of every object's existence, namely, its *to einai katholikon.* In this way transcendental historical thinking speaks with authority as an absolute objectivity, with a disinterest infinitely inconceivable to thinking in the past. Nor, indeed, is this authority to be circumscribed by the misunderstanding which takes its object to be in the past; it is of the essence of the historical object that it appears in its effect now.

That the historical object appears in its effect now is attributable not to what it is in terms of its own essence, but is due to the radical renewal of its essence effected by the essence of history at this time in thought. The essence of history itself, that is, the appearance of the transcendental essence of existence itself, issues at this time in the transcendental thought of this world's historical essence: the issuance of the formal invitation to the essence of this world to exist; or, the potential renewal of its essence (*ho logos katholikos*), related to the essence of history (*to einai katholikon*) through the historical perpetuation of time, that is, related to this appearance of the transcendental essence through the absolute novelty that history itself is in the world. The *novitas mundi* or absolute novelty of the world is the fact of its existence. This fact is *known to be in* its essence in the appearance with which previous thought begins, but it is *known in* its essence in this new thinking for which the appearance exists in absolute evidence, that is, in the absolute novelty that history itself is; *the absolute novelty of the world* (*novitas mundi*) *in the world as the novitas mentis*, the absolute novelty of this thinking now occurring. In the *novitas mentis* the identity of the storyteller with the story of what occurred to him during the course of his being in time is thought. The historical identity of the man is known to be the transcendental essence of existence itself (*nota bene*, the man's essence itself is known only in its absolutely evident *appearance* for thinking). This *new fact* is the essential fact of the *novitas mentis*. The man's historical essence is known to be his identity with existence itself in the *novitas mentis*. It is therefore clear that the appearance of the transcendental essence of existence itself is in no way able to be understood as *the end of transcendent existence* (as it appears only in the event of Nothing's overcoming thinking in the past), but rather *the absolute beginning of transcendent existence* in thought, before now in time. It is not the absolute beginning of existence itself in thought (*novitas mundi*); nor is it the absolute beginning of existence itself in time. But it is the absolute beginning of transcendent existence. This perception of the *novitas mentis,* that transcendence absolutely begins in the novelty of history itself, is the manifest evidence of the essential indifference of transcendence itself to the polarity of time-eternity. The transcendent itself is neither incapable, in its essence, of change, nor is its appearance in its essence assimilable to time so as to be forever beginning anew at the expense absolutely of the identity of every object. The absolute beginning of transcendent existence in the appearance of the transcendental essence of existence itself in the *novitas mentis* manifests the essential indifference of transcendental appearance to the dichotomy: transcendent-immanent. The absolutely immanent beginning of transcendent existence is the essential change in time whereby the identity of every object is perpetuated historically, that is, newly begun in essence by the perpetuation of the man's identity with existence itself. The truth of the new fact exists in the sheer authority of the fact of its existence: now occurring in thought. Existence itself is *essentially* historical. If existence itself were in fact anything less than historical, if it were not indifferent to the dichotomy, transcendent-immanent, and if, consequently, it were not indifferent to the polarity of time-eternity, if, indeed, it were known to be in its essence as it appears to that thinking which begins with the appearance of the transcendental essence of existence itself (which '*as*' is essentially the case with modern thought), then, in the event of Nothing at all, some reason would have arisen for being disappointed in existence itself. This world would have ended in the absolute gravity of its own essence, but time would have gone unredeemed, left to its own resource. Then modernity, true enough to its own essence, would, in the event of pure determination to exist in the face of Nothing, apprehend in that *appearance* the *reason* for its present state of affairs: *the end of the essence*. But modernity itself comprehends whatever reason arises for being disappointed in the final event in *appropriation*. And, although modernity essentially lacks that absolute objectivity now occurring for the first time in thought, so that it remains, in itself, totally unaware of the *novitas mentis* in which, in comprehension of the history of being, the essential priority of history to thought is recognized, and in which, therefore, being disappointed in existence is recognized to be a determination of the final result of the appropriation of the appearance of the transcendental essence of existence itself to thought, it, nevertheless, truly enough comprehends that beyond every ground of history or thought its ultimate resource is appropriation. This is modernity's candor. Indeed, this appropriation in Being beyond thought or history is modernity's candid admission of its radical inconsistency in being disappointed with existence itself. The promise of existence itself exists only in the absolute novelty of history itself in the world. An existence the essence of which, prior to thought, is not the essence of history, in its appearance promises nothing (neither does it disappoint); but an existence the very essence of which is historical appears absolutely evidently in its factuality as that truth now occurring in thinking taken up transcendentally with it so as to preclude essentially this *novitas mentis* to appropriation. Disappointment is precluded to this thinking that recognizes its essence as now fulfilling the promise of existence. The active contemplation of its historical essence reveals that modernity's candor cloaks its ignorance of the sheer fact of existence, known in that thinking now occurring. The *novitas mundi* in the world as the *novitas mentis,* this absolute novelty of history itself, this *new fact,* being the issuance of the invitation to the entire universe through the transcendental thought of this world's existence, daybreaks upon modernity's lingering candor. The latter takes refuge in appropriation's pale of appearances. The thinking that before now began with the appearance of the transcendental essence of existence itself knew in one way or another (not without the essential embarrassment of its own time) the promise to be in its essence in that appearance. By a formal abstraction of the appearance the promise of existence itself was known to be beyond itself in thinking; known in its essence in the appearance, if known, through ἔκστασις, or self-transcendence: a union of two lovers made sweet by being beyond themselves. This ascension of the self to participation in the promise of existence itself (at its extremities sheer joy or pain) is now able to be seen as the necessary form of comprehending the fact of existence, its promise itself, in the past. Indeed it was this betrothal of the mystic pair that modernity took to be in thought its own accomplished fact, when it resolved upon compounding this abstract form through a simplification in accord with its own determinations; with the final result, in the event of Nothing, that the clarity of the mystical distinction has been converted into the luminous mystery of difference itself. What previous ecstatic, self-transcending forms of comprehending the promise of existence comprehended not, in comprehending so (the necessity being the prior nonexistence of the *novitas mentis*), was the essential indifference of existence itself to the distinction of self-other in its appearance; this essential distinction exists in fact indifferently in its appearance as the historical essence of the appearance now occurring in its effect in thought. Now, for the first time, the essence of existence itself appears transcendentally to thinking itself. It renews the essence of thought by being transcendentally for this *novitas mentis* its historical essence (*to einai katholikon*). The new fact is that the fact of existence itself, its promise, is *essentially* historical. It exists now in the world in the form of a thinking for the first time conscious of its radical historicity (*ho logos katholikos*). In this absolutely evident factuality of creation now occurring in the *novitas mentis,* an absolutely active contemplation knows the essence of the promise in the historical existence of this world. Taken up with the transcendental essence, steadily it comprehends the fact of the essence of history; thus arises the authority of its sober judgment. The bittersweet ecstasy of the promise postponed in essence compares nothing with the knowledge of the promise appearing in essence now as history itself. This transcendental thinking comprehends through its essence the perpetual existence of every object on its own terms; in its objectivity it knows the *schema* of every object in its essential indifference to existence itself to be *holding to difference itself*. Therefore, without in the least disturbing the object, it knows it to be *as such essentially unhistorical,* while *at the same time* it knows the object to be *in existence historically*. It knows then the object to be in fact at variance with itself; every object will reflect this variance, under the steady gaze of the active contemplation of the *novitas mentis,* as the *inconsistency of its schema*. The inconsistency of the schema is radical. It is not a matter of objective self-actualization in time; nor is it the form of varying subjective intentions; nor, finally, is it simply to be set aside in the event of Nothing. This thinking now occurring is not the conversion of temporal to eternal being, nor of transcendent to transcendental existence; nor, finally, is it the conversion of Nothing to something. But, taken up as it is with the transcendental essence of existence itself (beyond every point of view in essence, being itself *ho logos katholikos*), it is the perpetual inversion of every object (indifferently to its being grounded or having lost its ground) to existence itself. *In this inversion the inconsistency of the schema is convertible with the intelligibility of the fact:* This is the absolutely untheoretical end of theoretical science, that is, of science theoretically founded. It is the *absolute end of reflection in existence*. In the *novitas mentis* all science is seen to be *essentially* historical, that is, science of the fact of existence. The radical inconsistency of the schema of the object of every science (beginning with the essential history of thought, as our first task) is potentially intelligible as a reflection at variance with the essential fact, a reflection of the object itself, *qua* object, that is, as it is in its own essence. The objective inconsistency of the schema is that with which science necessarily contends so long as it itself is transcendental subjectivity, essentially unaware of its historical appearance. But science is now the absolutely immediate beginning of a transcendent occupation with existence itself in thought, therefore, not being related to the object, *qua* object, in the object-subject duality, essentially not reflection, but a pure transparency to essence, liberated from the apparently irreducible formal contradictions of the objective appearance. Transcendental historical science, by its very essence, shares not in its object's schematic inconsistency whereby modern thought is encumbered by the self-imposed spatiotemporal limitations upon its absolutely weighty essence (the self-restricting motion of mass converting to energy). Nor does it share in the clarification of the intentional structures of an intersubjective monadology deriving from the absolute point of view of pure rational existence. Nor, finally, in the event of Nothing, is it diverted to its own end clinging to the thing in its saving embrace, in its horror of grace resounding to the thing, shining with the face of an other, reverting to primitive solicitude. Rather, the *novitas mundi* exists in the world as the *novitas mentis* inversely proportionate to this world's extension in whatever possible form of thinking (or Being beyond thinking) whose essence is that it belongs to the past. Its transcendent occupation with this world's existence now exists for the first time; in face of everything that has happened *in* the world, this thinking now occurring perceives what has happened *in* the world *to* the world, its irreducibly new essential appearance: the fact of history itself.

The appearance of history itself does not contest this world's Being; Being need not be at stake. To have it so is to deal with this world on its own terms. Indeed, such a contest of the *novitas mentis* with Being would be in this thinking now occurring for the first time in history unthinkable. It would be a manifest inconsistency of the schema of an object at variance with its own existence, by which this essentially new thinking would be in fact assimilated to thinking in the past. To contest this world's Being belongs not to this thinking's transcendent occupation with existence itself. It is but a transcendental preoccupation with this world's Being in the form of a self-affirming essence denying to Being existence itself. But this absolutely unhypothetical truth now occurring recognizes having a thesis to be its essential impossibility. Indeed, it regards it as its special temptation standing over against it in the form of absolute self-abstraction: its infinite tenuity. The appearance of history itself, that is, the absolutely evident appearance of the transcendental essence of existence itself, is the appearance of the fact of this world's historical essence, through which, in the presentation of the essential history of thought, this world is related in terms of its own essence to existence itself, in fact, demonstrably. That which is demonstrated in the essential history of thought is that *this world exists in fact*. Now this could hardly be construed as a thesis, since its contrary is essentially indemonstrable, while its contradictory (*this world exists in essence*) is recognized as thoroughly compatible, in the essential inconsistency of the schema of this world, with its existence in fact. The very language of the demonstration conveys transparently this world's essence to existence. It expresses the pure activity of contemplative judgment (*ho logos katholikos*) essentially liberated from the *logos* of *beginning with*—anything apart from its own historical essence, but then *it begins absolutely*. (The language of the essential history of thought is faithful to its transcendental essence in its appearance, even its appearance on the page. Being here an immediate expression of the *novitas mentis* it is essentially devoid of that schematic inconsistency of the language of past thought.) To demonstrate that this world exists in fact is, first, to recognize that creation's *factuality* accounts for the appearance of the essence of history (the *new* fact that history itself now appears in thinking is absolutely evident), second, to comprehend that this world's historical essence is compatible with its schema. The presentation of the essential history of thought is, precisely, this latter portion of the demonstration set out *formally* as potential knowledge; the prior portion of the proof belongs to the essence of the *novitas mentis*. Now to demonstrate that this world exists through the historical essence of this thinking now occurring is not first to have doubted its existence. Such a doubt belongs essentially to thinking in the past, to which the *novitas mentis* is related only by way of the essence of its potentiality. In its actuality that thinking now existing for the first time in history has no doubt about existence whatsoever. Its judgment is purely contemplative. Nor, indeed, did ancient, true metaphysics begin with any doubt of existence. Thinking in essence *prior* to the *appearance* of the transcendental essence of existence itself (beginning with which appearance existence itself is an issue in this world for thinking), *existence was in thought prior to reason* as an essential identity of knowledge with its object. The *absurd* was that *something thought to be, at the same time not be*. What was demonstrable was the *cause* of something being what it was. In that thinking that *begins with* the *appearance* of the transcendental essence in time it is possible to doubt this world's existence, but then only in the form of the *absurd* proposition that *nothing is now in existence*. By virtue of the formal understanding that the essence of existence itself was in the appearance, this thinking, embarrassed by time to one degree or another, immediately denies existence to nothing. What is demonstrable is the *reason* for this world's existence. *The fact of the existence of the world is self-evident in the appearance in time*. The self-evidence of the transcendental form of reason, unlike that of ancient metaphysics, did not derive from the *essence* of thought, nor was its object (the *existence* of this world) the same as the former's object (the *intelligible identity* of this world). Modern thought, through its appropriation of the transcendental form of faith (wherein it converted the *fact* of existence to the *thought* of existence, thereby demonstrating in its own way the actual *power* of the *apparent essence of the world's existence,* by simply *positing its identity therewith*), has effected the coming to pass of the absurd, namely, that Nothing is now in existence as the final event beyond the evacuation of the fact of existence from the history of thought. The self-evident fact of the existence of the world appeared before now *in time,* known to be in its essence in the appearance of the transcendental essence. In the course of time, not in fact independently of this appearance, history came to be assimilated to thought, time to thought, thought to the history of Being diverted to its own end. The fact of existence came to be assimilated to this world's schema. Now, *in thought,* occurs (in its effect absolutely evident) the appearance of the transcendental essence of existence itself. Herewith this essence is comprehended as that of history itself in its priority to thought. It brings with it, through its now-evident perpetuation of time, the historical essence through which the *fact* of the world's existence is to be demonstrated in purely active contemplative judgment, beginning with the essential history of thought. Taken up with the transcendental essence in this fact-evident appearance, the *novitas mentis* is the new fact, an essentially new thinking in history.

Section B

REFLECTION ON THE HISTORY OF BEING

Précis

PART I: A RETROSPECT: FAITH AND SELF-CONSCIOUSNESS

1. *ARISTOTLE: THE PARADOX OF GOOD SENSE.* Simply beginning distinguished from the beginning of thought. In Aristotle, pure thought thinks itself absolutely. The intellect, essentially divine, is specifically a man's true self, but a self-not-evident-to-reason. This peculiar structure of self-consciousness provides a referent from which to measure that dispersion in consciousness which has led at last to the intelligibility of the question of the history of being. The unintentional thought of God: its difference in being from this universe in which man is. The essence, or *what it was to be,* is evidence of the ultimate particularity of reality. The knowledge of God's existence grounded on the thoughtful experience of the intelligibility of things. Protagoras refuses to acknowledge the indemonstrable starting point of knowledge: the law of contradiction. It expresses intellect's priority to reason, which, *qua* reason, is the human *potential* for knowing. *Actual* knowledge is that of the individual, who, *qua* knower, transcends his own humanity. The real distinction of intellectual from practical life. The paradox: the essential priority of the individual in existence, in knowledge. Knowledge is being identically a particular essence. This priority to reason's understanding is the essence of a true metaphysics. Knowledge not at all an instrument. Allusion to other possibilities in the event this were not so.

2. *THOMAS AQUINAS: A NEW REALITY*. The universality of knowledge in Aristotle is inseparable from its *potentiality;* actual knowledge is of particulars. The law of identity is meant existentially. The absence of systematic unity in Aristotle. In Thomas metaphysics, in contradiction to its essence, is subordinated to revelation as an instrument of its understanding. Intellect is assimilated to reason. Nature is *formally* creation. Science for the first time is *knowledge of what is transcendent*. Kant's critique will presuppose a science departed in essence from sacred doctrine. The doubt. The transformation of metaphysical necessity into being provided. The ordering of all things, intellectual as well as moral, to a single transcendent end. Subsequent insubordination will mean desperate instrumentality. The principles of faith as the basis of science. Distinction from Neoplatonism. Essence becomes the form of a potential existence, actualized by participation in God's creative act. *Novitas mundi*. The formal understanding of simple existence. The transcendental form of natural reason comes to be explicitly in light of revelation. The principle of knowledge is existence in the form of the universal. In analogy to God's knowing, the object is *known in existence to be an other.* Sacred doctrine is, in essence, presuppositionless science. Aristotle's conception that 'only the form of the stone exists in the soul' contrasted with Thomas' understanding. The abstract employment of metaphysics to deny the *simply natural* tendency of the transcendental form. The self-evidence of the existing other.

3. *DESCARTES: A NEW THOUGHT.* Divine simplicity transforms, in its providence, merely natural forms into forms of its own activity. In light of revelation humanity is mercy or love of neighbor, informed by charity or love of God. Descartes creates a new science modeled in essence on the presuppositionless science of sacred doctrine, but without its *formal* presupposition of faith. The new science is not merely *formally* discontinuous with previous thought, but *materially* so. Contrast of

metaphysical with free natural reason. A self-employing agent for all objects explicitly contingent. The new principle of thought is a potentially universal *humanity*, superalternated to the principles of identity and contradiction. The universal middle: not for knowing *what is*, as in Aristotle, but *what is new*. The representative object exists in a purely formal subjectivity. It exists with respect to reason's absolute self-respect. The analogy to the thought of the God of faith. The *necessary* understanding of the *novitas mundi* comes about through *free resolve*. Science as an essentially moral enterprise dissembled in the *cogito*. This is the dissembling of history in modern thought's very inception. The implicit contingency of the *cogito* in the proof of God's existence in the Third Meditation, where the idea of Infinite Reason is understood to be innate to the presuppositionless phenomenon. The product of history appropriates that history as its birthright.

4. *THE INFINITE PRACTICAL I: KANT.* The priority of existence to truth in the immediacy of the *cogito;* along with God it exists in humanity's eventuality. In this eventuality existence itself is in doubt. The doubt is ultimately *provisional,* centered in the self as its reason for being. By contrast, Thomas recalls something of Aristotle's objectivity, but lays a groundwork for reason's acquiring an infinite power in the future. Kant sets the doubt against the reality *in itself* of the object. In answer to Hume, the object is known as the conceptualization of an appearance itself constructed by reason. Sensible intuition replaces rational intuition; Descartes' simple universal is, in light of the presupposition of the ideal totality of experience, particularized. The criterion of truth becomes a universally valid objective judgment. Law is the medium of the particularizing of the universal. The ultimate lawgiver is the original synthetic unity of apperception. Concretely, the validating agent is the corporation of scientific humanity, not the God of Descartes' abstract subjectivity. Feuerbach as an example of this essential modern idea looped back into Christianity. Modernity as revelation's subrogation. The exclusion of grace in the pure faith of reason. *Logical* in place of *factual* contingency: the avoidance of self-contradiction. Primordial being: the transcendental ideal of pure reason. God as symbol derived from a purely conceptual analogy to the unknown presupposition of science. The priority of pure practical reason to speculative reason, responding in absolute freedom to the moral imperative of its own nature. Descartes' intention now public policy.

5. *THE INFINITE PRACTICAL II: HEGEL.* In Kant the Ideal is unrealized. It is a purely empirical principle valid only for objects as they appear. Pure reason's essentially passive root: its inability to maintain itself face to face with its object's otherness. This passivity in Descartes compared with Aristotle, with Thomas. Modern thought occupies a position unto itself. It understands its relation to Christianity to be a *free* acceptance of the latter's 'conceptions' as its own, excluding in principle those elements that point to the history of being as being *unknowable,* therefore of no practical import. The *paránoia* of modern thought: the fruitful mistake. Neither mathematical method nor a renaissance of learning, but only an alteration in Being or in reason's self-conception accounts for this mistake. Pure reason's *pure potentiality:* its essential indifference to the distinction of matter and form is prepared for in sacred doctrine: *materia signata*. Matter as a reality for thought. The Hegelian essence is the Kantian ideal realized in matter. *The identification of two as such.* The appearance is the thing in light of the intuitive judgment that absolutely determines to existence what it determines to knowledge. Logic is metaphysics. Creation *ex nihilo* as the matter of a deeper *insight*. Christianity is the coming to pass of the idea of history in the process of time: the necessary comprehension of the essential identity of man with the Divine Spirit. The absolute elimination of the *novitas mundi* by the 'feeling of necessity.'

6. *KIERKEGAARD AND LESSING: THE LEAP OF FAITH.* In Hegel the individual has its substantial existence through participation in the totality of Absolute Reason that actualizes itself in history; concretely through the state. In essence Kant's means-end relation is superseded. Contrast with Aristotle. *What* *is actually is what ought to be.* The natural individual's formal subjectivity is the beginning of evil; the good is identification with universal Reason, man's divine essence. Christianity as an historical phenomenon manifesting an eternal truth. Kierkegaard opposes to Hegel's good or evil a third alternative, *faith,* whereby the individual, *qua* individual, is related immediately to God. Absolute inwardness: the *pure act of faith.* Being related *absolutely* to an *other.* The loss of essence, one's *very being for thought.* Transcendence to humanity. The novelty of faith itself. *Credo ergo sum.* The paradox in the way of existence. If God exists, faith exists. God's factual existence in time: the Eternal comes into existence. Faith believes that God exists as a matter of fact as this man. Incompatability of faith with knowledge necessitates God's providing it. By faith man becomes a sinner, *qua* man. Hegel and Lessing. Christianity as morality. Kierkegaard's conception of the Person of Christ. The decision made in time: an absolute difference made for eternity. Lessing's eternal self-assurance as a form of despair. Comparison with Thomas: act of being is act of faith. Lessing's abhorrence of the leap. The *novitas mundi* as a revelation to faith.

PART II: THE PROSPECT: INTRODUCTORY PRESENTATIONS IN THE ESSENTIAL HISTORY OF THOUGHT

7. *AUGUSTINE: THE KNOWLEDGE OF EXISTENCE.* No *coexistence* in Aristotle's universe. It is a mere potentiality. Knowledge transcends the infinite divisibility of time. The indivisible Now. Only a dimensional Now is a logical basis for an *actual* coexistence, where actuality is understood to consist in satisfying conditions of *reason,* not of *being,* where knowledge is *conviction* not *cognition.* Stoicism is this dissipation of the Now in time; time dissipated in motion. The actual, but not absolute perishability of the whole motion of the world. Dimensional consciousness as the limit. Time's perpetual reiteration. The *criterion* grounded in the self-guaranteeing identity of reason.

Sensation in Aristotle. Epictetus' epitome of epistemology. Meretricious identity. Virtue establishes truth. Stoicism's constitutional insincerity. The total obscurity of the residual self in Skepticism. Reason's absolute coincidence with its own possibility. Ancient and modern forms of scepticism point to the history of being; Kant's knowledge without existence. Probability without conviction. The prophetic spirit of Augustine. The coalescence of creation with conversion; its discontinuity with God. The infinite reluctance of reason to simply be itself is overcome by the fact of the conversion of the world, which *invites* it to return from the abyss. The significance of invitation. A new identity in Truth, the *very medium for existence.* The transcendental essence of existence in the form of conversion. Descartes' doubt is related to that of Augustine by way of contradiction.

8. *LEIBNIZ: THE IDEAL OF THE HISTORY OF BEING.* History is the identity of the storyteller with the story of what has occurred to him in the course of his being in the world. History is not a Kantian ideal; its matter is an occurrence to the essence of man. Kant's presupposition of time to change: the Moment. The abstract Now in Kant. Thomas' detachment of the Now from time. *The rift in time.* The discontinuous Nows endure on the ground of the transcendental essence of existence itself. The distinction of the transcendental form of reason from its *merely natural* state collapses in Descartes, so that natural reason takes the transcendental essence to be formally its own, witnessing through this peculiar form of misunderstanding to the history that made it *possible,* but *not necessary.* Under ban, the natural reason is cut off from the *fact* of existence. *Personal,* not *essential* immortality is history's possibility. Modern thought knows nothing of the former; it is destined to deny history only in the form of its purely abstract recollection of Aristotle. The clarification of the Aristotelian essence. Leibniz: reason's essential determination of creation *ex nihilo.* The thought of contingency distributed to the intelligible realm. The *potential* essence of the transcendental form becomes the *possible* essence of natural reason, under the domination of Sufficient Reason, or God in essence. The republic of being is the power of the essence of reason itself. The rift in time is retrospectively anticipated by the bridge of sufficient reason. The abstraction of time in the ideal of the history of being.

9. *HEGEL: THE ABSOLUTE TRUTH.* The transcendental essence as ground for the existence of the essential history of thought. The priority of existence to essence in Thomas *after the fact;* in Leibniz, *after the thought:* the confusion of creation with essence. Possible made actual through reason's necessity. Glimpse at Kierkegaard's essential limitation. The substantial continuum is constructed on analogies to the time and space of the transcendental form of reason; the monad: an analogy in substance to the Now of eternity in sacred doctrine. This is a recollection of the *potentiality,* not the *actuality* of the essence: an eternal region of essential possibilities. Comparison with Aristotle, with Thomas. (Kierkegaard's *transcendent passion to exist.*) The indiscernible identity of the individual with the universe in the concrete of matter. Kant *specifies* this generic intelligibility as *appearance* through the revocation of thought from matter. Hegel perceives the *essential possibility* in the *form* of pure reason: the movement culminates in the essential reason of matter itself; matter is, in its very essence, *weight. The differentiated unity of substance with function.* Absolute pathos. Kant's motion is a rearrangement of extrinsic relationships in a hypothetical moment. Hegel sees the 'higher continuity' by which this abstraction *lives.* Life *is* pure form. Absolute reason holds itself in the Notion of its essence. Oblivious to the essence of existence itself, it *lives* in the truth of appearances.

10. *CLARIFICATION OF THE ABSOLUTE I: KIERKEGAARD.* The interminable pathos of absolute reason reduces each moment of the thought of existence, in itself, to nothing. In the absence of knowledge of the essential history of thought no *intellectually* radical critique of this appropriation is possible; this potential knowledge belongs to God's wisdom. Augustine's in-dwelling Truth: the continuity of faith with knowledge. In Thomas their discontinuity grounded in the specific difference of divine from human intellect. For Kierkegaard faith is the absolute contradiction to knowledge; it is the deed of the individual, *qua* individual, in opposition to the absolute thought of *itself* as *nothing*, an opposition *within* that thought, coincidentally before God. In its transcendent passion to exist self-knowledge separates itself from thought; brought to this extremity by the essential history of which it is ignorant. The demonic preference for nonexistence is overcome by the absolute inwardness relating itself to the paradox of *the Eternal made historical*. Kierkegaard's modern *confusion of history with time*. The eternal enters time in the moment; Kierkegaard's eternity is substantially that of thought, appropriated by the individual, *qua* individual, who denies to thought *existence* apart from *this* appropriation. Time: the *consequence* of history presupposed as the *condition* of history. The antipathy to limitations. The individual, in himself, lacks an eternal determination; his faith is, *for us*, a measure of the extreme remove of modern thought from the *fact* of its origins.

11. *CLARIFICATION OF THE ABSOLUTE II: HUSSERL.* Kierkegaard's spiritual monadology. Under the law of grace the individual transcends his human self-determination, which, spiritually understood, is nothing but sin. *To be sacrificed* in place of *self-sacrifice*. Faith is the measure of the nothingness of worldly being. Man's free choice of God understood absolutely as God's love; this is actually what faith is in the event modernity's passing is not anticipated. In Husserl's phenomenology the species, *qua* species, denies to the truth of appearances *self-evidence*, the sole criterion of a *pure transcendental subjectivity*, or *purely rational existence*. The *possibility* of the world's existence. The *epoché* suspends the *substantiality* of the essence, that is, its *absolute self-relation;* the essence is now related to the pure formality of subjectivity existing in abstraction from absolute reason. Relation of substance and self-evidence in sacred doctrine, in Descartes; the preponderance of self-evidence in Husserl: *the end of reflection in essence*, or the self-reflection of the species, *qua* species, on its own finitude. Previous self-understandings contrasted with this definitive rational understanding of personality, grounded in the *pure ego*. The purely disinterested reappropriation of this world by a transcendental subjectivity purged of its *natural* element. In the 'essential history' of *this* world it seeks 'the origin of a possible world in general.' Knowledge of its own constitution. The *infinite a priori.* In the absence of *pre-existent* formality, it *originally* knows *in general.* Phenomenology's radical lack of clarity: its own essential historicity is not self-evident.

12. *CLARIFICATION OF THE ABSOLUTE III: HEIDEGGER.* The fragmentation of reality in opposition to absolute thought. In Heidegger the final moment of thought, the universe, *qua* universe, is beheld in *its immediacy* as the *visible actuality* of this world in its own Being independent of the thinking beholding it. The finitude of Being. Nothing. *Dasein's* essence is the revelation of Being's essential priority to thought in *absolute nearness* to man, overcoming the absolute distance of Kierkegaard, the relative distance of Husserl. The indispensability of man. In the dissipation of metaphysics in time man remembers that he *belongs* to Being. The reinterpretation of Leibniz' alternatives *within this world.* Heidegger's thinking presupposes the history of being, the *actuality* of which he knows *nothing*. The leap from metaphysics. The self-supporting contingency: the resource of man's survival. *Difference as such.* Hegel contrasted. *To begin with Nothing, not with the resultant thought.* In light of the essential history of thought, *we* see the *essentially derivative* enterprise of Heidegger in the *pure beginning with.* The shining forth of the *fact* of existence in the demonstration of the history of thought. The discontinuity of metaphysics with the transcendental form of reason: Heidegger's misconception of the *nothing* in Christianity. The *final* result: man's self-denial in essential communion with this world's Being. *Appropriation.* Mortals in the fourfold unity of the world.

Part I

A RETROSPECT: FAITH AND SELF-CONSCIOUSNESS

Chapter 1

ARISTOTLE: THE PARADOX OF GOOD SENSE

Beginning is an absolute. To begin is simple. One either begins or does not begin. "Well, begin!" I'm thinking about beginning, but, as I think about beginning, I think that beginning in itself is so simple that so long as I think about beginning I shall not begin. But, perhaps, thinking about beginning is beginning; in which event, without exactly beginning, I begin. Or, I began before thinking about beginning, so that what I'm thinking about now, at this moment, is not simply beginning, but the beginning of thought, that is, the beginning insofar as it is thought about. But a beginning thought about is not simple; it presupposes thought; it is a principle for thought. It is certainly not the simple beginning with which I began before thinking about beginning. To begin absolutely presupposes nothing; it is not a principle; it is not intelligible source or regulative original.

Beginning is a principle for a thought itself absolute, that is, simply present, together with nothing, but a thought that, as a matter of principle, thinks of itself not as absolutely present but as possessing a beginning together with other things. But this common principle is the principle of absolute thought. Nothing stands in the way of thought's being an absolute principle. No principle is absolute for thought except thought itself. Thought that as a matter of principle thinks of other things, for example, in science, thought that thinks of itself as a principle of order, or, in ethics, thought that thinks of itself as a principle of decision, thought, so conceived, is not itself, concerned as it is with other things. Thought in this instrumental mode concerned with other things as a *logos* or reason, as a principle or principles deployed in order, or in order to, this thought is only potentially itself. Although at times capable of thinking itself, it does not know itself. But pure thought thinks itself absolutely, not as a matter of principle; it is thought as light, constant, so intimately present, so intensively clear as to be invisible to reason save in its effects, that is, in its illuminating the forms of reason.

Pure thought is the *aither,* that divine fire of the upper region that makes my thinking shine, that makes this universe of reason 'to move, to breathe, to be.' This pure thought absolutely in upon itself is Life Itself to reason, which reason 'bodies' forth; it is, as it were, reason's Soul, its true Self. To analogize reason to this pure thought as body to soul points, in the context of an examination of Aristotle, to the unity of an actuality to a potentiality whose it is: a nexus of thought with reason within the identity of thought. It points to the paradoxical structure of self-consciousness. Since we come upon pure thought within our own horizon of self-consciousness, it is necessary to note, then, at the outset, that, although this pure thought is not at all unlike God's essence in its own essence, it is, nevertheless, directly or immediately reason's or human nature's principle. God for Aristotle is pure thought, but in God this essence is directly or immediately the principle, so to speak, of the *divine* nature. There is no direct communion between God and man through the principle of 'likeness' as, for instance, is to be found in Thomas Aquinas. In Aristotle, 'likeness' is a mirror that reflects each mind back upon its own essence. While, therefore, this absolute thought, which is reason's Life, is often taken as a god, it might be wise to take it also, especially in distinction from God himself, as reason's own 'invisible man.' In this way, attention is focused to *self-consciousness* as the object under consideration in general, but first, specifically in Aristotle, to reason's *raison d'etre,* that is, to the essential purity of thought which is, for Aristotle, the Intellect, reason's true Self, but a Self-not-evident to reason. By examining the peculiar, particular structure of Aristotelian self-consciousness, without regard either to interests of *faith* or interests of *reason,* it will be possible to delineate a space from which it is possible to measure to what extent there has taken place a dispersion in Western man's consciousness so that not only is perception of God or God's reality radically altered, even altered into nonbeing, but such that self-consciousness itself is altered in its very being, such that the question of the alteration of being itself is at last intelligible.

Hegel tells us that ". . . if we would be serious with Philosophy, nothing would be more desirable than to lecture upon Aristotle, for he is of all the ancients the most deserving of study." [1] But, says Hegel, although Aristotle "presses further into the speculative nature of the object," he does so in such a way "that the latter remains in its concrete determination, and Aristotle seldom leads it back to abstract thought-determinations. The study of Aristotle is hence inexhaustible, but to give an account of him is difficult, because his teaching must be reduced to universal principles. Thus in order to set forth the Aristotelian philosophy, the particular content of each thing would have to be specified." [2] But it is precisely this difficulty for us and for Hegel, namely, that in Aristotle the universal appears constantly in the particular, it is this that makes Aristotle that unique touchstone that he is when we set out to examine the question of a subsequent radical alteration in humanity, that is, in self-consciousness. In fact, this difficulty in dealing with Aristotle may be infinitely extended beyond what Hegel's conception takes it to be.

If it were possible to draw near to Aristotle's universe by coming upon it as an outsider (this possibility does not exist), it would be possible to occupy God's place. But this is twice impossible, first, because human nature presupposes this universe, and second, because God, who occupies his place outside this world absolutely, cannot draw near to, cannot conceive of this universe. Aristotle knows that God exists as *first* cause of this universe's ordered motions. But, in order that God be *sufficient*

---

[1] *Hegel's Lectures On The History of Philosophy* II (trans. E. S. Haldane and F. H. Simson, London, 1894), 134.

[2] Ibid.

cause, God must be *final* cause; that is, he himself must be an *unmoved* mover. Paradoxically, God's causality is an *in*direction; God's thought is, strictly, *un*intentional, that is, in itself. God exists simply as what he is; he lacks potentiality. In himself, or thought of by another in his relation to this world, he cannot be or think himself other than he is: that absolutely simple being that stands as other to this world's complex existence, the latter bound up with one degree or another of potentiality, with one kind or another of matter. The dynamic process of this universe, operating within itself through a multiplicity of causes, is for Aristotle, in its totality, an *eros,* a love, a passionate but self-circumscribed thrust reaching toward God, its Beloved Object. But it is a love perpetually revolving within its own potentiality, self-attaining, but unrequited by a God eternally his own object, a perfect, necessary being. God's Absolute Life, then, reflects consciousness back upon itself, points to this universe's existing differently, that is, outside forever of God's Life, enjoying a life of its own, proper to it. *A fortiori*, man, who is in this universe, who, as Aristotle says, is "in fact not the best thing in the universe," is possessed of his own proper nature, namely, his reason. Man is a living animal who has reason; or, man is a rational animal. It seems to us we understand this until we begin to think about it: wherein, in this complex, living-animal-possessing-reason, is man's essence? but this question immediately tows us into the undercurrent: 'wherein is essence?'

Aristotle's complex word for essence is three words *τί ἦν εἶναι*, *what it was to be*. The primary being of anything is *what* reason thinks *it was to be* that thing: Reason's understanding is retrospective. The primary being of anything is *what it was* for reason *to be* that thing: Reason's understanding is retrospectively identity. Further, the *to be,* or, the existence of anything, consists in its essentiality, that is, in its thinkability. The essence of a thing is to be found in that thing, or that thing does not exist, that is, its identity in being is its potential intelligibility. Essence bears kangaroo-like existence in its pouch; existence is borne from potentiality to actuality within the limits of essence. Reality is, if I may say it, a kangaroo court with no negative implications (except that there is no Adam, no Virgin Birth, etc.), its radical intelligibility being a foregone conclusion. This universe perpetually revolves through its varied motions; existence suffers its privations; rationality exercises its freedom to choose between contrary possibilities. But, finally, there is no doubt about either the possibility of knowledge or, what comes to the same thing, the actuality of existence. There is no universal or methodical doubt à la Descartes possible in this concrete universe. Such a doubt would be nonsense, not merely because it would deny experience (for there is a sense in which Aristotle's thought denies experience in reaching beyond it), but essentially because it would deny the ultimate particularity of reality. It would appear to Aristotle that one would have to be God, but, indeed, quite a different God than Aristotle knows, to entertain such a doubt; it would be not nonsense merely, but absolute nonsense.

If, in thinking about God, self-consciousness is reflected back upon itself, then it is instructive to note that Aristotle does not demonstrate that God exists without qualification, but that to be God is to be pure thought: God is, in himself, necessarily what he is. But God's necessity is not only in God. His necessity is in reason's demonstration. Aristotle, speaking of God as the unmoved mover in *Metaphysics* XII.7, says, "Since this is a possible account, and if it were not so the world would have proceeded out of night and 'all things together' and nonbeing, these questions must be taken as solved." [3] The necessity in reason for God's existence is, as indicated by Aristotle, that the contrary proposition contradicts the intelligibility of the universe whose ordered existence is not in doubt. This, combined with the fact that nothing in the account is self-contradictory, is, for Aristotle, the solution to the question of God's existence. Note that this is a two-legged proof. First leg: I can conceive without contradiction of God's existence. Second leg: God's existence is the sufficient reason of this universe. This two-legged proof belongs to a two-legged prover; science is reason's enterprise, but reason is the form of a psychosomatic being. The synthetic ordering of physics to metaphysics for example, that parti-

[3] *Aristotle: Metaphysics* (trans. R. Hope, U. Michigan, 1960), 258.

cularization of science so foreign to Hegel's ultramodern spirit, reflects the synthetic constitution of the human knower, which, in turn, is reflected in the two-legged proof ultimately because of the synthetic structure of all reality outside of God. If God's existence were merely possible (necessary in itself, or, if we could import into Aristotle's thinker such an abstract thought, necessary to thought), Aristotle would not know of it. But knowledge of God's existence is certain not because it is a necessity of rational thought nor because it is in itself necessary. It is certain because God's existence is *sufficient for thought;* that is, God's existence accounts for our intelligent experience, or, it is clear to our thoughtful experience of the intelligibility of things. There is no doubt that reason itself is a sufficient instrument, that is, that it experiences reality. On this condition, God's existence is a matter of intelligible fact. So that God's existence, on the one hand, is neither simply a necessity of my thought as will be the case with St. Anselm (where logical intuition determines existence), nor, on the other hand, is it simply a transcendental ideal of 'pure reason' as with Kant (where sensuous intuition determines appearances). At both of these later points in Western thought there emerges a radical subjectivity of reason foreign to Aristotle: For whether I claim to know existence directly (Anselm), or to know, but not to know existence (Kant), I assert that my reason is self-sufficient for religious purposes (Anselm), or for practical-moral purposes (Kant). But in Aristotle not only is reason not self-sufficient, but, consequently, the religio-practical, moral dimension, while necessary to human nature, is, in itself, finally insufficient. But Aristotle's scientific reason is sufficient for knowing God to exist on fundamentally the same ground on which it knows whatever it knows, namely, that the objective reality of the world informs subjectivity. 'Objectivity' is first receptivity. Aristotle compares his understanding to that of Protagoras: ". . . we say that science and sense measure things, because by them we get to know things; whereas they really do not measure, but are measured. We feel as though someone were taking our measure, and we get to know our size because the measuring tape is repeatedly applied to us. But Protagoras says, 'man is the measure of all things,' as if he had meant to say 'the man of science' or 'the man of sense'; for such men are measures because they possess science, or sense, which we know to be measures of whatever is submitted to them. Therefore, this saying, though it seems to say something, really says nothing." [4] As Aristotle makes clear, actual science is science of a concrete man who measures his subject matter only to the extent that he himself has been shaped by that section of reality. For Aristotle, 'man is the measure of all things' must, if it is to be interpreted into intelligibility, actually contradict the radical subjectivity it seems to announce. That man should be the arbiter of reality is perverse nonsense to Aristotle, so much so that he pretends to understand Protagoras' statement as a misleading superfluity. Such is Aristotle's contempt for this great Sophist.

Actually Protagoras belongs among those who seek a reason for everything; that is, there is more than one dimension in which reason is ultimate in sophism. Sophism, or the exaltation of reason, in which man places man at the center (apart from what it is morally: an ungoverned, or self-governing will) is scientifically, for Aristotle, the refusal to recognize the law of contradiction (that a thing cannot be and not be at the same time in the same respect) as the *indemonstrable* basis of all scientific demonstration. If reason is to be reason, then there is something it can *not* seek a reason for; reason stands on a limit: the law of contradiction. Reason abides by this law. In abiding by this law, it entertains no doubt about itself. It acknowledges its own essentiality, that is, that it actually exists transparently for pure thought, that it is in itself potentially intelligible. It is therefore in no position to establish its own existence (there is no 'I doubt, therefore I am' with St. Augustine, nor with Descartes, 'I think, therefore I am'). Its existence depends upon its essence, pure thought or intellect, together with which it is bound in the latter's identity. Reason in itself is not absolute or pure thought, but it knows the latter, that is, intellect, to be its own principle. It knows itself to possess, as an original possession, the cause of its own existence. The essential priority of

[4] Ibid., 203–204.

intellect to reason precludes doubt. At the same time, it guarantees independence in existence to every rational mind, *qua* intellectual. That is to say, *every rational animal possesses, above and beyond the sensible and rational conditions of its knowing, in the innermost essence of its being, that intellectual power which makes it to be itself, to be a rational animal, and which, in itself, is identically real.* This is the true speculative essence of Aristotle. No reason is to be sought for this.

Aristotle describes the nature of human reason in the *De Anima* III.4: "Concerning that part of the soul (whether it is separable in extended space, or only in thought) with which the soul knows and thinks, we have to consider what is its distinguishing characteristic, and how thinking comes about. . . . This part, then, must (although impassive) be receptive of the form of an object, *i.e.,* must be potentially the same as its object, although not identical with it: as the sensitive is to the sensible, so must mind be to the thinkable. It is necessary then that mind, since it thinks all things, should be uncontaminated, as Anaxagoras says, in order that it may be in control, that is, that it may know; for the intrusion of anything foreign hinders and obstructs it. Hence the mind, too, can have no characteristic except its capacity to receive. That part of the soul, then, which we call mind (by mind I mean that part by which the soul thinks and forms judgements) has no actual existence until it thinks."[5] First, it should be noted that mind is here treated *qua* reason, it thinks, reflects, intends, it judges, assumes, understands [ᾧ γινώσκει . . . φρονεῖ . . . . ᾧ διανοεῖται καὶ ὑπολαμβάνει].[6] This reason is for Aristotle the human mind *qua* human. It is specified as being what it is insofar as its existence is merely potential. Human reason is specifically differentiated from other possible minds by being *capable* of knowing an object. To grasp this distinction vividly, consider that God's mind knows an object (itself) without potentiality. It is clear, then, that for Aristotle mind includes reason, but that reason is specifically

[5] *Aristotle: On the Soul; Parva Naturalia; On Breath* (trans. W. S. Hett, Cambridge, Mass., 1957), 163, 165.
[6] Ibid., 162, 164.

the potentiality for knowing, which in and of itself does not include mind. The latter, for instance in God, exists in itself without reason, indeed, exists as perfectly actual knowledge. Reason is that difference by which man is distinguished from other animal species. But soul apart from reason is not human, nor, apart from an object, does reason actually exist. With this result (which I better appreciate as I keep in mind Aristotle's insight into the concrete particularity of the real), namely, that where I find a capacity for thought, that is, reason as such, I do not find a man, but I find a species. While he would be a man, he actually is not because he is not yet actually a thinker. I note that not only does reason in general not make a thinker, but that there is a conjunction of actual thought with humanity's concrete individual man. I am forced to conclude that it takes more than reason not only to make a thinker, but to make a man. (Of course, by more than reason I do not mean a body; the body comes together with reason in the species.) Since, *qua* rational, this man possesses the intelligible forms of thinking only potentially, it is evident that there must be a power in the intellect, as the essential identity in which reason is circumscribed, whereby thought can take place. As a matter of fact, Aristotle tells us something about this in a famous passage in Book III of the *De Anima;* but it is the locus of philosophic trials, where Aristotle has tied a tight knot of insight to 'tease us out of thought/As doth eternity.'

But before dealing directly with the question of the Active Intellect of the *De Anima*, let us first turn our attention to certain remarks of Aristotle in *Nicomachean Ethics* X: ". . . now activity in accordance with wisdom is admittedly the most pleasant of the activities in accordance with virtue: at all events it is held that philosophy or the pursuit of wisdom contains pleasures of marvellous purity and permanence, and it is reasonable to suppose that the enjoyment of knowledge is a still pleasanter occupation than the pursuit of it. Also the activity of contemplation will be found to possess in the highest degree the quality that is termed self-sufficiency. . . . the wise man . . . can also contemplate by himself, and the more so the wiser he is; no doubt he will study better with the aid of fellow-workers, but still he is the most self-sufficient of men. Also the activity of contemplation may be held to be the only activity that is loved for its own sake: it produces no result beyond the actual act of contemplation, whereas from practical pursuits we look to secure some advantage, greater or smaller, beyond the action itself." [7] Since we have just discovered ourselves that Aristotle's thought is that science resides actually in a concrete individual man, so intimately that it is the same thing 'to know and to be' a man, we, then, find ourselves in a position truly to appreciate Aristotle's words on scientific contemplation, located as they are at the conclusion of his treatise on *Ethics*. The actual man, *qua* actual, transcends humanity. Corresponding to this so thoroughly unmodern insight, Aristotle sharply distinguishes intellectual life from practical life. Moral virtues (practical wisdom, fortitude, justice and so on) need, as context for their proper exercise, an organized human community, what ancient Greece understood comprehensively as *political* life. And moral virtues need that species-life of the *polis* not only for their *acquisition* but also for their *exercise*. However, intellectual life, that life by which the individual transcends the limiting conditions of his own humanity, does not need other men for its exercise, although, incidentally, in acquisition of the materials of science, a man's *potential* knowledge is increased by the presence of fellow-workers. Within scientific life itself, therefore, we discern the same distinction between potential and actual as we see to exist between species and individual, or between reason and pure thought. That absolute intensity of intellect convergent with actual individuality manifests itself in Aristotle's understanding that contemplative activity is its own end, but that practical-moral activity due to its specific conditionality is extensively ordered beyond itself. Man, *qua* man, is *zoon politikon*, a living political animal. Therefore, the individual is not virtuous for his own sake, except incidentally, but for the good of the *polis*. With this distinction in mind we can better appreciate what Aristotle says in *Ethics* I: "For even though it be the case

[7] *Aristotle: The Nicomachean Ethics* (trans. H. Rackham, Cambridge, Mass., 1934), 613, 615.

that the good is the same for the individual and for the state, nevertheless, the good of the state is manifestly a greater and more perfect good, both to attain and to preserve. To secure the good of one person only is better than nothing; but to secure the good of a nation or a state is a nobler and more divine achievement."[8] At first glance it seems that the individual is submerged in the common good so thoroughly that, by analogy to modern times, he would need the Christian faith of a Kierkegaard to preserve himself from the totalitarian claims of the *spirit of man*. But what makes Kierkegaard intelligible is an event that is so primary in being that it could not occur in Aristotle's universe without destroying it totally. Therefore, the analogy is inappropriate. But, that it is inappropriate is instructive in understanding not only Aristotle but ourselves, for our worlds are not at all the same. Here is a structural tension so characteristic of Aristotle's thought, touching on human being, by extension on epistemological questions—here it is at a fairly tractable point. Aristotle understands intellectual activity to excel moral activity by reason of its self-sufficiency. Within intellectual life itself research is to contemplation as acquisition is to exercise, as politics is to knowledge, as insufficient reason is to self-sufficient intellect. But reason knows its 'invisible man,' insofar as it is self-sufficient, to occupy a state like God's. In fact, for Aristotle, the individual, *qua* individual, is like God. He is especially Godlike, especially the individual, in contemplation. But Aristotle in his words just cited, exalting the common good over the good of one man, states that to "secure the good of one person only is better than nothing; but *to secure the good of a nation or a state is a nobler and more divine achievement*." There's the rub! Godlike individual against Godlike *polis*. How is it both ways? But it is only one way: Aristotle's thought is relentlessly synthetic. The common good excels the good of one man insofar as only one leg of two-legged reality is being considered, namely, an order of sufficient causality, that is, this universe. Here a man taken in isolation is by definition insufficient. For this perspective only God is a sufficient individual, precisely

[8] Ibid., 7.

because he is outside this universe. But within this universe God's sufficient causality is approximated to varying degrees by eternal movements of heavenly magnitudes. On this earth sufficient cause has it locus in a species; that is, it takes a man *qua* man to make a man. Consequently, it is perfectly true that in an order of sufficient causality common good is more divine. But the order of sufficient causality does not affect God's thought. The latter is absolutely self-identical. Therefore, on the other leg, God's existence tells Aristotle not only that self-sufficiency is individuality, but that individuality is prior, in reality, to the conditioned existence of the universe. As a result, an order of sufficient causality not only does not submerge individuality, but is, as a matter of fact, ordered to it. *The individual,* qua *individual, engaged in what he alone can do, active contemplation, is most divine.*

Now it must also be clear that although we speak of two legs, they are not related to one another indifferently. Rather, there is an *order* relating species to individual, what is more divine to what is most divine, as matter to form. Order denotes at the same time subordination, but real unity. So, for example, the life of an organism is this particular organism's life. The life of an individual man is that he is not either an individual or a man, but, organically, one is an individual man. This psychosomatic synthesis, this living man, exists because soul is united to body not indifferently but in an intelligible order. This is to say that the man himself, so considered, is a universe of sufficient causality. But the further implication is that this man's *individuality* must itself be the sufficient cause of this *order.* Note well: not of his existence in the sense that he has come into existence; the sufficient cause of this man's existence, insofar as he is one-among-men in this universe, is the species. If we say of a man that he is a synthesis of body and soul, of form and matter, of sensation and reason, still we do not account for his being *this individual man.* If his individuality is attributed to his matter, then it is understood in and of itself to be nonexistent; if to his form, then his existence is some other man's existence, or he is simply a species. But for Aristotle this man must possess within himself his own principle of individuality or ordered existence;

if not, no knowledge is possible, because reason, in itself, is an insufficiency. In Book II of the *De Anima,* Aristotle defines form as that "in virtue of which [καθ' ἥν] individuality [τόδε τι] is directly attributed [ἤδη λέγεται]." [9] Within the horizon of form it is said, here and now, there is an individual present. But what is known to reason, what is able to be spoken about, here and now, namely, individuality, does not by virtue of that fact or form of reason *exist.* That would be to attribute to reason, in its sheer transparency, a creative power it simply does not possess (actually a creative power that exists nowhere in Aristotle's universe). Reason is insufficiently ordered to existence (that is, to individuality). The rational soul or form of man communicates existence not actually but potentially; individuality is in itself incommunicable existence. Forms existing for reason exist potentially. It is sufficient for scientific reason that this man exist by virtue of his form, that is, sufficient for science as potential knowledge, as reason, research, as *logos,* or that cooperative gathering of materials for what finally is to be an individual knower's actuality. Such an understanding is sufficient when science is taken to be directed to another man. But it is radically insufficient to reality *per se,* and therefore to science undertaken for its own sake, for the sake of the knower. Science is not yet truly itself so long as existence is known bounded by an horizon of forms: so long as self-existence is merely science's *implication.* This is the heart of Aristotle's criticism of Plato. What the dialectical formalism of Platonism gains in scope, in universality, it loses in intensity, in power; as Aristotle says in *Metaphysics* IV: ". . . philosophy differs from dialectic in degree of power. . . . For dialectic puts questions about matters which philosophy knows. . . ." [10]

And, of course, science is only pure in isolated moments, in isolated men; this is a function of man's imperfect nature, that is, his rationality. Aristotle says in *De Anima* III: ". . . when the mind has become the several groups of its objects, as the learned man when active is said to do (and this happens, when

---

[9] *On The Soul,* op. cit., 67.
[10] *Metaphysics,* op. cit., 65.

he can exercise his function by himself), even then the mind is in a sense potential, though not quite in the same way as before it learned and discovered; moreover the mind is then capable of thinking itself."[11] In a pure act of scientific contemplation, all preparations having been made, the army having been marshalled in order, intellect transcends formal judgments, committed to act it knows through identity the incommunicable existence which is an object's essence: Pure intellectual knowledge is identical in existence with its object. This is the absolutely uncommon life of contemplative science. It is a sabbath between two evenings; not union but identity. It is Aristotle himself who compares the intellectual act to battle in *Metaphysics* XII: "We must also inquire in which way the nature of the whole enjoys its good or highest good: whether as something separate and by itself, or as its own order, or in both ways, as does an army. For an army's good lies both in its order and in its commander, more especially the latter; for he is not the result of the order, but it results from him."[12] The Unmoved Mover, the Divine Intellect, by analogy, the human intellect in relation to the universe this man is, is a still point for the turning world, but in itself, like a general's being, it is *to act.* After all, a general is not a general if he is merely a model for demonstration purposes; the maneuvers of reason exist for knowing. Intellect is not a toy soldier. But knowing is knowing a particular essence; so that there is no danger of human intellect in its identity with its object becoming God: that would be to forget that being in Aristotle is everywhere particular being. But knowing is knowing what it is to be this thing; *afterwards* reason *remembers* this thing's incommunicable essence as *τί ἦν εἶναι*, what it *was* to be this thing.

Essence, then, is, in reality, individual existence. Reason names it according to its own formal principles. In so doing it betrays its passivity. It is then the intellect, as essential individuality, that shapes this human soul (common in kind) to its particular difference in matter. Proximate matter is the principle of

[11] *On The Soul,* op. cit., 167.
[12] *Metaphysics,* op. cit., 267.

sensible individuality, but this is derivative. Rational form is the principle of a *remembered* (conscious) individuality, but it too is derivative. But pure intellect is the principle—no, the *very individuality,* the 'invisible man.' If we think with Aristotle it is impossible to take either matter or form as a principle of individuality. Listen to Aristotle in *Metaphysics* XII: "And even the explanatory factors of things in the same kind are different, not in kind, but because those of different individuals are different [οὐκ εἴδει ἀλλ' ὅτι τῶν καθ' ἕκαστον ἄλλο]: your matter and form and mover differ from mine [ἥ τε σὴ ὕλη καὶ τὸ εἶδος καὶ τὸ κινῆσαν καὶ ἡ ἐμή]; but they are the same insofar as they have a common formula [τῷ καθόλου δὲ λόγῳ ταὐτά]." [13] It is only *logos* or reason which does not reach to that difference in existence which descends from intellect (mover) to form, to matter. This is the Aristotelian essence. Whatever passages might be cited to show that matter or form are principles of individuality must be understood as significant for logical or methodological purposes, but not for the act of knowing. Aristotle's understanding of intellect, or this man's true self, is meant to be a perfect contradiction of Plato's *idea* of man, indeed, to all ideas of reality. This man's essence is his own consummate difference in existence; he belongs to himself. He is a person. To put it contextually: This man, essentially a mover or intellect, is self-moved in existence; as such, neither God nor man is his end; nor need this man be moved by God or man except by virtue of the potentiality of his own being.

Now let us turn our attention to *De Anima* III.5, where Aristotle discusses the so-called Active Intellect: "Since in every class of objects, just as in the whole of nature, there is something which is their matter, *i.e.,* which is potentially all the individuals, and something else that is their cause or agent in that it makes them all—the two being related as an art to its material—[ἕτερον δὲ τὸ αἴτιον καὶ ποιητικόν, τῷ ποιεῖν πάντα, οἷον ἡ τέχνη πρὸς τὴν ὕλην πέπονθεν] these distinct elements must be pres-

[13] *Metaphysics,* op. cit., 255. I have rearranged the translator's 'and mover and form' to correspond with Ross' text: *Aristotle's Metaphysics* (Oxford 1924), 1071a, 27–29.

ent in the soul also. Mind in the passive sense is such because it becomes all things, but mind has another aspect in that it makes all things; this is a kind of positive state like light; for in a sense light makes potential into actual colors. Mind in this sense is separable, impassive, and unmixed, since it is essentially an activity; for the agent is always superior to the patient, and the originating cause to the matter. Actual knowledge is identical with its object. Potential is prior in time to actual knowledge in the individual, but in general it is not prior in time. Mind does not think intermittently. When isolated it is its true self and nothing more, and this alone is immortal and everlasting (we do not remember because, while mind in this sense cannot be acted upon, mind in the passive sense is perishable), and without this nothing thinks."[14] The first thing to be noted here is that intellect (active mind) is related to reason (passive mind) not as form is to matter, but, precisely, as skill or *technē* is to matter. Intellect is not a static form, but a creative energy. It is the skill which brings reason's forms into being. Therefore, it is not itself a form of reason; it is not matter acted upon. Also, it is not a form, because it would then be a form of forms: a redundant magnification of reason; or, what is the same thing, an efficient cause. But intellect is not compared to a sculptor's face, as if reason were a studio filled with likenesses of their maker. But, by way of contradiction, intellect is compared to a sculptor's skill or *technē:* It is that energetic trick so much more itself as it does not, itself, appear in the forms it brings into being. Pure intellect is a disappearance of the true cause of reason's forms; it contrives its effects to appear that much more real. So that, when reason speaks formally of the essence of anything as τί ἦν εἶναι, *what it was to be,* it, strictly, knows not whereof it speaks. Intellect is neither same as nor different than reason; these terms are reason's. Intellect, in itself, is finally discontinuous with reason. *Reason is potentially its object, but, intellect is actually, self-identically, its object's existent individuality, or essence.* In itself, it is immortal as this universe's very being, everywhere individual existence, is immortal. Now, since Aristotle says in *Metaphysics*

[14] *On The Soul,* op. cit., 171.

XII: ". . . wherever things are immaterial the mind and its object are not different, so that they are the same; and knowing is united with what is known," [15] it might be objected that intellect is not one with its object if that object exists in this universe. But matter apart from form does not exist; form apart from individualizing essence does not exist; in other words, that division that matter introduces into knowing is relevant to reason, not to intellect; it is relevant to science so long as science is logical or methodological arrangement. Reason is formally proportioned to natures, but only potentially to essences.

With regard to immortality, it follows that, while I can have no experience of God's existence apart from that likeness to it that is my own essence, my reason is able to demonstrate that the Divine Nature exists. Further, since reason is disproportionate to my essential identity, I can know it only in moments, insofar as reason is still. Then I think I see this Light (it disappears), but I do not see it; I am it identically. This absolute incommunicable certitude in existence is ground to nature, to science. Individual immortality is for Aristotle a substantial fact. This is not to say that it is in itself what reason takes it to be. There is not a reason for reason's being more real after death than it is in this life. But then, reason's idea of its ultimate reality is inadequate. But the naturalistic-logical interpretation of Aristotle, for example, that of Pietro Pomponazzi (1462–1525), whose understanding fundamentally anticipates the modern reception of Aristotle, whereby intellect is taken insubstantially, that is, merely functionally, or logically, as descriptive of an operation of human reason, is also reason's idea. It is, perhaps, noteworthy that Pomponazzi's arguments for intellect's actual mortality were directed against Thomas Aquinas' understanding of the immortality of the soul. Here it can be noted that in sacred doctrine reason must already have moved to the center, if Pomponazzi's procedure is to be intelligible.[16] It will be left to subsequent analysis to examine that logical con-

[15] *Metaphysics,* op. cit., 266.

[16] P. Pomponazzi, *On The Immortality of The Soul* (trans. W. H. Hay II in *The Renaissance Philosophy of Man,* ed. Cassirer, et al., Chicago, 1948), 280–381.

tinuity whereby modern science takes its point of departure from the Queen of Sciences.

In the meantime, it is clear to us that Aristotle stands in perfect discontinuity with all possible ideas of reason insofar as they are mere ideas, directed practically to human interests as if to man's highest good, with all understandings of man which take man to be finally a religious, a moral, or a technical being. On the contrary, Aristotle speaks of man's intellectual life in *Nicomachean Ethics* X in this way: "Such a life as this however will be higher than the human level [*κρείττων ἢ κατ' ἄνθρωπον*]: not in virtue of his humanity will a man achieve it, but in virtue of something within him that is divine; and by as much as this something is superior to his composite nature [*διαφέρει τοῦτο τοῦ συνθέτου*], by so much is its activity superior to the exercise of the other forms of virtue. If then the intellect is something divine in comparison with man, so is the life of the intellect divine in comparison with human life. Nor ought we to obey those who enjoin that a man should have man's thoughts and a mortal the thoughts of mortality, but we ought so far as possible to achieve immortality, and do all that man may to live in accordance with the highest thing in him; for though this be small in bulk, in power and value it far surpasses all the rest. It may even be held that this is the true self of each, inasmuch as it is the dominant and better part; and therefore it would be a strange thing if a man should choose to live not his own life but the life of some other than himself." [17] Actually, what Aristotle states here is *nothing other than a true metaphysics' indispensable condition: that reason possess within itself as its absolute priority essential being.* Only on this condition is it possible for a knower to be *identical* with a reality other than itself. Only under intellect's light will reason submit to this condition beyond its understanding. Only on this condition is science disinterested: when it possesses in itself all that it might desire: when it is essentially itself. Then it is not at all an instrument. To depart from this condition is to take up reason's perspective on metaphysics,

[17] *Nicomachean Ethics,* op. cit., 617, 619.

namely, that it is naturally an instrument with these possible epistemological consequences: either, that a knower is not identical but *like* what is known; or, that a knower is identical with what is known, but it is none other than himself, or, finally, what is known is other than knowledge, but knowledge *determines it to exist.*

But what is incomprehensible to ungoverned reason is to reason under intellect's guidance understood as good sense (γνώμη). It is nothing other than good judgment on the part of a truly metaphysically conditioned reason to direct its attention to particulars. Scientific reason is metaphysical insofar as it judges every object considerately (εὐγνώμων), that is, takes into consideration that particular object's own essential identity, its individuality in existence. Indeed, this science goes so far as to take this object's side in the judgment. That is, it recognizes the object's own essential identity; it identifies itself with it, forgives it (συγγνώμη), that is, absolves it of being as it appears to it. Reason recognizes its limits, acknowledges that what it knows is an independent substance. This conciliatory spirit of metaphysical science is rooted in the soil of experience, but, essentially, flourishes in the light of intellect. Since reason judges nature by intellect, not by its own universalizing tendency, it judges well (εὐγνώμων) of nature. That is, it judges that in nature there exist substantial differences in being; consequently, it is in little danger of taking itself for God (since it and God are proportioned to different natures), nor is it in danger of a phenomenological solipsism (since it is essentially a mediator, not an agent).[18]

While intellect shines, reason restricts its scientific enterprise to making whatever preparations might be necessary for the act of knowing what something is. To do otherwise would simply be willful.

---

[18] For Aristotle's original use of the Greek terms in this paragraph, see *Nicomachean Ethics,* op. cit., 358, 360, 362.

Chapter 2

THOMAS AQUINAS: A NEW REALITY

'It is not possible for a thing to be and not to be at the same time in the same respect.' The law of contradiction is for Aristotle the indemonstrable presupposition of scientific *logos* or reason. But reason for Aristotle is, in itself, distinguished from intellect as matter is from the skill that shapes it. Reason itself is passive mind, simply receptive, or transparent. The law of contradiction means for reason, then, that it be determined to one thing or another in particular. When Aristotle says in *De Anima* III. 5 that "mind in the passive sense is such because it becomes all things," [19] it is to be understood that this universality of reason is inseparable from its potentiality. That is, its actual knowledge is that of particulars. Reason's particular competence is to determine by formal principles that *this thing* (*τόδε τι*) exists. But that it exists for reason is due, in the first place, to intellect's constant activity. But intellect itself, while it shines on particular objects in such a way as to make them to be for reason, while at the same time it makes clear to reason that the law of contradiction is its indemonstrable beginning, is, itself, secretly concentrating on what it regards above everything in these particulars, or in the law of contradiction, identity. Identity is the law of intellect's own being: 'It is not possible for a thing to be and not to be,' or, *a thing actually is what it is.* While, therefore, intellect makes things to exist *for reason,* it does not make things to exist simply. Even when intellect is considered as

[19] Above, pp. 49–50, note 14.

the ultimate principle or very individuality of this man, still, it presupposes his rational animality. In fact, intellect is immediately certain that simple existence is *bound up with being what it is,* that essence is existent individuality. For intellect to know is *to be identically what* is known. That is, it is not *simply* to be either itself or another, for simply to be is not to be known. But if *what* is known is not self-identical *in existence* then intellect *is simply* another, not itself, nor, therefore, is there knowledge. Knowledge requires difference in existence. To know is *to be what* another is, but for another to be it is to be *what* it is. *What,* then, is the actuality of existence. Also, *what* is the actuality, for intellect, of knowledge. *What* is not, metaphysically, a question, but an answer. Metaphysics is a happy child's joke: "D'ya know what?" "What?" "That's what!" *What* is that metaphysical identity of an actual existent *in what* is known to intellect. *What* is the absolute limit upon which intellect stands: The law of identity is to intellect, as the law of contradiction is to reason. Metaphysical joy consists in the knowledge that *what* is the solution to existence, that what is there to be known is known, since essence or identity is meant existentially. Science's consummate pleasure is beholding not simply a nature but an existing individual known through a nature. In this situation a distinction is to be made between nature, on the one hand, with its universality and degrees of being, and, on the other hand, essence, with its self-identical individuality. Natures exist in order, or as mediums of order, or as results of order; but essences originally cause order. Differences attributed to essences from the order they cause do not touch them, except that ultimate difference known as identity, but this difference is touched by intellect which, in the act of knowing, prescinds from order. Insofar as essence is known from that complexity of its attributes it is that it touches them, that it draws out of reason those shapes that reason woos. But it abides by itself apart, accessible only to intellect which leaves everything else behind to come to it. For Aristotle essence is a distinction in being, beyond nature. Thus that particularity in being that informs the particularity of Aristotle's science. Because he could not appreciate the former from his vantage point Hegel complained often of the latter:

"In the Aristotelian teaching the Idea of the self-reflecting thought is thus grasped as the highest truth; but its realization, the knowledge of the natural and spiritual universe, constitutes outside of that idea a long series of particular conceptions, which are external to one another, and in which a unifying principle, led through the particular, is wanting. The highest Idea with Aristotle consequently once more stands only as a particular in its own place and without being the principle of his whole philosophy." [20] But Aristotle's Divine Intelligence thinks what it is; nor can it think what other things are without ceasing to be what it is, namely, God. The systematic unity that Hegel seeks to complement what he calls the 'highest Idea' in Aristotle is missing precisely for the reason that, even if God's identity could be preserved in his knowing other essences (which it could not), other essences would need to be comprehended through a natural medium such as *simple being.* But this is impossible as has been indicated. Metaphysics did finally achieve that systematic unity in Neoplatonism, but in so doing its metaphysical conscience understood the principle of the whole to be not a 'highest Idea' or 'self-reflecting thought' or Divine Intelligence, but an Absolute One, prior to, beyond Divine Intelligence. [21]

But what is not possible to metaphysics is to sacred doctrine its very fount. To turn from Aristotle's crystalline reality to Thomas Aquinas' infinite reality is, in the first instance, to enter into a church. It is not a transit lasting 600,000 days, but it is to submit; it is to enter the cave in God's service, to stand in God's hollow, to cave in, to worship His child. Form of mind is form of church is form of universe is form in God's mind; universe, church, mind is God's will working: it is God's creation. The Spirit, which is Love, is God's Spirit. Here, in perfect contradiction to metaphysics, *to know is not to be, but to be is to be known by Him.* And the proper principle of the science of sacred doctrine is not light of intellect, but light of revelation. The possibility of

[20] *Hegel's Lectures On The History of Philosophy* II, op. cit., 229.

[21] Proclus, *The Elements of Theology* (trans. E. R. Dodds, 2nd ed., Oxford, 1963), 23: ". . . while all things, whatsoever their grade of reality, participate unity . . . , not all participate intelligence . . ."

its reception is not transparent reason, but a pure heart. Not the law of identity, but the presence of the Trinity; not the law of contradiction, but the fact of incarnation. Not a metaphysics, in its good sense, forgiving reality, but a theology itself forgiven its necessity. Thomas writes in *Summa Theologica* 1,1,5 ad 2; "This science can in a sense depend upon the philosophical sciences, not as though it stood in need of them, but only in order to make its teaching clearer. For it accepts its principles not from other sciences; but immediately from God, by revelation. Therefore it does not depend upon other sciences as upon the higher, but makes use of them as of the lesser, and as handmaidens: even so the master sciences make use of the sciences that supply their materials, as political of military science. That it thus uses them is not due to its own defect or insufficiency, but to the defect of our intelligence [*propter defectum intellectus nostri*], which is more easily led by what is known through natural reason [*per naturalem rationem*] (from which proceed the other sciences) [*ex qua procedunt aliae scientiae*], to that which is above reason, such as are the teachings of this science."[22] If the principal science proceeding from natural reason is for Aristotle metaphysics, then Thomas subordinates metaphysics to sacred doctrine as matter to form, so that, in light of revelation, intellect is itself seen as defective, that is, it is seen. It itself becomes an object; intellect is assimilated to reason. As sacred doctrine knows reality, intellect is no longer self-identically real; it is no longer self-sufficient. It is for sacred doctrine *in*essential, that is, merely instrumental. The teachings of sacred doctrine exist above reason. But what is above *metaphysical* reason, as opposed to Thomas' *natural* reason, is, as its absolute priority, essential being or knowledge of being bound up with what it is. It is therefore clear that sacred doctrine is related to metaphysics disjunctively: it is *either* faith *or* knowledge. But since grace does not destroy, but rather perfects, nature, this disjunctive conjunction means as a matter of fact, on the one hand, knowledge's transformation from essen-

[22] Thomas Aquinas, *Summa Theologica* (trans. Fathers of the English Dominican Province, New York, 1947), 3.

tial being to being natural, to simply being, or to being something that comes to be. On the other hand, it means nature's transformation into actuality. But since on neither hand is there any potentiality for these respective transformations, namely, of essence to nature, of nature to actuality, we find ourselves *formally* in creation's presence. As in form, so in substance: sacred doctrine is an absolute novelty in science. It is *knowledge of what is transcendent.* But since, as Thomas makes clear, this knowledge is *explicitly* grounded as a matter of principle in faith or revelation, no Kant can question this possibility of knowing a transcendent reality, nor would such a critique as Kant undertakes of science's foundations be at all intelligible to Aristotle. In other words, it must be expected that Kant's critique is intelligible if science itself, outside of sacred doctrine, is understood to be also a novelty. But if this were so science would take its point of departure *implicitly* from sacred doctrine; its perpetual departure its way of being itself; its doubt purely a matter of method. Its will to be itself would then be intelligible, but being will, still frightening.

But for Thomas sacred doctrine is a matter of explication, not implication. He writes in *Summa Theologica* I,1,5: "Now one speculative science is said to be nobler than another, either by reason of its greater certitude, or by reason of the higher worth of its subject-matter. In both these respects this science surpasses other speculative sciences; in point of greater certitude, because other sciences derive their certitude from the natural light of human reason, which can err; whereas this derives its certitude from the light of the divine knowledge, which cannot be misled: in point of the higher worth of its subject-matter, because this science treats chiefly of those things which by their sublimity transcend human reason; while other sciences consider only those things which are within reason's grasp. Of the practical sciences, that one is nobler which is ordained to a further purpose, as political science is nobler than military science; for the good of the army is directed to the good of the state. But the purpose of this science, in so far as it is practical, is eternal bliss; to which as to an ultimate end the purposes of every practical science are directed. Hence it is clear that from every standpoint it is nobler than other sciences."[23] If the essential light of intellect is become in Thomas the natural light of reason, then, sacred doctrine illuminated by God's light is explicitly *alone* possessed of certitude, while other sciences operate in the *doubtful* light of nature. The doubt of modern science undoubtedly derives without ado from sacred doctrine; but the immediate condition is the alteration of intellect itself to nature, that is, a reduction of man to being merely man. Man sees himself as he simply is, inessentially existing, not possessing within himself a sufficient cause of his own existence. Identity is no longer *existentially* significant; in God's light man's essence is confounded with nature. Like Adam, man in light of creation is reduced to a naked humanity. So starkly rational is this animal, that Aristotle's call 'to be like gods' stings like snake bite, or, like Yahweh's parting shot in *Genesis* 3:22: "See, the man has become like one of us, with his knowledge of good and evil. He must not be allowed to stretch his hand out next and pick from the tree of life also, and eat some and live for ever."[24] With the disappearance of essence as the self-sufficiency of particular individuals prescinding from natural order as a necessity of being, sacred doctrine is able to unify in itself what otherwise would be the disparate domains of speculation and practical science. If sacred doctrine were not a practical science, but only speculative, then no divine light would pursue naked man into his natural state ordering his practical science to eternal beatitude. But it is just this that God's humiliation of man brings into being: an ordering of man's practical life, *qua* practical, to eternity. If man is become a simply natural being through the creative light of God's revelation, then nature, *qua* nature, human nature, *qua* human, is forgiven its imperfection. The metaphysical forgiveness of a being substantially independent, that is, itself indifferent to intellectual light's good sense, is shown to be what it is: absolute self-centeredness. This luminous opacity of metaphysical self-love is shown up by the light of God's science that forgives what *needs* forgiveness. It is nature

[23] Ibid.

[24] *The Jerusalem Bible: O.T.* (Garden City, 1966), 18.

together with natural man, this metaphysical insubstantiality, this impoverished here and now, to which God's light extends His redemptive resources in absolute freedom; the need is wholly on the part of man or nature. In divine light self is absolutely transparent so that what is seen is simply that *what is necessary is provided.* But this new reality is not nature or man in themselves. It is, therefore, not in advance of faith, not in advance of hope, not in advance of charity. As St. Paul says in his *Letter to the Romans* 8:19–25: "The whole creation is eagerly waiting for God to reveal his sons. It was not for any fault on the part of creation that it was made unable to attain its purpose, it was made so by God; but creation still retains the hope of being freed, like us, from its slavery to decadence, to enjoy the same freedom and glory as the children of God. From the beginning till now the entire creation, as we know, has been groaning in one great act of giving birth; and not only creation, but all of us who possess the first-fruits of the Spirit, we too groan inwardly as we wait for our bodies to be set free. For we must be content to hope that we shall be saved—our salvation is not in sight, we should not have to be hoping for it if it were—but, as I say, we must hope to be saved since we are not saved yet—it is something we must wait for with patience." [25] Paul's statement is not metaphysical. The practical end of Thomas' sacred doctrine is the practical end of *created* nature, nature *explicitly* in light of revelation. It belongs to men of God's Spirit. It is not the metaphysical self-love by which man non-identifies with nature, but it is precisely men who love God because He first freely forgave them their essential necessity who express solidarity with created nature. Paul does not say 'we wait to be set free from our bodies'; he says 'we wait for our bodies to be set free.' Man is totally, essentially a natural entity; his spirituality is not an alternative to this. But this fact exists only in the disjunctive union of grace with nature. It is explicitly in a circle of grace that man embraces nature: in this union nature suffers with man; into this union God alone penetrates as Identity. Thomas writes in *Summa Contra Gentiles* III.153: ". . . if by

[25] *The Jerusalem Bible: N.T.* (Garden City, 1966), 280.

grace man is made a lover of God, there must be produced in him a desire for union with God, according as that is possible. But faith, which is caused by grace, makes it clear that the union of man with God in the perfect enjoyment in which happiness consists is possible. Therefore, the desire for this fruition results in man from the love of God. But the desire for anything bothers the soul of the desirer, unless there be present some hope of attainment. So, it was appropriate that in man, in whom God's love and faith are caused by grace, there should also be caused a hope of acquiring future happiness." [26] This practical-natural reality created by grace is necessarily a realm of action, but not of a self-sufficient virtue as Thomas says in *Summa Theologica* I–II,69,1: "*We are saved by hope.* Again, we hope to obtain an end, because we are suitably moved towards that end, and approach thereto; and this implies some action. And a man is moved towards, and approaches the happy end by works of virtue, and above all by the works of the gifts, if we speak of eternal happiness, for which our reason is not sufficient, since we need to be moved by the Holy Spirit, and to be perfected with His gifts that we may obey and follow him." [27] In action, in the light of revelation, virtue is subordinate to spiritual gifts, self-reliance to obedience, just as, in thought, metaphysics is subordinate to sacred doctrine. But this subordination is transformative of virtue or metaphysics, in light of intellect ends in themselves, into instruments directed, by faith, to a single transcendent end. A subsequent insubordination of metaphysics or virtue to spiritual form would be not only, in Paul's terms, that impatience that thinks its salvation is in sight, but it would be the despair of a transformed consciousness: a consciousness transformed into a desperate instrumentality. It would be a final loss of innocence.

If sacred doctrine is founded not on the principle of identity or contradiction, but on faith, this means that in place of the metaphysical principles, this single science of transcendent real-

---

[26] Thomas Aquinas, *Summa Contra Gentiles,* Bk III, Pt 2 (trans. V. J. Bourke, Garden City, 1956), 238.

[27] *Summa Theologica,* op. cit., 886.

ity is grounded in its own proper principles. As Thomas points out in *Summa Theologica* II-II,1,7: ". . . all the articles of faith are contained implicitly in certain primary matters of faith, such as God's existence, and His providence over the salvation of man, according to Heb. xi: *He that cometh to God, must believe that He is, and is a rewarder to them that seek him.*" [28] Sacred doctrine is not only to be distinguished from Aristotelian metaphysics, but also from Neoplatonism's systematic unity; God for sacred doctrine is not an Absolute One necessarily beyond being or knowledge; but sacred doctrine identifies God's Infinity with being so that God knows being's particularity, otherwise metaphysical, to be his own creation. This absolute positivity of God's being is His own revelation; this fact is vigorously expressed by Dionysius in his *Divine Names* IV: ". . . we must dare to affirm (for 'tis the truth) that the Creator of the Universe Himself, in His Beautiful and Good Yearning towards the Universe, is through the excessive yearning of His Goodness, transported outside of Himself in His providential activities towards all things that have being, and is touched by the sweet spell of Goodness, Love, and Yearning, and so is drawn from His transcendent throne above all things, through a super-essential and ecstatic power whereby He yet stays within Himself. Hence Doctors call Him 'jealous,' because He is vehement in His Good Yearning towards the world, and because He stirs men up to a zealous search of yearning desire for Him, and thus shows Himself zealous inasmuch as zeal is always felt concerning things which are desired, and inasmuch as He hath a zeal concerning the creatures for which He careth." [29] God empties Himself of metaphysical reserve absolutely. Insofar as creation is related to Him, God is altogether outside of Himself. He is beside Himself. The principle of identity in sacred doctrine is become love of another. In each *what it is to be* God is Another's Love; God is become infinitely, intimately indwelling. It is not God, Being Himself, Who for sacred doctrine is abstract; rather is metaphysics, together with its depen-

[28] Ibid., 1174.

[29] Dionysius, *On the Divine Names* (trans. C. E. Rolt, London, 1940), 106.

dents, potentialized. It is intellect, not God, that suffers abstraction. It becomes a power of man's soul. Man's soul, in turn, actually exists only as God's creation. Thomas says in *Summa Theologica* I,90,2 ad 1: "The soul's simple essence is as the material element, while its participated existence [*esse participatum*] is its formal element; which participated existence necessarily co-exists [*ex necessitate simul est cum essentia*] with the soul's essence, because existence naturally follows the form [*quia esse per se consequitur ad formam*]." [30] In Aristotle's metaphysics essence includes existence in its pouch. In Thomas essence is become form or nature. Considered formally, existence continues to follow upon nature. But nature itself is brought into being by God's creative act. Therefore, that existence, formally nature's, is actually, by participation, God's gift. Essence reduced to form is therefore in itself a potential existence, a matter to be actualized by participated existence.

There is a special question as to whether or not it is an article of faith that the world began, to which Thomas replies in *Summa Theologica* I,46,2 as follows: "By faith alone do we hold, and by no demonstration can it be proved, that the world did not always exist, as was said above of the mystery of the Trinity. The reason of this is that the newness of the world (*novitas mundi*) cannot be demonstrated on the part of the world itself. For the principle of demonstration is the essence (*quod quid est*) of a thing. Now everything according to its species (*secundum rationem suae speciei*) is abstracted from *here* and *now;* whence it is said that universals are everywhere and always." [31] Although we might pass this over as a self-evident statement, perhaps we should think carefully about it. For Aristotle this universe did not begin; moreover, this is demonstrably clear for Aristotle. Thomas tells us that Aristotle's reasons are only relatively demonstrative (*ST* I,46,1) [32] since his arguments intend to contradict only particularly impossible *modes* of this world's coming into being advanced by Empedocles, Anaxagoras, and others.

---

[30] *Summa Theologica,* op. cit., 460.
[31] Ibid., 243.
[32] Ibid., 241.

But Aristotle, as a matter of fact, is not engaged in a dialectical dispute for purposes of clarification; on the contrary, as he says in *Physics* VIII.1: ". . . we must look into this question of whether there was a beginning of motion; for it is worthwhile to get at the truth [ἰδεῖν τὴν ἀλήθειαν] not only with a view to our speculations as to nature, but also for its bearing on our study of the first principle." [33] The impossibility of this universe's beginning is for Aristotle a necessary truth of not only physics but metaphysics. The never beginning never ending movement of this world is, for Aristotle, unmoving God's eternal effect. It is clear here how far away in reality Thomas' *natural* reason is from Aristotle's *metaphysical* reason. For Thomas that the world began is not to be known because the principle of scientific demonstration is a universal abstracted from time and space. That is, the eternity of this universe is a necessary form of reason, indeed merely that. No species, for Thomas, by virtue of its being reason's abstraction, could be anything but eternal *for reason.* Sacred doctrine is, among other things, a critique of reason. In face of revelation *natural* reason becomes conscious of itself as operating by a process of abstraction and having for its proper objects the universal essences of things. Aristotle, for whom essence was not synonymous with form, knew, in knowing things essentially, not abstract universals but particulars. Aristotle rejected a beginning for this universe not because of the limitations of the abstract formality of reason, but because it contradicts the definitive structure of the world of his experience to the reality of which his intellect was able to penetrate immediately. Not only is the operation of intellect changed in Thomas from what it was in Aristotle, but yet another facet of this change is that intellect's light is explicitly infused by God himself as Thomas says in his treatise *On Truth* 11: "Now the light of reason by which [self-evident] principles are evident to us is implanted in us by God as a kind of reflected likeness in us of the uncreated truth. So, since all human teaching can be effective only in virtue of that light, it is obvious that God alone

[33] Aristotle, *Physics* II (trans. P. H. Wicksteed and F. M. Cornford, Cambridge, Mass., 1934), 271.

teaches interiorly and principally, just as nature alone heals interiorly and principally." [34] Intellect is here not simply identity in existence, but similitude to uncreated existence itself: What was a self-reflecting mirror is now a transparency to otherness. Through likeness I see myself in God, God in myself.

Now, analogy is not possible between two absolutes considered absolutely, nor between a nonabsolute and an absolute so considered. What analogy requires is two nonabsolutes or comparable structure: For example, *x* (structured *a:b*) is analogous to *y* (structured *c:d*). Aristotelian metaphysics strictly does not analogize nature to God because God is absolutely simple. Metaphysics knows itself to be structured process, but that God is not. It sees however an analogy within itself (intellect:reason) to ultimate causality (divine intellect:universe). Likewise Neoplatonism's Absolute One beyond Intelligence corresponds to that system's perception that divine intellect in knowing itself lacked unity. But the Absolute, whether it be Aristotle's Divine Intellect or Plotinus' One, is incomparable. But for sacred doctrine this universe began to be by God's creative act; that is, it sees this world *in* God's Absolute Existence. The *novitas mundi,* this universe's novelty, is indemonstrable to metaphysics actually because metaphysics does not see except for that condition known as a *blind spot.* That is, metaphysical eyes take root in a synthetically structured reality: Metaphysics knows nothing of *simple* being which in sacred doctrine is God's formative causality of the totality of substances. To see participated existence as the actuality of nature is to see creation; to see by an absolute receptivity to God is to see in absolute simplicity the Trinity, to see creation through this medium of simple being in its likeness to God. This sight requires not two, but an infinity of eyes. This is to say that sacred doctrine is not only in principle a transcendence to metaphysical causality, for example, in its doctrine of *novitas mundi;* it is *natural* reason's *transcendental form.* In light of God's revelation

[34] Thomas Aquinas, *The Teacher; The Mind* (trans. V. McGlynn, Chicago, 1962), 18.

reason's blind spot is healed over. Instead of darkness, in God's light it sees its structure to be sheer transparency; it actually sees through itself. It sees itself potentiated to God's creative act, its new existence. It sees in its natural existence God's intention to be together with nature its freedom. But nature's new existence, ultimately free from the necessity of essence, is not to be understood as a reduction unqualified to natural potentiality, although it so might appear to metaphysical eyes. It is not left in a state of radical unintelligibility, but, to the contrary, Christianity exalts nature through creation to a most perfect intelligibility, perfect, that is, in a new order of faith. This order is constituted not by exclusive metaphysical distinctions but by infinite differences known to God, Who is *a new reason for things being what they are.* As Thomas writes in *Summa Theologica* I,47,1: ". . . we must say that the distinction and multitude of things come from the intention of the first agent, who is God. For He brought things into being in order that His goodness might be communicated to creatures, and be represented by them; and because His goodness could not be adequately represented by one creature alone, He produced many and diverse creatures, that what was wanting to one in the representation of the divine goodness might be supplied by another. For goodness, which in God is simple and uniform, in creatures is manifold and divided; and hence the whole universe together participates the divine goodness more perfectly, and represents it better than any single creature whatever. And because the divine wisdom is the cause of the distinction of things, therefore Moses said that things are made distinct by the word of God, which is the concept of His wisdom; and this is what we read in Genesis (i.3,4): *God said: Be light made. . . . And He divided the light from the darkness.*" [35] That distinction that resides in forms is attributed to creatures not because they are *per se* actual, that is, not because *to know is to be,* but because of their *relation* to God's infinite goodness, or, because *to be known is to be.* It is participated existence that is Thomas' principle of intelligibility. God is Infinite Reason, not metaphysics' God, not pure intellect, in itself,

[35] *Summa Theologica,* op. cit., 246.

unrelated, unrepresentable essence. Rather, God is sacred doctrine's God, knowing himself, but in that act knowing things *other* than himself, contained, by similitude, in His essence or absolute species. Thomas says in *Summa Theologica* I,14,5 ad 2: "The object understood is a perfection of the one understanding not by its substance, but by its image, according to which it is in the intellect, as its form and perfection, as is said in *De Anima* iii. For *a stone is not in the soul, but its image.* Now those things which are other than God are understood by God, inasmuch as the essence of God contains their images as above explained; hence it does not follow that there is any perfection in the divine intellect other than the divine essence." [36] Now, if you know what's what in metaphysics, you recall that *what* is what intellect knows itself to be identically when it knows what it is to be something. But God, as sacred doctrine's Infinite Reason, *knows* things *to be without* identifying himself with *what* it is to be them. That is, *God knows to be without what it is to be.* God knows things other than himself through that absolute being that is his own creative essence. If difference in existence is the obverse of the principle of identity, and, as such, is the principle of intelligibility, then, sacred doctrine, seeing that God knows these differences immediately in His infinite reason, sees that the principle of identity is obviated. The principle of intelligibility for reason is directly otherness. What reason knows is not what it is to be, but *that it is another's to be.* Likeness, which is an obstacle to metaphysical intellect, is to be the very medium of natural reason.

In infinite reason sacred doctrine sees a simultaneous conception of all things together, in which conception effects are perceived immediately in their cause; that cause is God's will. This conception without time or space is an absolute species or reason: created nature is its trace, man its image. Thomas writes in *Summa Theologica* I,93,1: ". . . it is manifest that in man there is some likeness to God, copied from God as from an exemplar; yet this likeness is not one of equality, for such an exemplar infinitely excels its copy. Therefore there is in man a

[36] Ibid., 76.

likeness to God; not, indeed, a perfect likeness, but imperfect."[37] To get an insight into how absolutely unhypothetical is the circle of grace: Thomas is able to argue to God's nature by *starting* with human nature, for example, in *Summa Contra Gentiles* II.6.7 as follows: ". . . the more perfect is the principle of a thing's action, to so many more and more remote things can it extend its action. . . . But pure act, which God is, is more perfect than act mingled with potentiality, as it is in us. But act is the principle of action. Since, then, by the act which is in us we can proceed not only to actions abiding in us, such as understanding and willing, but also to actions which terminate in things outside of us, and through which certain things are made by us, much more can God, because He is in act, not only understand and will, but also produce an effect. And thus He can be the cause of being to other things."[38] Now it might be objected that sacred doctrine after all *presupposes* what it then concludes to by analogy, namely, God's creative activity. But it is impossible to presuppose creation. *Metaphysical* reason presupposes essential being's absolute priority; that is, it presupposes its own necessity. But if *natural* reason begins with its own contingency, it presupposes nothing. To begin science with an article of faith is a radical alteration of science to being presuppositionless. But a presuppositionless reason is in its simplicity a perfect instrument for knowing something other than itself. No such concept of human mind exists for Aristotle except insofar as he understands reason, *qua* reason, or passive mind assimilated to soul to be instrumental. In *De Anima* III.8 Aristotle says: "Now summing up what we have said about the soul, let us assert once more that in a sense the soul is all existing things. . . . The sensitive and cognitive faculties of the soul are potentially these objects, *viz.,* the sensible and the knowable. These faculties, then, must be identical either with the objects themselves or with their forms. Now they are not identical with the objects; for the stone does not exist in the soul, but only the

[37] Ibid., 469.

[38] *Summa Contra Gentiles* II (trans. J. E. Anderson, Garden City, 1956), 37–38.

form of the stone. The soul then acts like a hand; for the hand is an instrument which employs instruments, and in the same way the mind is a form which employs forms, and sense is a form which employs the forms of sensible objects. But since apparently (ὡς δοκεῖ) nothing has a separate existence, except sensible magnitudes, the objects of thought—both the so-called abstractions of mathematics and all states and affections of sensible things—reside in the sensible forms. And for this reason as no one could ever learn or understand anything without the exercise of perception, so even when we think speculatively, we must have some mental picture of which to think; for mental images are similar to objects perceived except that they are without matter."[39] Remembering that Aristotle's understanding of the consummate act of knowledge is identity with its object, we clearly perceive that soul for Aristotle is, as he says, a hand, or better, a right hand, scientific reason, a left hand, sensation, in turn employing images. It is a hand that holds an object formally before intellect so that that essential insight of intellect that knowledge actually is ensues. It is in this act of intellect that those sensible magnitudes, which as Aristotle says *seem* to exist separately, disappear, exhausted in an object's essential identity with what it is to be. Now Thomas often quotes Aristotle to the effect that the 'stone does not exist in the soul.' But this idea that 'only the form of the stone' exists in the soul depending upon whether it is said of a metaphysical reason (Aristotle) or a natural reason (Thomas) means different things. For Thomas' presuppositionless reason, a perfect instrument for knowing things other than itself, a material object is known apart from the individuating principle of its matter. That is, it knows by abstraction of a species the existence of another as an object for itself. But, for Aristotle's metaphysical reason, presupposing essential identity, to possess an object's form apart from its matter, as if the latter were the principle of individuality, is not yet to know actually, or, identically, what it is. Reason so far and no further is for Aristotle merely potential knowledge, merely instrumental. The objects of such knowledge

[39] *On The Soul,* op. cit., 179, 181.

would exist only potentially. For Thomas, on the contrary, such an object is known to exist actually; that is, a materially individuated object *exists* as such. It is this actual *existence per se* that is known through the species. It is not the particular, *qua* particular, that is known but *its existence as a universal.* For Thomas the actually existing stone is actually intelligible; that is, external stone is actually known through its form in the soul. But for Aristotle the actually existing stone is potentially intelligible; that is, it is actually intelligible to an intellect which prescinds from its externality, its natural medium. Perhaps our perception of this crucial but subtle distinction will be sharpened by recalling that Thomas carefully distinguishes between *mode* of knowing and *mode* of existing, so that an individual may exist in one way but be known to exist in another way. In Aristotle such a flexibility of perspective is ultimately to give way to reason's presupposition of the *essential* identity of knower and known. The natural reason of man created to the image of God, Whose Infinite Reason is, in turn, ground of the *novitas mundi,* is in its being without presupposition possessed of a freedom of movement unknown to Aristotle.

But this new freedom of natural reason toward its object bestirs a question as to whether or not its knowledge might not be simply its own invention, which takes the form in *Summa Theologica* I,85,2 of the question 'Whether the Intelligible Species Abstracted from the Phantasm Is Related to Our Intellect As That Which Is Understood?' To this Thomas responds: "According to this theory, the intellect understands only its own impression. This is, however, manifestly false for two reasons. First, because the things we understand are the objects of science; therefore if what we understand is merely the intelligible species in the soul, it would follow that every science would not be concerned with objects outside the soul, but only with the intelligible species within the soul. . . . Secondly, it is untrue, because it would lead to the opinion of the ancients who maintained that *whatever seems, is true*, and that consequently contradictories are true simultaneously." [40] Now this question arises because sacred doctrine in its disjunctive conjunction of

[40] *Summa Theologica*, op. cit., 433.

grace with nature transforms reason into a transcendental form. That is, it simplifies metaphysical reason by de-essentializing intellect, thereby permitting reason, on the one hand, sensation, on the other, to specialize in knowing: reason, the universal; sensation, the individual. This is natural reason; but Thomas here proceeding negatively rules out any possibility that reason knows only its own impression by saying that such a theory would violate, in effect, in the first place, the law of identity, that is, different sciences would be one science, in the second place, the law of contradiction, that is, each man would in Protagoras' language 'be the measure of all things.' Confronted with an interpretation of reason that is not at all unnatural, by which human nature would simply take its head in its hands, Thomas applies, after the fact of creation, the 'good sense' of identity and contradiction. This turning of the keys of metaphysics, in advance, against that pure intentionality of modern reason is sacred doctrine's agreement with metaphysics that for man there is a limit: for Thomas that limit is not essence, but it is grace or God's revelation. While outside of grace to be in God's image would mean that man is God, that theology is actually anthropology, within grace, in sacred doctrine explicitly, being in the image of God is an intention quite contrary to man's thinking his own thoughts. In *Summa Theologica* I,93,4 Thomas says: "Since man is said to be to the image of God by reason of his intellectual nature, he is the most perfectly like God according to that in which he can best imitate God in his intellectual nature. Now the intellectual nature imitates God chiefly in this, that God understands and loves Himself. Wherefore we see that the image of God is in man in three ways. First, inasmuch as man possesses a natural aptitude for understanding and loving God; and this aptitude consists in the very nature of the mind, which is common to all men. Secondly, inasmuch as man actually or habitually knows and loves God, though imperfectly; and this image consists in the conformity of grace. Thirdly, inasmuch as man knows and loves God perfectly; and this image consists in the likeness of glory."[41] Beginning, middle, and end likeness to God is thinking first of

[41] Ibid., 471.

God. God is the limiting identity of natural reason, but this identity, unlike that of metaphysics, is absolutely received; therefore by virtue of its innermost being it is related to another beyond itself, both theologically (God) and epistemologically (world). This natural reason waits, hopes, serves.

To return to the epistemological question, Thomas' positive response is as follows (*ST* I,85,2): ". . . the intelligible species is related to the intellect as that by which it understands . . . . as the form from which proceeds an act tending to something external is the likeness of the object of the action . . . . so the form from which proceeds an action remaining in the agent is the likeness of the object. Hence that by which the sight sees is the likeness of the visible thing; and the likeness of the thing understood, that is, the intelligible species, is the form by which the intellect understands. But since the intellect reflects upon itself, by such reflection it understands both its own act of intelligence, and the species by which it understands. Thus the intelligible species is that which is understood secondarily; but that which is primarily understood is the object, of which the species is the likeness."[42] In reflection reason knows itself; in knowing itself as distinctly not its intelligible species, it knows the latter as its instrument by which it knows universals outside of itself, or else no distinction would exist between it and its species; also by reflection, in a less strict sense, it knows singulars by its phantasm.[43] That its knowledge terminates in an actually existing object other than itself is self-evident to natural reason. This reason is presuppositionless; or else God goes first. But this points to a critical decision.

[42] Ibid., 434.
[43] Ibid. (I,86,1), 440–441.

Chapter 3

DESCARTES: A NEW THOUGHT

There is power in simplicity. God's power is simply His existence, its simplicity. By bringing into being this universe's totality God manifests His absolute existence. It is that simple. Another's existence reveals itself in transparent simplicity. Simplicity's is the Sabbath, its message that what is necessary is provided. The Law's perfection is simplicity itself. In *Matthew* 12:1–8 we are told: "At that time Jesus took a walk one sabbath day through the cornfields. His disciples were hungry and began to pick ears of corn and eat them. The Pharisees noticed it and said to him, 'Look, your disciples are doing something that is forbidden on the sabbath.' But he said to them, 'Have you not read what David did when he and his followers were hungry—how he went into the house of God and how they ate the loaves of offering which neither he nor his followers were allowed to eat, but which were for the priests alone? Or again, have you not read in the Law that on the sabbath day the Temple priests break the sabbath without being blamed for it? Now here, I tell you, is something greater than the Temple. And if you had understood the meaning of the words: *What I want is mercy, not sacrifice,* you would not have condemned the blameless. For the Son of Man is master of the sabbath.' "[44] The mercy of God's simplicity pierces through *what things are* to their very being, that is, to their being without qualification provisions for another's necessity; that necessity itself forgiven

[44] *The Jerusalem Bible: N.T.,* op. cit., 33.

by Divine Providence. If, in light of revelation, man loses essential self-sufficiency, if he is reduced to a naked humanity, to a state of natural reason, he, nevertheless, is in that same light God's creature, so that, as Jesus says at *Matthew* 7:7–11: "Ask, and it will be given to you; search, and you will find; knock, and the door will be opened to you. For the one who asks always receives; the one who searches always finds; the one who knocks will always have the door opened to him. Is there a man among you who would hand his son a stone when he asked for bread? Or would hand him a snake when he asked for a fish? If you, then, who are evil, know how to give your children what is good, how much more will your Father in heaven give good things to those who ask him!" [45] In his *Summa Theologica* Thomas Aquinas considers both sacrifice as well as prayer as acts of religion, under the category of moral virtue, specifically annexed to the cardinal virtue of Justice. [46] This placing makes clear that religious acts, *qua* religious, belong to the order of nature, *qua* nature. For sacred doctrine, nature, otherwise considered metaphysically or in itself finally imperfect, is perfected by grace. That is, nature is substantially altered to being created from being what it is. Or, more properly, there is believed to be a *novitas mundi* or novelty of this universe, specifically, that it came into being after not being absolutely, therefore, by God's omnipotent will. As a result, it must appear subsequently to metaphysical eyes necessary to suppose what it is not possible to suppose, namely, that there is an alteration of everything's being, in itself, unrelated to God (or, related indirectly through natural order) to everything's being, in itself, related directly to God as to another, without, nevertheless, simply prescinding from natural order. In the light of faith, *metaphysical* reason is made over into *natural* reason. Man's religious acts, then, likewise undergo a new formation. The new form of sacrifice is mercy; prayer's new form is charity or love of God. These forms of that simplicity by which God is related to His creation

[45] Ibid., 25.

[46] Cf. *Summa Theologica*, op. cit., II—II,80,81,83,85.

are discussed by Thomas in *Summa Theologica* II-II,30,4: "In itself, mercy takes precedence of other virtues, for it belongs to mercy to be bountiful to others, and what is more, to succor others in their wants, which pertains chiefly to one who stands above. Hence mercy is accounted as being proper to God: and therein His omnipotence is declared to be chiefly manifested. On the other hand, with regard to its subject, mercy is not the greatest virtue, unless that subject be greater than all others, surpassed by none and excelling all: since for him that has anyone above him it is better to be united to that which is above than to supply the defect of that which is beneath. Hence, as regards man, who has God above him, charity which unites him to God, is greater than mercy, whereby he supplies the defects of his neighbor. But of all the virtues which relate to our neighbor, mercy is the greatest, even as its act surpasses all others, since it belongs to one who is higher and better to supply the defect of another, in so far as the latter is deficient." [47] If mercy is sacrifice's true form, then, applying Thomas' distinctions, only that sacrifice is pleasing to God that is properly undertaken, that is, by Himself as subject. Not God, but man, is the proper object of man's sacrifices. That is, man is to show mercy not to God, but to his neighbor. But man's mercy toward his neighbor is neither his neighbor's greatest possible good, since that is God's own self-sacrificing mercy immediately present in his neighbor's very being, nor is it man's greatest good to be merciful to his fellowman, for he himself is directly related to God in that circle of mercy received, charity returned, in which God is loved as absolutely provident Father. Mercy is to sacrifice as form to matter; so too, charity is to prayer; but charity is to mercy, love of God is to love of neighbor, as participated existence is to essence. Therefore, in grace, humanity essentially is mercy or love of neighbor, but this mercy or humanity is actual only in prayerful communion with God; but again, since nature is not destroyed, since existence naturally follows form, charity is *simultaneously* love of neighbor.

[47] Ibid.,1320.

When Christianity came to grips, intellectually, with God's simplicity, human reason, like Jacob wrestling with Yahweh, [48] sought a blessing; but like Jacob, at daybreak, it was not reason that identified God, rather reason itself suffered a new identity: " 'What is your name?' 'Jacob', he replied. He said, 'Your name shall no longer be Jacob, but Israel, because you have been strong against God, you shall prevail against men.' " The enterprise of sacred doctrine freed reason from presupposing its own necessity, that is, from essential being's priority. Its encounter with God left it radically conscious of its own contingency, but this loss was reason's gain. It came to be a perfect instrument for knowing an object other than itself, *qua* other, as actually existing together with it in that totality of intelligibility that is God's creation. Its only presupposition was divine revelation, what, metaphysically, it is not possible to presuppose.[49]

With this universe of grace in mind we turn our attention to Descartes who writes as follows in *Discourse On The Method* I: "I honoured our Theology and aspired as much as anyone to reach to heaven, but having learned to regard it as a most highly assured fact that the road is not less open to the most ignorant than to the most learned, and that the revealed truths which conduct thither are quite above our intelligence, I should not have dared to submit them to the feebleness of my reasonings; and I thought that, in order to undertake to examine them and succeed in so doing, it was necessary to have some extraordinary assistance from above and to be more than a mere man. I shall not say anything about Philosophy, but that, seeing that it has been cultivated for many centuries by the best minds that have ever lived, and that nevertheless no single thing is to be found in it which is not subject of dispute, and in consequence which is not dubious, I had not enough presumption to hope to fare better there than other men had done." [50] Descartes is no Jacob. He considers theological learning, in the

[48] *The Jerusalem Bible: O. T.*, op cit. (*Genesis* 32:25–30), 54.

[49] Cf. above, pp. 68–72.

[50] *The Philosophical Works of Descartes*, I (trans. E. S. Haldane and G. R. T. Ross, Cambridge, 1970), 85–86.

first place, not to be necessary to salvation. In the second place, he thinks that revelation is not able to be reasoned about, except through a special grace, by a man, *qua* man. Descartes, quite content to be no more than a man, is, nevertheless, not able to devote himself to philosophy, or to any other of those sciences that derive their principles from it, since nothing in it is beyond doubt. Descartes says that after several years "I one day formed the resolution of also making myself an object of study and of employing all the strength of my mind in choosing the road I should follow." [51] While Descartes is no Jacob he will come to grips with himself; this man's introspective resolution will radically alter science's foundation, no, he will create a new science. But it may fairly be inquired how this is possible, especially to one who is explicitly nothing other or more than a man, that is, in his intention, merely human, in his resolution, merely one. For example, Aristotle, who was other than merely human, who was essentially divine, nevertheless thought that scientific reason, that dealt with objects, developed in continuity with previous thought. So Aristotle, who otherwise certainly appreciated his own originality in science, entertained no thought as to his science's absolute novelty, such as is implicit in Descartes' resolution. Aristotle's continuity with his predecessors is one of subject matter. That is, reality or an object potentially knowable exists actually in itself prior to being known. Aristotle's discontinuity with what went before is not material but essentially a matter of form. This is to say that disagreement within reason itself is understood *not* to affect an object's substantial reality, on Aristotle's basic assumption that an actually existing stone, composed of form and matter, is potentially intelligible, so that an object's form in reason is that reality's potential, not actual, existence. Actual knowledge of an actual object is that consummate act of intellect, beyond reason, in which knowledge is identical with its object, that is, in which the distinction between knower and object is lost sight of, or, in which otherness, *qua* otherness, vanishes totally. To put it summarily, for Aristotle, actuality is *finally* incompatible with other-

[51] Ibid., 87.

ness. Or, to distinguish Aristotle in advance from Hegel, otherness exists insofar as actuality is not simply present. Demonstration is for Aristotle the method of scientific knowledge because the actuality of an object is not self-evident to reason. But Descartes' discontinuity with other thinkers is not merely formal but material. It presupposes no prior reality or object actually existing outside of being known. He, Descartes, is not only subject but object of his meditations. *Cogito ergo sum,* I think therefore I am, is Descartes' perfect isolation in the thought of his own existence. This is Descartes' rock; if there is to be a rock actually existing outside of Descartes' mind, its existence will first necessarily be grounded in that self-evident existence of Descartes' rational intuition. It is not therefore Descartes' intention in doubting to call into question without qualification the existence of other things as objects of knowledge, as he says in the *Discourse On The Method* III: "Not that indeed I imitated the sceptics, who only doubt for the sake of doubting, and pretend to be always uncertain; for, on the contrary, my design was only to provide myself with good ground for assurance, and to reject the quicksand and mud in order to find the rock or clay." [52] Descartes' resolution to doubt is then his design for assurance; implicit in his doubt is his assurance that his doubt will reach to that rock, *cogito ergo sum;* his scepticism is circumscribed by this faith, so that he tells us again in the *Discourse On The Method* IV: ". . . immediately afterwards I noticed that whilst I thus wished to think all things false, it was absolutely essential that the 'I' who thought this should be somewhat, and remarking that this truth '*I think, therefore I am*' was so certain and so assured that all the most extravagant suppositions brought forward by the sceptics were incapable of shaking it, I came to the conclusion that I should receive it without scruple as the first principle of the Philosophy for which I was seeking." [53] *Metaphysical* reason knew on the basis of experience its necessity to be in intellect's essential priority. *Natural* reason knew in light of revelation its contingency, but its contingency meant an as-

[52] Ibid., 99.
[53] Ibid., 101.

similation of what was intellect's former essential priority to reason itself, so that, *in the elaboration of sacred doctrine, reason came to know not itself on the basis of experience, but rather to know its experience on the basis of a reflection upon itself.* Reason came from God's hand its own agent; a perfectly transparent self in God's light it saw its object known to it as an actually existing other. Now it is this reason acting as its own agent that Descartes would employ outside that realm of explicit grace that created it. But whereas this presuppositionless natural reason had in sacred doctrine at least a formal presupposition in the articles of faith, it is in Descartes' hands simply without presuppositions whatsoever. If God is not presupposed, then neither is created nature presupposed. Only the phenomenon of its existence self-evidently presents itself to reason. Descartes accepts sacred doctrine's judgment that natural reason errs, but cannot doubt that since reason errs it exists. It exists as a self-employing agent. Descartes describes it in the *Discourse On The Method* V as a "universal instrument which can serve for all contingencies." [54] But it is to be noted that the object which was *implicitly* contingent when reason knew it as an actually existing other in sacred doctrine is in Descartes' new science *explicitly* contingent. This points to the fact that Descartes' doubt which is intended to be merely provisional, to be resolved in the *cogito ergo sum*, is, nevertheless, in natural reason operating in its own light permanently present.

This perdurable doubt, far-reaching in its consequences in modern thought, is to be further examined in Descartes. But first let us look at another aspect of his thought as he enunciates it in *The Search After Truth By The Light Of Nature:* "In this work I propose to show what these means are [by which a man may carry his knowledge to the highest point it can possibly attain], and to bring to light the true riches of our souls, by opening to each one the road by which he can find in himself, and without borrowing from any, the whole knowledge which is essential to him in the direction of his life, and then by his study succeed in acquiring the most curious forms of knowledge that the human

[54] Ibid., 116.

reason is capable of possessing." [55] Descartes' procedure is that of *a self employed in the service of;* this is reason's form as it was shaped by sacred doctrine's employment of reason in the clarification of revelation. But here reason is in the service of man; Descartes' thought is an ideal thought or model useful to others which every man is understood to be capable of appropriating for his own personal enrichment. Descartes thinks of his intellectual activity as a focus of a universal human potential. His method is open to every man, *qua* man. It is curiously like salvation, offered to all without distinction. It is not a matter of learned definitions or the like, it presupposes only a *sanus sensus*, a sound mind. The new principle of the new science is *humanity*. Listen to Descartes in *The Search After Truth:* "What, Epistemon, do you think of what Polyander has just said? [He has just by doubt reached the *cogito ergo sum.*] Do you find his argument to be halting or inconsequent? Should you have thought that an unlettered man, and one who had not studied, would have reasoned so well and followed out his ideas so rigorously? Here, if I do not mistake, you must begin to see that he who knows how properly to avail himself of doubt can deduce from it absolutely certain knowledge, better, more certain, and more useful than that derived from this great principle which we usually establish as the basis or centre to which all other principles are referred and from which they start forth, viz. *it is impossible that one and the same thing should both be and not be.*" [56] The principle of humanity takes up within its own self-directed instrumentality, into its immediate perception of its simply being a thinking thing, the law of identity; it superalternates doubt to identity; by doubt it establishes identity for itself. The principle of humanity is absolute instrumentality, so that doubt itself springing as it does from reason's contingency is become reason's very own instrument for establishing its existence. Sacred doctrine subordinated metaphysics to its clarification of God's revelation; it substituted God's existence for

[55] Ibid., 305.
[56] Ibid., 322.

metaphysical identity. Descartes' new science, shaped by reason's service to revelation, nevertheless substitutes humanity's existence for identity. Thereby, in a wholly internal way it reshapes science, not implicitly as in sacred doctrine, but explicitly, resolutely science is taken up into humanity's interiority, into human freedom. Interest centers on this concrete, factual freedom that reason is become in time, in space, but, as it appears to itself, free from history. Only this freedom is, for Descartes, humanity's palpable stuff; this is his new subject matter, his new science's radical discontinuity with everything past: *Humanity*. It is a universal object, equally each man's property. Descartes uses school language to make this last point clear in the *Discourse On The Method* I: "For myself I have never ventured to presume that my mind was in any way more perfect than that of the ordinary man; I have even longed to possess thought as quick, or an imagination as accurate and distinct, or a memory as comprehensive or ready as some others. And besides these I do not know any other qualities that make for the perfection of the human mind. For as to reason or sense, inasmuch as it is the only thing that constitutes us men and distinguishes us from the brutes, I would fain believe that it is to be found complete in each individual, and in this I follow the common opinion of the philosophers, who say that the question of more or less occurs only in the sphere of the *accidents* and does not affect the *forms* or natures of the *individuals* in the same *species*." [57] Every man is, through the universality of the species, humanity; that is, essentially himself the form or nature of man, freedom or self-employing reason. So each man insofar as he transcends his individuality can by doubting come to identify himself as actually existing in the freedom of a universal reason.

Descartes writes in the *Discourse On The Method* II as follows: "Among the different branches of Philosophy, I had in my younger days to a certain extent studied Logic. . . . I observed in respect to Logic that the syllogisms and the greater part of the other teaching served better in explaining to others

[57] Ibid., 82.

those things that one knows . . . than in learning what is new."[58] Descartes sees that syllogism relates one to others. It must be of little or no use to a self-relating reason, just as the principle of identity is less useful to reason than its own originality. That reason which employs itself in learning what is new is absolutely formal; humanity immediately conceiving its own existence. This free reason is self-relating, representative thought; it excludes, as a matter of its simple priority, others. Descartes' *cogito ergo sum* as his new science's rock is rational intuition without identity or contradiction; it is a *universal medium*. For Aristotle demonstrative syllogism was science's method because reason's object was not self-evidently actual; that is, reason knew its object as other in actuality. So reason abided by the law of identity or contradiction; it acknowledged in its subordination to intellect an other's priority in existence. Aristotle's knower, not being a universal medium, is this individual man, *qua* individual, not an ideal or species man (paradoxically, therefore, not a mere man in Descartes' sense), but a man synthesized of body and soul, a two-legged intellect. Aristotle handles the question of learning what is new by distinguishing new knowledge from old in this way in the *Posterior Analytics* I.i: ". . . I presume there is no reason why a man should not in one sense know, and in another not know, that which he is learning. The absurdity consists not in his knowing in some qualified sense that which he learns, but in his knowing it in a certain particular sense, *viz.*, in the exact way and manner in which he learns it."[59] For Aristotle an individual knower does not know except his knowledge is first potential; that is, knowledge is not unqualifiedly new, but new in particular. There is no universal science for discovering what is universally new. Such a science is inconceivable to Aristotle: *A universal science would require a universal middle term*. In Aristotle's scientific universe that middle term through which knowledge is

[58] Ibid., 91.

[59] *Aristotle: Posterior Analytics* (trans. H. Tredennick, Cambridge, Mass., 1960), 29.

acquired is other than reason; intellect, as reason's essential priority, lays it down as Aristotle says in *Posterior Analytics* I.ii "that the cause from which the fact results is the cause of that fact." [60] It is, not the indifferent, universal middle 'reason itself.' But for Descartes, who appropriated from sacred doctrine a natural reason essentially altered from previous metaphysical identity, reason is that 'universal instrument which can serve for all contingencies,' that is, for knowing not *what is,* but *what is new*. But what is new in sacred doctrine is ultimately this entire universe. What is universally new, absolutely without prior potentiality, what began to be after simply not being, what comes perfectly fresh from God's omnipotent hand is this universe. But *novitas mundi,* this universe's novelty, creation *per se,* is not properly, that is, metaphysically, *caused*.[61] We see that it is faith's form that Descartes' universal middle, free reason, patterns itself upon implicitly. Natural reason is implicitly itself a novelty; it is uncaused, or it is its own cause. That is, it takes itself to be a cause for its own purposes; it is itself its own solution to *what is new.*

This new science of what is new begins at that moment when it is resolved in Descartes' person to make itself representatively an object of its own study. Universal science is free humanity; but free humanity exists, *qua* principle, as a representative object. Aristotle's concretely individualized scientific subjectivity is no more, knowing particulars, *qua* particulars; Descartes' new science is a purely formal subjectivity, a universal object. In his *Rules For The Direction Of The Mind* Descartes writes, Rule IV: "*There is need of a method for finding out the truth.*" [62] Rule V: "*Method consists entirely in the order and disposition of the objects towards which our mental vision must be directed if we would find out any truth. We shall comply with it exactly if we reduce involved and obscure propositions step by step to those that are simpler, and then starting with the intuitive apprehension of all those that are absolutely*

[60] Ibid.
[61] Cf. above, pp. 63–64.
[62] *The Philosophical Works of Descartes,* op. cit., 9.

simple, attempt to ascend to the knowledge of all others by precisely similar steps."[63] Rule VI: "*In order to separate out what is quite simple from what is complex and to arrange these matters methodically, we ought, in the case of every series in which we have deduced certain facts the one from the other, to notice which fact is simple, and to mark the interval, greater, less, or equal, which separates all the others from this.*"[64] Of Rule VI Descartes says: "Although this proposition seems to teach nothing very new, it contains, nevertheless, the chief secret of method, and none in the whole of this treatise is of greater utility. For it tells us that all facts can be arranged in certain series, not indeed in the sense of being referred to some ontological genus such as the categories employed by Philosophers in their classification, but in so far as certain truths can be known from others; and thus, whenever a difficulty occurs we are able at once to perceive whether it will be profitable to examine certain others first, and which, and in what order."[65] Descartes tells us that reason's vision sees its object, *qua* object, only in order, that is, only in relationship to other objects. The principle of order is not metaphysical causality, not natural necessity, for this method is the self-employment of humanity, free reason, transcendent to natural causality, itself a novelty, a self-relating causality: *cogito ergo sum, I think, therefore I am.* The principle of humanity is a self-reflecting absolute; it is reason's simple self-respect. Its principle is *likeness*. (In its simple immediacy free reason is only implicitly aware of its history, namely, that it came into being in sacred doctrine through God's creative will which made it to exist through communicating to it a *likeness* of His own existence; were it to be aware of this history of its being, *qua* history, it would cease to be what it is for itself, namely, a *new* science's *new* principle.) Reason creates its objects' order in its own image: Reason, Descartes tells us, searches among its objects for the *simplest*, the *absolute;* to this, when found, it relates each other

[63] Ibid., 14.
[64] Ibid., 15.
[65] Ibid.

object—to this absolute. Everything is what it is with respect to this absolute; but this absolute is implicitly the order, so that from it can be deduced by careful observation of relationships a remote, or complex, object. But Descartes' rational intuition itself is that original simplicity, that absolute rock, *cogito ergo sum,* upon which is to be erected, since a building is implicit in its foundation, modern science's complex structure, or this universe's natural order. Descartes distinguishes solitude from simplicity; solitude is, we may understand, that incommunicable essence of metaphysics; but simplicity, like Thomas Aquinas' God, is of its very being communicative. It is an analogical order's object. An incommunicable essence is nothing for Descartes. In his comment following Rule XIV of the *Rules For The Direction Of The Mind* Descartes writes: "The figure of a silver crown which we imagine, is just the same as that of one that is golden. Further this common idea is transferred from one subject to another, merely by means of the simple comparison by which we affirm that the object sought for is in this or that respect like, identical with, or equal to a particular datum. Consequently in every train of reasoning it is by comparison merely that we attain to a precise knowledge of the truth. Here is an example:—all *A* is *B*, all *B* is *C*, therefore all *A* is *C*. Here we compare with one another a *quaesitum* and a *datum,* viz. *A* and *C,* in respect of the fact that each is B, and so on. But because, as we have often announced, the syllogistic forms are of no aid in perceiving the truth about objects, it will be for the reader's profit to reject them altogether and to conceive that all knowledge whatsoever, other than that which consists in the simple and naked intuition of single independent objects, is a matter of the comparison of two things or more, with each other. In fact practically the whole of the task set the human reason consists in preparing for this operation; for when it is open and simple, we need no aid from art, but are bound to rely upon the light of nature alone, in beholding the truth which comparison gives us." [66] Just as in sacred doctrine

[66] Ibid., 55.

Thomas tells us "that the distinction and multitude of things come from the intention of the first agent, who is God,"[67] so, likewise, this free intentionality of scientific reason understands that its vision of its objects by which it relates them one to another in light of its own understanding constitutes, together with simple intuition, their entire intelligibility. That is to say, objects become knowable by being included in human reason's order, so, in Rule XVII of the *Rules For The Direction Of The Mind* Descartes says: "When a problem is proposed for discussion we should run it over, taking a direct course, and for this reason neglecting the fact that some of its terms are known, others unknown."[68] Or, even more radically, Descartes states in his third law in the *Discourse On The Method* II ". . . to carry on my reflections in due order, commencing with objects that were the most simple and easy to understand, in order to rise little by little, or by degrees, to knowledge of the most complex, assuming an order, even if a fictitious one, among those which do not follow a natural sequence relatively to one another."[69] Such is this reason's simplicity: It prescinds from natural order as Aristotle's intellect, but unlike that intellect not to the solitude of identity, but to a new, reasoned order of its own in which it in absolute freedom relates itself to reality. Human reason is that presuppositionless phenomenon, resting on the rock of the *cogito ergo sum,* by which I know reality.

In Aristotle what is known is a particular through identity. In Thomas Aquinas what is known is a particular's existence as a universal through abstraction of its species. In Descartes what is known is a particular's potential existence through reason's universal species. This shadowlike existence of particulars in Descartes' science is mirrored in his own existence, that is, in that very existence whose resolution it is to doubt. Descartes' enterprise is a severe, if secret, discipline, an intellectual asceticism, bread and water unrelieved, if I may say it, by the wine of Aquinas. Descartes speaks of nine years spent employing his

---

[67] Cf. above, p. 66.
[68] *The Philosophical Works of Descartes*, op. cit., 70.
[69] Ibid., 92.

method in the *Discourse On The Method* III: ". . . without living to all appearance in any way differently from those who, having no occupation beyond spending their lives in ease and innocence, study to separate pleasure from vice, and who, in order to enjoy their leisure without weariness, make use of all distractions that are innocent and good, I did not cease to prosecute my design, and to profit perhaps even more in my study of Truth than if I had done nothing but read books or associate with literary people." [70] And again, in the same place, of eight years spent in search of a foundation for philosophy: "And it is just eight years ago that this desire [the 'honest' desire not to be esteemed as different than what he was] made me resolve to remove myself from all places where any acquaintances were possible, and to retire to a country such as this [Holland], where the long-continued war has caused such order to be established that the armies which are maintained seem only to be of use in allowing the inhabitants to enjoy the fruits of peace with so much the more security; and where, in the crowded throng of a great and very active nation, which is more concerned with its own affairs than curious about those of others, without missing any of the conveniences of the most populous towns, I can live as solitary and retired as in deserts the most remote." [71] Descartes' honesty is such that he cannot be a hypocrite; he is, nevertheless, a dissembler. Disguised with an indifferent worldliness he attends, with the solitary passion of a desert monk, to Truth. Likewise that indifferent middle that is the method, presuppositionless phenomenal intuition (the objective principle of humanity), eclipses particular individuals. But this intuitive dimension is entered into by Descartes' personal resolve; likewise by personal resolve in everyone who does, as everyone is able to, this that Descartes does explicitly as a mere man. *It is through his free resolve that this man contemplates himself as nothing other than pure, universal humanity*. This is *what is new*. This is, for reason, what it is to be a new man. This is the *necessary* understanding of the *novitas mundi,* the createdness of

[70] Ibid., 99.
[71] Ibid., 100.

this universe. Free reason perceives in its freedom a moral imperative. Descartes' science is a new science for a new reality in which the moral imperative is implicit. Both in origin, resolve, and in form, pure universal humanity, modern science is a moral enterprise. In Aristotle's metaphysical science contemplative self-transcendence is that essential necessity science is subject to; science is disinterestedly pursued by a man knowing his ultimate truth not to be in nature or human nature. For Aristotle, while man was a psychosomatic synthesis, his intellectual life was possessed of its own end, apart from moral interest. Whereas in Descartes, although he is famous for radically distinguishing soul from body, science is clearly a practical-moral endeavor.

Sacred doctrine in its disjunctive conjunction of grace with nature transformed both speculative and practical reason into instruments ordered to a single end beyond themselves, eternal union with God. Sacred doctrine is a unified science *par excellence;* but to it humanity's practical-moral activity is, in light of grace, in light of the end to which it is ordered beyond reason's sufficiency, explicitly *mercy.* For sacred doctrine humanity, or love of neighbor, is actual only if it is informed by charity, or love of God. Absolute human reason is in itself an abstraction of humanity from grace; it is for its own sake, for man's sake; it is the eclipse of grace. But since self-employing human reason is explicitly a method, its moral-religious dimension is perfectly dissembled. That is, personal resolution is hidden from us in Descartes' *cogito ergo sum* (necessarily so since this reason will be presuppositionless). *A fortiori* personal will's conformity to God's will, or that grace requisite to willing absolutely to be God's instrument—this absolute self-forgetfulness is altogether invisible; it is as if it did not exist. This dissemblance of methodical reason is so complete that, although one might discover its moral nature from an historical inspection (which does not touch it in itself), one is absolutely precluded from discovering its spiritual valence with certitude, since spirituality is a particular man's. (But considering free reason as a power, or principality, it does not acknowledge God's eternal law. It lacks the Spirit of God, being apart from God. It is oblivious of sin; that is, it is in principle unaware of its radical otherness from God. Free reason is this power *explicitly* as *pure* reason.)

For Descartes this new science's practical significance, its pure utility, is not at all in doubt. He writes in the *Discourse On The Method* VI as follows: ". . . it is possible to attain knowledge which is very useful in life, and . . . instead of that speculative philosophy which is taught in the Schools, we may find a practical philosophy by means of which, knowing the force and the action of fire, water, air, the stars, heavens and all other bodies that environ us, as distinctly as we know the different crafts of our artisans, we can in the same way employ them in all those uses to which they are adapted, and thus render ourselves the masters and possessors of nature. This is not merely to be desired with a view to the invention of an infinity of arts and crafts which enable us to enjoy without any trouble the fruits of the earth and all the good things which are to be found there, but also principally because it brings about the preservation of health, which is without doubt the chief blessing and the foundation of all other blessings in this life. For the mind depends so much on the temperament and disposition of the bodily organs that, if it is possible to find a means of rendering men wiser and cleverer than they have hitherto been, I believe that it is in medicine that it must be sought." [72] Here Descartes speaks of the medical significance of his new science, but notice, he supposes a most intimate dependence of mind upon body. This raises the question of Descartes' dualism, his exclusion of his body from the rational intuition, *cogito ergo sum*, in which he identifies himself as simply a *res cogitans,* a thinking thing, a soul. *Cogito ergo sum* is that rock or indubitable intuition reached in the course of a universal doubt but especially a doubt of what is learned from or by means of the senses. As Descartes says in the *Meditations On The First Philosophy* II, "What of thinking? I find here that thought is an attribute that belongs to me; it alone cannot be separated from me. I am, I exist, that is certain. But how often? Just when I think; for it might possibly be the case if I ceased entirely to think, that I should likewise cease

[72] Ibid., 119–120.

altogether to exist. I do not now admit anything which is not necessarily true: to speak accurately I am not more than a thing that thinks, that is to say a mind or a soul, or an understanding, or a reason, which are terms whose significance was formerly unknown to me. I am, however, a real thing and really exist; but what thing? I have answered: a thing which thinks." [73] This is Descartes' presuppositionless phenomenon. It is to be noted that Descartes notices that his essential existence is not able to be supposed. Thus, while he exists, his existence is implicitly contingent. Descartes' absolute proposition then is 'I think, therefore I am,' but his next thought which follows as an arch through which he will pass out from his simplicity to a universe of other existents, even to the reality of his body, is 'I exist, therefore God is.' This second most simple element in Descartes' chain of meditation reveals to us that conversion of faith's proposition 'God exists, therefore I am' which renders revelation a natural possession of human reason. Descartes' proof for the existence of God in the Third Meditation is a double one: Its first part relates to the 'I think' of the *cogito ergo sum;* its second part to the 'I am' or existence. In the first part Descartes says: ". . . there remains only the idea of God, concerning which we must consider whether it is something which cannot have proceeded from me myself. By the name God I understand a substance that is infinite [eternal, immutable], independent, all-knowing, all-powerful, and by which I myself and everything else, if anything else does exist, have been created. Now all these characteristics are such that the more diligently I attend to them the less do they appear capable of proceeding from me alone; hence, from what has been already said, we must conclude that God necessarily exists. For although the idea of substance is within me owing to the fact that I am a substance, nevertheless I should not have the idea of an infinite substance—since I am finite—if it had not proceeded from some substance which was veritably infinite." [74] In the second part of his proof Descartes says: ". . . [were I inde-

[73] Ibid., 151–152.
[74] Ibid., 165–166.

pendent of every other and] were I myself the author of my being, I should doubt nothing and I should desire nothing, and finally no perfection would be lacking to me; for I should have bestowed on myself every perfection of which I possessed any idea and should thus be God . . . . it is quite evident that it was a matter of much greater difficulty to bring to pass that I, that is to say, a thing or a substance that thinks, should emerge out of nothing, than it would be to attain to the knowledge of many things of which I am ignorant . . . . But though I assume that perhaps I have always existed just as I am at present, neither can I escape the force of this reasoning, and imagine that the conclusion to be drawn from this is, that I need not seek for any author of my existence. For all the course of my life may be divided into an infinite number of parts, none of which is in any way dependent on the other; and thus from the fact that I was in existence a short time ago it does not follow that I must be in existence now, unless some cause at this instant, so to speak, produces me anew, that is to say, conserves me."[75] So far Descartes is speaking the language of faith. God is, on the one hand, the only possible cause of Descartes' idea of an Infinite Being, Descartes obviously finite. On the other hand, God alone is powerful enough to have brought Descartes into being from not being, indeed to conserve him. But now listen to Descartes, immediately afterwards in the Third Meditation: "It only remains to me to examine into the manner in which I have acquired this idea from God; for I have not received it through the senses, and it is never presented to me unexpectedly, as is usual with the ideas of sensible things when these things present themselves, or seem to present themselves, to the external organs of my senses; nor is it likewise a fiction of my mind, for it is not in my power to take from or to add anything to it; and consequently the only alternative is that it is innate in me, just as the idea of myself is innate in me."[76] Further Descartes says: ". . . from the sole fact that God created me it is most probable that in some way he has placed

[75] Ibid., 168.
[76] Ibid., 170.

his image and similitude upon me, and that I perceive this similitude (in which the idea of God is contained) by means of the same faculty by which I perceive myself—"[77] It is when Descartes raises the question of the *acquisition* of the idea of the Infinite Reason that created him that we are sharply reminded that we are listening to the meditations of a presuppositionless phenomenon, or a being not extended in space, a being aware of time, but apparently not of history, but itself a product of history here appropriating that history as its birthright.

---

[77] Ibid.

Chapter 4

THE INFINITE PRACTICAL. I: KANT

Descartes in his *Meditations On The First Philosophy* founds a new science which rests neither on the metaphysical principles of identity or contradiction, nor on the principles of sacred doctrine, God or God's providence, but upon that presuppositionless phenomenon, *cogito ergo sum,* or the principle of humanity. In Descartes' rational intuition humanity thinks its own existence, directly, primarily. It ought therefore to be carefully noted that what Descartes thinks *immediately* is not truth but simply existence. *Cogito ergo sum* is for Descartes an unshakable truth. It is the first principle of his philosophy.[78] But it is so only when it is considered, so to speak, externally, when it is thought about by another thought, by a traditional thought that takes it to be a principle for its thinking. Or, *cogito ergo sum* is truth when viewed from the perspective of Descartes' resolution to begin a new science. But then truth is explicitly a means to an end, relative to Descartes' resolution, its unshakableness a matter of will, not vision. In fact the more I think about the truth of Descartes' *cogito ergo sum* so much more doubtful does this truth become, born, as it is, of doubt turned about upon doubt. '*I think, therefore I am*' becomes for me a pristine declaration of truth's secondariness: its being related finally to universal humanity's contingent existence. It is this universal medium of its own existence that humanity thinks through in Descartes' *cogito ergo sum;* that is, free reason thinks through its own contingency

[78] Cf. above, p. 78.

to God first, thence to truth. Neither God nor truth is presupposed; both exist only in humanity's eventuality, but in humanity's eventuality existence itself is in doubt. *Cogito ergo sum* as certitude is a product of thinking as doubting; this certitude is an obverse face of that dark thought: nothing is certain. So that nothing is certain but existence, or, in certitude existence alternates with nothing: *'I think, therefore I am'* but the next moment this very existence is in doubt. Were it not that God creates me anew each moment I would fall into nothing. I cannot doubt that I exist, but I doubt that I can continue to exist. That is, I perceive that this existence of which I am certain is nevertheless existence in time. I see that I am a finite creature; therefore the proof of God's existence in the Third Meditation. But doubt continues to be present not only as a result of finite existence's temporality, but also, of course, because things extended, known through sensation, were not able to be known without doubt. As a result, Descartes only established his immediate certitude of his existence (in that absolute moment of the *cogito ergo sum*) on condition that thought itself is not extended; but subsequently Descartes requires that things corporeal might be known. So he then distinguishes between corporeal nature as known to mathematics on the one hand, on the other hand to sensation. For knowledge of both kinds God is necessary. Descartes says at the end of the Fifth Meditation: "And so I very clearly recognize that the certainty and truth of all knowledge depends alone on the knowledge of the true God, in so much that, before I knew Him, I could not have a perfect knowledge of any other thing. And now that I know Him I have the means of acquiring a perfect knowledge of an infinitude of things, not only of those which relate to God Himself and other intellectual matters, but also of those which pertain to corporeal nature in so far as it is the object of pure mathematics [which have no concern with whether it exists or not]." [79] Now as Descartes says at the beginning of Meditation VI: "Nothing further remains but to inquire whether material things exist." [80] This is treated

[79] *The Philosophical Works of Descartes,* op. cit., 185.
[80] Ibid.

as follows in the Sixth Meditation: "There is certainly further in me a certain passive faculty of perception, that is, of receiving and recognising the ideas of sensible things, but this would be useless to me [and I could in no way avail myself of it], if there were not either in me or in some other thing another active faculty capable of forming and producing these ideas. But this active faculty cannot exist in me [inasmuch as I am a thing that thinks] seeing that it does not presuppose thought, and also that those ideas are often produced in me without my contributing in any way to the same, and often even against my will; it is thus necessarily the case that the faculty resides in some substance different from me in which all the reality which is objectively in the ideas that are produced by this faculty is formally or eminently contained . . . . this substance is either a body, that is, a corporeal nature in which there is contained formally [and really] all that which is objectively [and by representation] in those ideas, or it is God Himself, or some other creature more noble than body in which that same is contained eminently. But, since God is no deceiver, it is very manifest that he does not communicate to me these ideas immediately and by Himself, nor yet by the intervention of some creature in which their reality is not formally, but only eminently, contained. For since He has given me no faculty to recognise that this is the case, but, on the other hand, a very great inclination to believe [that they are sent to me or] that they are conveyed to me by corporeal objects, I do not see how He could be defended from the accusation of deceit if these ideas were produced by causes other than corporeal objects. Hence we must allow that corporeal things exist." [81] Here Descartes' knowledge of nature depends altogether upon his knowledge of what he calls the true God, by which we are to understand God the Creator who has made Descartes as well as the objects of his knowledge to exist; specifically, Descartes' knowledge depends on God's veracity, but God's veracity, for all practical purposes, merges with nature which for Descartes is God or God's creation, so that finally he trusts nature when it clearly teaches him that he is

[81] Ibid., 191.

most intimately united to his body.[82] God, for Descartes, is that arch through which he passes out to knowledge of created nature, but God is *not* a *deus ex machina* for knowledge's sake. God first appears not for knowledge's sake but for existence's sake. Likewise, that absolute *cogito ergo sum,* immediate thought of existence, precedes truth as that simplicity with respect to which truth exists. Precisely because God's existence, His veracity, is implicit here, for the very reason that God guarantees Descartes' knowledge in his *Meditations, nothing else* happens here but that it is made explicitly clear that free reason knows reality through the medium of its own representations. In the end doubt is overcome as it was in the beginning, in Descartes' *cogito ergo sum*. This presuppositionless phenomenon is seen to be, through and through, simple receptivity to reality. This receptivity is that rock with respect to which Descartes' doubt is ultimately, even reappearing, provisional.

Descartes tells us in the *Discourse On The Method* II that the first law of thinking he resolved to abide by "was to accept nothing as true which I did not clearly recognise to be so: that is to say carefully to avoid precipitation and prejudice in judgments, and to accept in them nothing more than what was presented to my mind so clearly and distinctly that I could have no occasion to doubt it." [83] And in the Fourth Meditation he says: ". . . I recognise that the power of will which I have received from God is not of itself the source of my errors—for it is very ample and very perfect of its kind—any more than is the power of understanding; for since I understand nothing but by the power which God has given me for understanding, there is no doubt that all that I understand, I understand as I ought, and it is not possible that I err in this. Whence then come my errors? They come from the sole fact that since the will is much wider in its range and compass than the understanding, I do not restrain it within the same bounds, but extend it also to things which I do not understand: and as the will is of itself indifferent to these, it easily falls into error and sin, and chooses

[82] Ibid., 192.
[83] Ibid., 92.

the evil for the good, or the false for the true."[84] But the resolve to restrain the will from affirming as true anything other than what God causes to appear distinctly and clearly to a purely receptive mind is anticipated in Descartes' original resolve to think of himself as pure universal humanity, that is, in his setting aside his particularity in favor of that universal simplicity of reason whose very constitution is communicated existence. How then could it occur to this free reason to doubt that reality appears in its distinct ideas? In sacred doctrine, as elaborated by Thomas Aquinas, natural reason knew its object, by means of an abstracted intelligible species, to actually exist as an other together with reason in the light of God's creation; natural reason knows itself to see directly to its object's existence. In Descartes instead of beginning with the objectivity of a created universe in which natural reason has its place beside other things, free reason begins with its own existence, swallowing up, as it were, Thomas' entire universe, making of it an implication of reason's existence. Reason becomes the representative object of reality, the glass in which, somewhat more darkly, reality mirrors itself. This self-centeredness of reason is the reason of its doubt. But although the *novitas mundi,* the novelty of this universe, is, in Descartes, poured down into reason's funnel, so that faith is naturalized, so that creation becomes an innate idea, still it flows through: Descartes' doubt is provisional, reality is understood to represent itself in reason's ideas. Nevertheless, for the first time, the object's existence is potentialized to its existence in reason's universal medium. (Notice how different this is from Aristotle for whom the actuality of an object's existence did not depend on reason; also how, in this one respect, for a moment, so to speak, Thomas Aquinas could look back to Aristotle, but also, how natural reason's transformation in light of revelation set up a possibility of reason's acquiring an infinite power over reality, in terms of which modern man saw, first with Descartes, reason's face set forever toward its own future.)

In the thought of Immanuel Kant just such a doubt as is

[84] Ibid., 175–176.

inconceivable to Descartes is institutionalized. In place of the latter's open, provisional doubt, Kant sets down at reason's frontier a closed, permanent doubt, a doubt directed against objects as things in themselves (*noumena*). This Kant did in order to better secure scientific knowledge a permanent dwelling place within an horizon of appearances (*phenomena*), which, in turn, he took to be the proper objects of knowledge. But to introduce doubt as a final closedness to the reality of an object in itself, except that it is understood that the thing in itself is real *per se,* is to darken the glass not quite totally. Actually Kant's procedure is a reaction to Hume's scepticism. He speaks as follows in the *Critique of Pure Reason:* "The illustrious Locke, failing to take account of these considerations [namely, Kant's considerations to the effect that pure concepts of the understanding cannot be deduced from but merely illustrated by experience], and meeting with pure concepts of the understanding in experience, deduced them also from experience, and yet proceeded so *inconsequently* that he attempted with their aid to obtain knowledge which far transcends all limits of experience. David Hume recognised that, in order to be able to do this, it was necessary that these concepts should have an *a priori* origin. But since he could not explain how it can be possible that the understanding must think concepts, which are not in themselves connected in the understanding, as being necessarily connected in the object, and since it never occurred to him that the understanding might itself, perhaps, through these concepts, be the author of the experience in which its objects are found, he was constrained to derive them from experience, namely, from a subjective necessity (that is, from *custom*), which arises from repeated association in experience, and which comes mistakenly to be regarded as objective. But from these premises he argued quite consistently. It is impossible, he declared, with these concepts and the principles to which they give rise, to pass beyond the limits of experience. Now this *empirical* derivation, in which both philosophers agree, cannot be reconciled with the scientific *a priori* knowledge which we do actually possess, namely, *pure mathematics* and *general science of nature;*

and this fact therefore suffices to disprove such derivation." [85] In other words, Kant is able to assume what does not even occur to Hume, namely, that human reason itself, specifically pure understanding, is, through its own concepts *a priori*, author of its own experience in which its object appears, that appearance then totally conditioned by reason, manifesting, therefore, nothing of what the object is in itself. But if the object in itself is not known to reason, but only its appearance, then that intuition that apprehends the object is not Descartes' *rational* intuition but it is, in Kant, *sensible* intuition, of which, in turn, the two pure *a priori* forms are time and space. To go from Descartes to Kant is to go from a rational intuition of existence to a sensuous intuition of experience; but to go from a reason, as with Descartes, negatively conditioned through time to a dependence on God for its existence, through space to a reliance on God's veracity, to a reason, as with Kant, *pure,* that is, containing in itself *a priori* those sensible forms of time and space which alone make possible the actual appearance of an object, upon which understanding reflects, subsuming it, in turn, under its *a priori* conceptualization so as to make it able to be known. Descartes dissociates free reason from space, therefore time from space, inner from outer, so that the criterion of truth is for Descartes 'clear and distinct' ideas or representations. By this dissociation Descartes maintains knowledge of reality itself, though this reality is potentiated to its existence in reason's universal medium, though, therefore, it is already somewhat more remote than that of Thomas Aquinas. But Kant dissociates pure reason from reality itself in order to acquire, in principle, a perfect knowledge of experience by deriving it, experience, formally-materially from the transcendental unity of self-consciousness. In this way the criterion of truth for Kant is no longer a subjective perception's clarity but is able to become the universal validity of an objective judgment. Descartes' presuppositionless phenomenon, *cogito ergo sum,* that simplicity of perception,

---

[85] Immanuel Kant, *Critique of Pure Reason* (trans. N. K. Smith, New York, 1965), 127–128.

gives way in Kant to complexity, to the presupposition of the totality of all possible experience, or, to state it negatively, to the presupposition of the unknowability of the thing in itself (*noumenon*). Pure reason presupposes totality: the simple universal of Descartes' thought is, in light of the presupposition of the totality of experience, particularized. That is, particularity, which is, by virtue of Descartes' resolution to think himself pure universal humanity, excluded from the patient receptivity of scientific thought so as to be merely implicit, is by virtue of Kant's irresolution (seen in the fact, among others, that the *noumena* continue to hang in the sky out of reach of knowledge) itself universalized. That is, the particular, *qua* particular, is universalized. The universal objectivity of science penetrates to bone's marrow: This man, *qua* time, *qua space,* is science's object. Descartes' rock, his absolute moment, his *cogito* Kant raises to second power; Descartes' ideal is concretized in universal laws of nature. Law is pure reason's medium in which the universal is explicitly particularized. Descartes' free reason, a self-employing agent, nevertheless took up a simply receptive attitude toward nature, or God's creation, or things in themselves; but in so doing Descartes is involved in a contradiction, only heightened by the absence of a theological justification, since his intuition is the rock upon which a new science is to be built. It is Kant who perceives the contradiction, moves with a precipitate motion to preclude reality in itself from knowledge, thus raising pure reason to a simply self-consistent agency for the production of all possible representations. Listen to Kant in the preface of the second edition of the *Critique of Pure Reason:* ". . . what necessarily forces us to transcend the limits of experience and of all appearances is the *unconditioned,* which reason, by necessity and by right, demands in things in themselves, as required to complete the series of conditions. If, then, on the supposition that our empirical knowledge conforms to objects as things in themselves, we find that the *unconditioned cannot be thought without contradiction,* and that when, on the other hand, we suppose that our representation of things, as they are given to us, does not conform to these things as they are in themselves, but that these objects, as appearances, conform to our mode of representation, *the contradiction vanishes;* and if, therefore, we thus find that the unconditioned is not to be met with in things, so far as we know them, that is, so far as they are given to us, but only so far as we do not know them, that is, so far as they are things in themselves, we are justified in concluding that what we at first assumed for the purposes of experiment is now definitely confirmed." [86] Now if we attend carefully to what Kant says here we are rewarded by seeing that not only is there a contradiction in a conditioned thought thinking the unconditioned or reality in itself, not only does the thought of the unconditioned contradict reason's freedom insofar as reason conforms to it, but most directly Kant tells us that the unconditioned itself is what 'reason, by necessity and by right, demands in things in themselves,' the very idea of unconditioned *noumena* is reason's. The *noumena* hang above the horizon in Kant because truly they bear silent witness to the absolutely self-sufficient agency of pure reason.

Whereas for Descartes thought immediately perceived its own existence, further, thought knew reality itself in its representative ideas, in Kant, on the contrary, that source of unity beyond reason itself but implicit in reason's thought of its own existence, namely, God, God who both unifies creation and creates anew each moment the continuity of Descartes' consciousness, is not available. The simplicity of Descartes' thought is the other side of its being presuppositionless; but Kant's thought is complex: It presupposes that pure reason contains within itself a source of that unity of thought with intuition which is knowledge. Kant identifies this source as the 'original synthetic unity of apperception.' [87] He speaks of it in the *Critique of Pure Reason* as follows: ". . . in the synthetic original unity of apperception, I am conscious of myself, not as I appear to myself, nor as I am in myself, but only that I am. This *representation* is a *thought,* not an *intuition.* . . . Accordingly I have no *knowledge* of myself as I am but merely as I appear to myself. . . . I exist as an intelligence which is conscious solely

[86] Ibid., 24.
[87] Ibid., 151–175.

of its power of combination; but in respect of the manifold which it has to combine I am subjected to a limiting condition (entitled inner sense), namely, that this combination can be made intuitable only according to relations of time, which lie entirely outside the concepts of understanding, strictly regarded. Such an intelligence, therefore, can know itself only as it appears to itself in respect of an intuition which is not intellectual and cannot be given by the understanding itself, not as it would know itself if its *intuition* were intellectual." [88] Kant adds in a footnote: ". . . I cannot determine my existence as that of a self-active being; all that I can do is to represent to myself the spontaneity of my thought, that is, of the determination; and my existence is still only determinable sensibly, that is, as the existence of an appearance. But it is owing to this spontaneity that I entitle myself an *intelligence*." [89] The transcendental unity of self-consciousness is for Kant a necessary, universal presupposition for contingent consciousness, or consciousness according to time. When I think I am, my existence as thought is temporally conditioned, my thinking is, as it appears to me, receptive. This transcendental self-sufficient agency of pure reason's understanding is a one or unity beyond knowledge because beyond that sensible intuition of existence which is the limit of reason's experience. To state it simply, human thinking for Kant is not itself intuition (were it so, thought would at once bring into existence its own objects [90]). In fact, were thought intuitive, it would know itself as something other than synthetic unity, other than the necessary presupposition of human knowledge; but, as it is, this abstract principle of unity is the only possible content for human consciousness of self. On the one hand, pure reason knows of itself nothing but that it exists; on the other hand, it contains within itself forms for the knowledge of all possible experience, including sensible intuition, imagination, and synthetic apperception. Human reason is then, while not at all competent scientifically for self-

[88] Ibid., 168–169.
[89] Ibid., 169.
[90] Ibid., 157.

knowledge, most perfectly equipped for knowledge of nature; or contingent consciousness ultimately understands itself to be knowable only as nature, that is, as particular universalized within a totality of human knowledge, which radical reduction of this particular to conformity with all possible particulars is implicit in the transcendental unity of self-consciousness. For all scientific purposes I exist totally as an appearance under a dome of universal human identity; my knowledge of myself is such that my very being or substance is contextual. My existence, as I know it, is simply in accord with universal laws of nature; I judge myself to be, then, both as knower (subject) and known (object), a nature. Anything I might be in myself is forever beyond my experience. Therefore my knowledge bends necessarily along a transcendental horizon, likewise my very being, unless I am dreaming.

The applicability of this transcendental structure of pure reason to specific knowledge is clearly stated by Kant in his *Prolegomena To Any Further Metaphysics* II: "All our judgments are at first merely judgments of perception [i.e., requiring no pure concept of understanding]; they hold good only for us (that is, for our subject), and we do not till afterward give them a new reference (to an object) and desire that they shall always hold good for us and in the same way for everybody else; for when a judgment agrees with an object, all judgments concerning the same object must likewise agree among themselves, and thus the objective validity of the judgment of experience signifies nothing else than its necessary universal validity. And conversely when we have ground for considering a judgment as necessarily having universal validity . . . we must consider that it is objective also—that is, that it expresses not merely a reference of our perception to a subject, but a characteristic of the object. For there would be no reason for the judgments of other men necessarily agreeing with mine if it were not the unity of the object to which they all refer and with which they accord; hence they must all agree with one another. Therefore objective validity and necessary universality (for everybody) are equivalent terms. . . . By this judgment we know the object (though it remains unknown as it is in itself) by the universal and necessary connection of the given perceptions."[91] In Descartes an object exists, *qua* object, in an order of objects as perceived by what Descartes takes to be a rational intuition. Kant concretizes Descartes' principle of universal reason by incorporating subjective judgments of perception themselves in an order of judgments. But such an order of judgments must be one of particular judgments of experience, so that the condition of objectivity is subsumption of perception under a pure concept of understanding, that is, ultimately, under pure reason's presupposed transcendental unity. This means that it, pure subjective judgment, becomes itself explicitly an appearance; therefore, its object is subject to universal validation. So, in conforming my judgment to pure understanding's conceptualization I enter into connection with other possible judgments, thus, together with them, realizing universal objective humanity, or nature. If, for Kant, there is to be any future metaphysics, or science purely theoretical, it must be this universal objectivity, capable of determining *a priori* all possible experience, of subsuming all subjectivity under a strict law of necessity. Only the objective validity of a judgment satisfies that desire implicit in judgment to be perpetuated; but for Descartes it was God who perpetuated his consciousness, whose existence assured Descartes of the perpetuity of his judgment, which was therefore implicitly universal; Descartes says in the Fifth Meditation: ". . . even though my nature is such that as soon as I understand anything very clearly and very distinctly I cannot help but believe it to be true, nevertheless, because I am also of such a nature that I cannot always confine my attention to one thing and frequently remember having judged a thing to be true when I have ceased considering the reasons which forced me to that conclusion, it can happen at such a time that other reasons occur to me which would easily make me change my mind if I did not know that there was a God. And so I would never have true and certain knowledge concerning anything at

[91] Immanuel Kant, *Prolegomena To Any Future Metaphysics* (ed. L. W. Beck, Indianapolis, 1950), 46.

all, but only vague and fluctuating opinions. . . . But after having recognized that there is a God, and having recognized at the same time that all things are dependent upon him and that he is not a deceiver, I can infer as a consequence that everything which I conceive clearly and distinctly is necessarily true." [92] If Descartes took the true God of sacred doctrine to be an innate idea of his scientific consciousness so as to quiet the perdurable doubt which threatens to corrupt scientific knowledge at its core, namely, at the point of judgment itself, then it is illuminating to see that Kant, particularizing, as he does, Descartes' intuition, transforming it into an appearance, translates God, as he appears in Descartes' subjectivity, into that ideal scientific corporation, *humanity,* or the objective community of judgments. God is to subjectivity as humanity is to objectivity; it is retrospectively clear, at least, that when Descartes speaks of God in the implication of that presuppositionless reason he appropriated from sacred doctrine what is significant is that his knowledge of God is *innate,* that is, that universal humanity is itself a self-sufficient agent of knowledge. The explication of this latter truth of modern science is Kant's work. Or the situation might be formulated: God answers to an individual knower's deficiencies, *qua* individual, at a more primitive stage of modern science's development; but, then, at a more perfect stage of self-conscious concretion scientific humanity recognizes that it itself, *qua* all possible judgments, answers to those same deficiencies: The idea of God becomes, for science, a superfluous abstraction compared to its own rigorous internal process of self-validation. If we pause here a moment, reflecting on science's emergence out of grace taking with it as its birthright this principle of humanity which rises to explicit independence in Kant, this might be an appropriate juncture to look at how this scientific ideal of humanity's self-validation is able to be superimposed upon that very theology which is its origin, so that we get a result essentially like this one which I quote from Ludwig Feuerbach's *The Essence of Christianity:* "Be-

[92] R. Descartes, *Philosophical Essays* (trans. L. J. Lafleur, Indianapolis, 1964), 124–125.

cause Christianity thus, from exaggerated subjectivity, knows nothing of the species, in which alone lies the redemption, the justification, the reconciliation and cure of the sins and deficiencies of the individual, it needed a supernatural and peculiar, nay, a personal, subjective aid in order to overcome sin. If I alone am the species, if no other, that is, no qualitatively different men exist, or, which is the same thing, if there is no distinction between me and others, if we are all perfectly alike, if my sins are not neutralised by the opposite qualities of other men: then assuredly my sin is a blot of shame which cries up to heaven; a revolting horror which can be exterminated only by extraordinary, superhuman, miraculous means. Happily, however, there *is* a natural reconciliation. My fellow-man is *per se* the mediator between me and the sacred idea of the species. *Homo homini Deus est.* My sin is made to shrink within its limits, is thrust back into its nothingness, by the fact that it is only mine, and not that of my fellows." [93] From this we perceive that modern human self-consciousness, shaped to science's transcendental form, is by no means indifferent to orthodox Christianity's doctrine of revelation; in fact, modern self-consciousness, which as Kant makes clear lays claim to human knowledge's total possible compass, is revelation's subrogation. This is quite directly stated in Kant's *Religion Within The Limits of Reason Alone,* in which work ecclesiastical, historical faith based on revelation is displaced by what Kant calls the 'pure faith of reason.' Or a universal morality, so-called, displaces grace.[94] Of Descartes we have said [95] that his free reason is an eclipse of grace, but since particularity is excluded from his simple universal medium, it remains an open question, dissembled by his method, what accommodation this man might make with God's will; but in Kant, where pure reason is become self-satisfying desire for perpetuity through humanity's infinite future possibility, then grace appears as an inadmissible contradiction to

[93] L. Feuerbach, *The Essence of Christianity* (trans. G. Eliot, New York, 1957), 159.
[94] Cf. Kant's *Religion Within The Limits of Reason Alone* (trans. T. M. Greene and H. H. Hudson, rprt. New York, 1960).
[95] Cf. above, p. 88.

pure reason's moral self-determination. At that point how thoroughly obscured is any question of an alteration of self-consciousness in history, since even where facts of history relevant fall under consideration it appears that they are inessential or related only by chance to reason's freedom. But if Thomas Aquinas is right in saying that *novitas mundi,* this universe's novelty, is affirmed by faith alone, by no demonstration [96] to be known, then how could it be otherwise? How might reason, more itself, not be more oblivious?

If we turn our attention back to the realm of scientific reason, we see that the idea of God serves reason as a purely regulative principle; as Kant says in the *Critique of Pure Reason:* "In other words, it must be a matter of complete indifference to us, when we perceive such unity [the seemingly purposive arrangement of the world], whether we say that God in his wisdom has willed it to be so, or that nature has wisely arranged it thus. For what has justified us in adopting the idea of a supreme intelligence as a schema of the regulative principle is precisely this greatest possible systematic and purposive unity—a unity which our reason has required as a regulative principle that must underlie all investigations of nature . . . . we cannot, without contradicting ourselves, ignore the universal laws of nature—with a view to discovering which the idea was alone adopted—and look upon this purposiveness of nature as contingent and hyperphysical in its origin." [97] Keeping in mind that the 'laws of nature' are prescribed by understanding itself to appearances ultimately grounded in reason's own pure intuition of time and space, it is then evident that, as Kant says, to look upon nature as contingent would be pure reason's contradiction of itself. It is now noticeable that no longer are we dealing with three distinct entities as in Aristotle's metaphysics, namely, God, universe, and self, nor, as in sacred doctrine, with God, self-reflecting reason, and actually existing world, but that, since Descartes' invention of that universal middle, free reason, only two, at first glance in Descartes, remain: I and God, at second glance in

[96] Cf. above, p. 63ff.

[97] *Critique of Pure Reason,* op. cit., 567–568.

Kant, only one: pure reason. So that, as scientific consciousness grows more determinate within its own horizon, it becomes, vis-à-vis its object increasingly a simplification of reality in the direction of disappearance; witness Kant's *noumena.* If it were a mystery, it would be entitled: 'The Case of The Disappearing Other.' Or to state it another way, with science's increasing objectivity, as the object grows more concrete it grows less real.

For Kant understanding knows objects in experience; as distinguished from understanding, pure reason's principle is, as he says in the *Critique of Pure Reason,* "to find for the conditioned knowledge obtained through the understanding the unconditioned whereby its unity is brought to completion." [98] The unconditioned is pure reason's supreme principle; its form is threefold, corresponding to the three forms of judgment: *categorical,* the subject; *hypothetical,* the series; *disjunctive,* the system.[99] Kant tells us then in the *Critique* that "Pure reason thus furnishes the idea for a transcendental doctrine of the soul, for a transcendental science of the world, and, finally, for a transcendental knowledge of God." [100] These ideas of pure reason, since reason itself is possessed of no intuition, refer to no objects, but serve merely as regulative principles for understanding's scientific ordering of its experience. We have had occasion to refer to Kant's understanding of the regulative use of the idea of God. These ideas as forms of the supreme principle of the unconditioned serve to bring to completion, that is, to unity, the manifold organized by understanding which for this purpose is dependent upon the logical priority of reason. Kant tells us further in the *Critique:* "In conformity with these ideas as principles we shall, *first,* in psychology, under the guidance of inner experience, connect all the appearances, all the actions and receptivity of our mind, *as if* the mind were a simple substance which persists with personal identity (in this life at least), while its states, to which those of the body belong only as outer conditions, are in continual change. *Secondly,* in cosmology, we

[98] Ibid., 306.
[99] Ibid., 316.
[100] Ibid., 323.

must follow up the conditions of both inner and outer natural appearances, in an enquiry which is to be regarded as never allowing of completion, just *as if* the series of appearances were in itself endless, without any first or supreme member. . . . *Thirdly,* and finally, in the domain of theology, we must view everything that can belong to the context of possible experience *as if* this experience formed an absolute but at the same time completely dependent and *sensibly* conditioned unity, and yet also at the same time *as if* the sum of all appearances (the sensible world itself) had a single, highest, and all-sufficient ground beyond itself, namely, a self-subsistent, original, creative reason."[101] As Kant points out in the *Prolegomena To Any Future Metaphysics,*[102] with the third transcendental idea, namely, God, reason breaks altogether with experience. That is to say that while the idea of personal identity would, if taken as referring to an object, be an unwarranted extension of inner sense (time), nevertheless, there is an inner sense. Likewise with the idea of natural necessity, there is an outer sense (space). But in the case of the idea of God there is no sense whatsoever. As a result, the idea of God is not related to the thinking subject nor to the natural world, but it is related to pure reason itself as its transcendental ideal, insofar as reason is itself the transcendental unity of self-consciousness. Kant describes reason's ideal in the *Critique* in this way: "The object of the ideal of reason, an object which is present to us only in and through reason, is therefore entitled the *primordial being* (*ens originarium*). As it has nothing above it, it is also entitled the *highest being* (*ens summum*); and as everything that is conditioned is subject to it, the *being of all beings* (*ens entium*). These terms are not, however, to be taken as signifying the objective relation of an actual object to other things, but of an *idea to concepts.* We are left entirely without knowledge as to the existence of a being of such outstanding pre-eminence."[103] So God is thought by analogy not to any possible object of knowledge but to the original synthetic unity

[101] Ibid., 551.
[102] *Prolegomena To Any Future Metaphysics,* op. cit., 96.
[103] *Critique of Pure Reason,* op. cit., 492.

of apperception itself, that is, to the transcendental horizon itself: self-relating relation, that purely logical identity of reason itself, which identity, like God, is beyond, absolutely, being known as an object. Kant says in the *Prolegomena:* ". . . we then do not attribute to the Supreme Being any of the properties in themselves by which we represent objects of experience, and thereby avoid *dogmatic* anthropomorphism; but we attribute them to the relation of this Being to the world and allow ourselves a *symbolical* anthropomorphism, which in fact concerns language only and not the object itself. . . . Such a cognition is one of analogy and does not signify . . . an imperfect similarity of two things, but a perfect similarity of relations between two quite dissimilar things. By means of this analogy, however, there remains a concept of the Supreme Being sufficiently determined *for us,* though we have left out everything that could determine it absolutely, or *in itself;* for we determine it as regards the world and hence as regards ourselves, and more do we not require."[104] In Thomas Aquinas the analogy was between creature and creator, it was an analogy of specific likeness between a finite and an infinite existence, both objects, that is, knowable, even if in God's case only by an extraordinary act on his part. For Descartes God is related to man simply through causality, which relation is, nevertheless, rationally intuitable for Descartes. In Kant the relation of causality is merely a concept itself unknowable, so that God is no object whatsoever for pure reason but a term in a logical relationship analogous to that of reason itself in its relation to its own understanding. God as symbol is pure reason's judgment that it itself, that is, reason, is bound for purposes of knowledge to conditions of understanding, that it is finally *technē,* hypothetical or purely logical in nature.

But pure reason beyond its employment as the ultimate logical ground of scientific understanding of nature is also a practical reason. Kant recognizes that there is in pure practical reason consciousness of a pure transcendental idea of freedom. Indeed, freedom is pure reason's supreme practical idea of itself.

[104] *Prolegomena To Any Future Metaphysics,* op. cit., 105–106.

It is reason as absolutely unconditioned causality of events. Kant says in the *Critique:* "Reason is the abiding condition of all those actions of the will under [the guise of] which man appears. Before ever they have happened, they are one and all predetermined in the empirical character. In respect of the intelligible character, of which the empirical character is the sensible schema, there can be no *before* and *after;* every action, irrespective of its relation in time to other appearances, is the immediate effect of the intelligible character of pure reason. Reason therefore acts freely; it is not dynamically determined in the chain of natural causes through either outer or inner grounds antecedent in time. This freedom ought not, therefore, to be conceived only negatively as independence of empirical conditions. The faculty of reason, so regarded, would cease to be a cause of appearances. It must also be described in positive terms, as the power of originating a series of events." [105] It is as if pure reason is for Kant a two-way mirror; reason sees itself looking at itself in a world of scientific reality, in which sensibly conditioned world of appearance its freedom is totally invisible. Indeed all events are knowable to understanding only insofar as they are considered *as if* members of an infinite chain of natural causality or necessity; but precisely because reason's transcendental idea of totality is merely a regulative principle of science, not able to be verified in experience, Kant concludes in his *Critique* that freedom "is at least *not incompatible with* nature," [106] so that reason stands this side of the mirror causing events in time that are simultaneously explainable by natural law. But the law that reason obeys this side of the mirror is a law of its own pure practical being, the moral law. Kant writes in the preface to the *Critique of Practical Reason:* "The concept of freedom, in so far as its reality is proved by an apodictic law of practical reason, is the keystone of the whole architecture of the system of pure reason and even of speculative reason. All other concepts (those of God and immortality) which, as mere ideas, are unsupported by anything in speculative reason now attach

[105] *Critique of Pure Reason,* op. cit., 476.
[106] Ibid., 479.

themselves to the concept of freedom and gain, with it and through it, stability and objective reality. That is, their possibility is proved by the fact that there really is freedom, for this idea is revealed by the moral law. Freedom, however, among all the ideas of speculative reason is the only one whose possibility we know a priori. We do not understand it, but we know it as the condition of the moral law which we do know. The ideas of God and immortality are, on the contrary, not conditions of the moral law, but only conditions of the necessary object of a will which is determined by this law, this will being merely the practical use of our pure reason." [107] The moral law is defined as follows: "So act that the maxim of your will could always hold at the same time as a principle establishing universal law." [108] In science subjectivity entered into objectivity through its incorporation into the universality of judgments by means of the conceptualization of pure understanding. Here, in its moral freedom, reason acts only in such a way that the principle of its act might be legislated as a universal law. The moral law itself is absolutely underived: It is the original fact of pure reason, which, as Kant says, "by it proclaims itself as originating law." [109] Reason discovers its freedom simply because it is the necessary condition of obedience to its own law. Pure reason is ultimately a moral-practical reason. In its freedom it obeys itself, not God; the idea of God and the idea of immortality being secondary, but necessary, postulates of the operation of its will, specifically related to the end or happiness.

Finally pure practical reason is to speculative or scientific reason as the unconditioned to the conditioned, the in itself complete to the in itself incomplete. Kant writes in the *Critique of Practical Reason:* ". . . in the combination of pure speculative with pure practical reason in one cognition, the latter has the primacy. . . . Without this subordination, a conflict of reason with itself would arise, since if the speculative and the practical reason were arranged merely side by side (co-ordinated),

[107] *Critique of Practical Reason* (trans. L. W. Beck, Indianapolis, 1956), 3–4.
[108] Ibid., 30.
[109] Ibid., 31.

the first would close its borders and admit into its domain nothing from the latter, while the latter would extend its boundaries over everything. . . . Nor could we reverse the order and expect practical reason to submit to speculative reason, because every interest is ultimately practical, even that of speculative reason being only conditional and reaching perfection only in practical use." [110] Thus science is become technique; Aristotle is stood on his head; the sanctity of the realm of grace has become the virtue of human freedom; and Descartes' subjective intention is explicitly declared to be pure reason's public policy.

[110] Ibid., 126.

Chapter 5

THE INFINITE PRACTICAL. II: HEGEL

In Kant the Ideal is unrealized. Pure reason's transcendental Ideal, namely, that there is an absolute individual existent possessing in itself the totality of reality, a being (God) in which all existents exist determinately, that is, that there is an individual being which is the ground for that systematic disjunction whereby each particular exists in its own right, with its own particular share of the totality of being, absolutely determined to exist in itself by reference to this totality, this transcendental Ideal is unreal as Kant says in his *Critique of Pure Reason:* "It is obvious that reason, in achieving its purpose, that, namely, of representing the necessary complete determination of things, does not presuppose the existence of a being that corresponds to this ideal, but only the idea of such a being, and this only for the purpose of deriving from an unconditioned totality of complete determination the conditioned totality, that is, the totality of the limited. The ideal is, therefore, the archetype (*prototypon*) of all things, which one and all, as imperfect copies (*ectypa*), derive from it the material of their possibility, and while approximating to it in varying degrees, yet always fall very far short of actually attaining it." [111] For Kant the function of the idea of God in science is to serve as a regulative principle for the organization of experience, that is, sensibly conditioned appearances. But because the organization of experience is the work of *understanding,* which in turn presupposes the transcendental unity of self-consciousness, God is conceived of as not

[111] *Critique of Pure Reason,* op. cit., 491–492.

merely the *being of beings,* but also as an *intelligence* or personality. This idea of God is for Kant not a being but a seeming, a symbol of pure reason's relationship to the world of its experience, of understanding's dependence upon transcendental apperception, or, a presupposed unity of consciousness, and a symbol of the necessarily presupposed totality of *empirical* reality. This last, the realm of empirical reality, is the proper place for the employment of the transcendental ideal, but as Kant observes in the *Critique:* ". . . owing to a natural illusion we regard this principle, which applies only to those things which are given as objects of our senses, as being a principle which must be valid of things in general. Accordingly, omitting this limitation, we treat the empirical principle of our concepts of the possibility of things, viewed as appearances, as being a transcendental principle of the possibility of things in general." [112] But such an idea is for Kant a pure fiction.

We understand Kant's pure reason to be like a two-way mirror.[113] On its scientific side reason sees itself bound to the conditions of its understanding. But on its moral side reason sees itself in its freedom as the cause of events in the world of time and space which events are otherwise explainable in terms of natural laws; this freedom of pure reason is discovered as the condition of its obedience to the moral law ('act only on that maxim through which you can at the same time will that it should become a universal law'). The moral law itself is absolutely underived: It is pure reason's ultimate fact. The freedom of reason is its moral autonomy; but, as such, pure practical reason is the unconditioned to which is related conditioned or scientific reason as subordinate to pure reason's supreme practical interests. Pure scientific understanding proceeds without interference in its own realm, beholding its image reflected in conditions of sensible intuition, on condition that it is itself incomplete, but completed by that practical-moral interest that pure reason's unconditional freedom is. Throughout science is instrumental; freedom directive; as the original legislative

---

[112] Ibid., 494.
[113] Cf. above, p. 111.

source of order pure reason lays it down that understanding organizes its experience in accordance with laws, as Kant says in the *Prolegomena To Any Future Metaphysics:* "Nature is the existence of things, so far as it is determined according to universal laws." [114] If we permit ourselves a biblical analogy, then pure reason in relating scientific understanding to its own unconditioned autonomy is *pharisaical,* in that it places all things under a strict law of necessity while exempting itself; but, in this, pure reason bears its own peculiar burden, namely, *keeping up appearances.* But in this keeping up appearances which is the active enterprise of Kant's science is to be discovered a passive root: *Human reason's inability to maintain itself face to face with its object's otherness.* In Aristotle reason's passivity is in evidence in its understanding essence according to its own formality, but in Aristotle this passivity of reason is subordinated to intellect's ultimate identity with its object wherein, its otherness presupposed, its existent individuality is known. [115] In Thomas Aquinas, where God is become natural reason's limiting identity, reason knew directly its actually existing object through the universal medium of created existence; here, in light of God's revelation, natural reason looked upon its object known to it as simply an other. [116] In Descartes, however, free human reason, no longer operating in light of an explicit creation, comes to know its object's separate existence as potentiated to reason's own simple self-reflection in which that object's reality is established. In Descartes' *cogito ergo sum* the object's reality is subjected to a radical doubt, as a result of which the object is centered in Descartes' intention which in turn is implicitly guaranteed by a veracious God. [117] Whereas in Thomas reason is explicitly infused with God's light [118] so as to be, at least in principle, not simply left to its own passivity, as it would otherwise most certainly have been when metaphysics was subordinated to sacred doctrine, in Descartes reason is completely on

[114] *Prolegomena To Any Future Metaphysics,* op. cit., 42.
[115] Cf. above, pp. 48, 54–55.
[116] Cf. above, p. 65ff.
[117] Cf. above, p. 89ff., p. 92ff.
[118] Cf. above, pp. 64–65, note 34.

its own, so that its passivity, which prior to revelation is limited by a superior active intellect, is now, after revelation, unlimited altogether except, in Descartes, by its own idea of God; later, in Kant, by its own idea of freedom. Or doubt itself, the doubt in which existence is involved [119] is the measure of reason's passivity, so that in Kant where doubt is institutionalized in the form of the doubt of things in themselves, in the *noumena,* modern thought is most explicitly passive. If, using Aristotle as our touchstone, an active science, purely so, is a science for the sake of knowing, subordinated to no moral-practical interest, or human interest, then modern science, which in Kant's words understands that "every interest is ultimately practical," is purely passive.[120] Or to state it in terms of principles: *The principle of a philosophy purely passive is the principle of humanity.* The principle of humanity is the invention of Descartes; in its invention Descartes appropriated an abstraction from sacred doctrine as something in its own right, namely, humanity; but in sacred doctrine not humanity but God is the principle of science. It was therefore our intention in introducing the new thought of Descartes to distinguish the *principle of humanity* of his new science from humanity as it is in a circle of grace where it is *mercy or love of neighbor,* and where, as mercy, humanity is actual only in relation to charity or love of God, where it is subordinated to charity as essence to participated existence.[121] We are now, perhaps, in a better position to see that that distinction indicates that *sacred doctrine is to be compared to modern science as an active to a passive thought.* At the same time sacred doctrine might be compared to ancient thought as perfect to imperfect. At this juncture it can be seen at a glance that while there exists an opposition between ancient thought and sacred doctrine it is, in some sense at least, a matter of degree. Sacred doctrine would be 'folly' to the Greeks because its content is a perfection impossibly known. It was this that Paul understood in his speech to the Athenian Areopagus in *Acts* 17:22–23 when

[119] Cf. above, p. 94.
[120] Cf. above, p. 113, note 110.
[121] Cf. above, pp. 74–75.

he said: "Men of Athens, I have seen for myself how extremely scrupulous you are in all religious matters, because I noticed, as I strolled round admiring your sacred monuments, that you had an altar inscribed: To An Unknown God. Well, the God whom I proclaim is in fact the one whom you already worship without knowing it." [122] But on the other hand between sacred doctrine and modern thought it might be conceived that there is no common ground, except as Kant indicates in the matter of language [123] (which phenomenon would itself indicate something of the emptiness of words in modern thought), but that between an active science and a passive one only an opposition of contradictory positions is possible, since modern thought, unlike Paul's audience, recognizes no God, not even an unknown God, that is, in its pure passivity it excludes absolutely the possibility of worship. Modern thought occupies a position unto itself whereby, beginning with Descartes' formal-material discontinuity with previous science, it is totally its own form, its own content.[124] For it, worship (which Kant identifies with ecclesiastical historical faith in his *Religion Within The Limits of Reason Alone* [125]) would be in cultus what knowledge of transcendent reality would be in sacred doctrine, or what, epistemologically, perception of an object's otherness would be: a perfect contradiction of its resolution to be itself.[126] Modern thought places God in a subordinate clause to this resolution, as Kant says in the *Critique of Judgement* II: "Objects that must be thought *a priori,* either as consequences or as grounds, if pure practical reason is to be used as duty commands, but which are transcendent for the theoretical use of reason, are mere *matters of faith.* Such is the *summum bonum* which has to be realized in the world through freedom. . . . This effect which is commanded, *together with the only conditions on which its possibility is conceivable by us,* namely the existence of God and the immortality of the soul, are *matters of faith* (*res fidei*) and, moreover, are of

[122] *The Jerusalem Bible: N.T.,* op. cit., 230–231.
[123] Cf. above, pp. 109–110, note 104.
[124] Cf. above, pp. 76–79. Also p. 107f.
[125] Op. cit.
[126] Cf. above, p. 72.

all objects the only ones that can be so called . . . Faith as *habitus,* not as *actus,* is the moral attitude of reason in its assurance of the truth of what is beyond the reach of theoretical knowledge." [127] Kant adds in a footnote: "It is a confidence in the promise of the moral law. . . . The very word *fides* expresses this; and it must seem suspicious how this expression and this particular idea get a place in moral philosophy, since it was first introduced with Christianity, and its acceptance might perhaps seem only a flattering imitation of the language of the latter. But this is not the only case in which this wonderful religion has in the great simplicity of its statement enriched philosophy with far more definite and purer conceptions of morality than morality itself could have previously supplied. But once these concepts are found, they are *freely* approved by reason, which adopts them as conceptions at which it could quite well have arrived itself and which it might and ought to have introduced." [128] For Kant, Christianity introduces, as a matter of fact, conceptions which pure reason, in its passivity, adopts freely, that is, takes on its own terms, or, morally or practically understands only in a manner compatible with its, pure reason's, original unconditioned freedom. In the light of pure reason, itself autonomous, Christianity, insofar as it is historical faith, insofar as its faith points to a history of being, indeed, insofar as its doctrine is one of grace, redemption, man's absolute need of God's intervention, of God's mercy, of sin—Christianity in its sacramental form, as Church, is, since in every way it is dependence of reason upon an other, *qua* other, unacceptable to pure reason, except that reason appropriates *freely* those elements of Christianity that are able to be assimilated to its moral autonomy. So, for example, the Logos or Word of the Prologue of John's Gospel is to Kant the "archetype of the moral disposition in all its purity." [129] Kant writes in *Religion Within The Limits of Reason Alone:* "Now if it were indeed a fact that such a truly godly-minded man at some particular

[127] I. Kant, *The Critique of Judgement* (trans. J. C. Meredith rprt., Oxford, 1957), 142–145.
[128] Ibid., 146.
[129] *Religion Within The Limits of Reason Alone,* op. cit., 55ff.

time had descended, as it were, from heaven to earth and had given men in his own person, through his teachings, his conduct, and his sufferings, as perfect an *example* of a man well-pleasing to God as one can expect to find in external experience (for be it remembered that the *archetype* of such a person is to be sought nowhere but in our own reason), and if he had, through all this, produced immeasurably great moral good upon earth by effecting a revolution in the human race—even then we should have no cause for supposing him other than a man naturally begotten. (Indeed, the naturally begotten man feels himself under obligation to furnish just such an example in himself.) This is not, to be sure, absolutely to deny that he might be a man supernaturally begotten. But to suppose the latter can in no way benefit us practically, inasmuch as the archetype which we find embodied in this manifestation must, after all, be sought in ourselves (even though we are but natural men)." [130] We recall that Kant informs us that there are for pure reason only three *matters of faith:* that there is a *summum bonum,* together with those conditions upon which alone it is conceivable to us, firstly, God's existence, secondly, immortality of soul. Beyond these facts of human freedom there exist no matters of faith but only matters of fact, as Kant says in the *Critique of Judgement* II: ". . . although we have to believe what we can only learn by *testimony* from the experience of others, yet that does not make what is so believed in itself a matter of faith, for with *one* of those witnesses it was personal experience and matter of fact, or is assumed to have been so. In addition it must be possible to arrive at knowledge by this path—the path of historical faith; and the Objects of history and geography, as, in general, everything that the nature of our cognitive faculties makes at least a possible subject of knowledge, are to be classed among matters of fact, not matters of faith." [131] The upshot of Kant's either/or is that Jesus' divinity is a question not of faith but of history, or, of knowledge; but it is quite clear that in a Kantian universe the very first witness is one whose own experi-

[130] Ibid., 57.
[131] *The Critique of Judgement,* op. cit., 143.

ence is sensibly conditioned, so that if one were able to follow a path of historical witness to Jesus himself, nevertheless, his testimony would extend only so far as he knows himself as an appearance. For Kant orthodox faith is neither faith nor knowledge; quite consistently Kant takes it to be symbolic of reason's own moral freedom. Just as the principles of Thomas Aquinas' sacred doctrine become Descartes' innate idea of God, in Kant (who in concretizing Descartes' abstract universal of reflection extends explicitly reason's province over man's entire experience) religion itself, in particular Christianity, is purged of its incredible elements with the result that the Child of Faith becomes a "man naturally begotten". One is reminded of Jesus in the synagogue at Nazareth, his home town, where, as recounted in *Luke* 4:16–30, Jesus understood the question "This is Joseph's son, surely?" as the form of rejection, by which he was led, in turn, to denouncing his fellow townsmen, so that a day which began rather well with a reading from Isaiah ended with an attempt to throw Jesus off a cliff.[132]

There is, perhaps, yet another direction in which pure reason's passive root, that is, its inability to maintain itself face to face with its object's otherness, might be explored, a clue to which is given us by the ancient Greek word, *paránoia,* meaning madness, more basically (*παρανοέω*) 'to think amiss,' 'to misconceive,' 'to misunderstand.' If one looks at Greek literature in order to discover instances of *paránoia* or madness, one will soon see that they abound in Athenian tragedy; Oedipus, Ajax, Heracles—each suffered *paránoia,* which, in Greek tragedy, is practically spiritual sickness' universal form, the 'mistake' of madness being everywhere a substitution of appearance for reality. This *paránoia* is further associated with, for ancient Greeks, such morally dubious qualities as disobedience of the gods, ignorance of self, along with, as the idea in which these preceding qualities possess a proper context, a belief in the inessential or simply accidental arrangement of this universe. By this not, perhaps, entirely playful analogy modern thought is revealed, for the moment, not least in Kant's formulation, as

[132] *The Jerusalem Bible: N.T.,* op. cit., 98.

structurally a *paránoia* in process whereby there takes place, as a logical extension of Descartes' original misconception (wherein he mistook the humanity or love of neighbor of sacred doctrine as a principle of humanity for a new science), a progressive displacement of reality itself (*noumena*) by appearances (*phenomena*), so that reason perceives itself beside itself, perceives beside things intelligible in themselves of which it knows nothing (*noumena*) appearances (*phenomena*) of which alone it has knowledge. Think of that two-way mirror which pure reason is, wherein pure practical reason as *noumenon* possessed of its own freedom (to which are subjoined the ideas of God and immortality) beholds itself, without interfering, on the other side of the mirror of itself completely immersed in a realm of appearances or natural necessity, on which outer side of the mirror alone *knowledge* is possible, but where reason understands itself *to know not itself* but *only that it is* as an original synthetic unity of apperception, that is, a transcendental self-consciousness or abstract logical presupposition. While on the inner, through-seeing side, the side of its freedom, it must simply *act,* although insofar as it proposes as an ultimate end of its moral action a *summum bonum,* it acts on faith, but a faith entirely derived out of its necessity to do its duty. That is, it acts on a faith in its own self; indeed, it might be said that it acts heroically, or, at least, that its task is of an heroic dimension. Pure reason is beside itself in a structural schism by which it is objectively divided from itself by that infinite indifference to particularity, *qua* particularity, that constitutes its transcendental unity. (To whatever extent is possible, Kant attempts to join these disparate realms of, on the one hand, moral freedom, on the other hand, natural necessity, to join, that is, understanding to freedom by means of the reflective judgment's principle of finality. But this pure reflection is neither knowledge nor faith, neither scientifically nor morally objective, but purely subjective. Its value therefore is ultimately determined by its usefulness to either scientific or moral enterprise, the objectivity of both of which realms is original to each respectively, i.e., independently of each other, but also of subjective reflection.) [133] So

[133] Cf. *The Critique of Judgement,* op. cit., passim.

within pure reason itself is reflected that external distinction between *noumena* and *phenomena* by which, through its peculiar 'mistake,' modern science dissociates knowledge from reality itself. Nor ought it to be objected that modern thought could not possibly be so radically a 'mistake' since it would not then have borne those fruits of invention which, on the contrary, testify to its palpable knowledge—such an objection ignores our certain experience that mistakes quite often bear fruit.

In order to relate Hegel to Kant properly it is necessary to pursue further pure reason's transcendental unity, and its infinite indifference to particularity. In scientific thought the particular is, as it were, in exteriority to itself, related immediately to the universal by its subsumption under law. Through law's universality, sensibly conditioned appearances become intelligible, but this immediate relationship presupposes an essentially relational or logical being for objects in respect of which *mathematics* is modern science's natural language. Ernst Cassirer in his *Determinism And Indeterminism In Modern Physics* calls attention to this novelty of modern natural science: "Even in the natural philosophy of antiquity the demand arises not to think of the basic elements out of which the cosmos is built up as separated from one another and as it were 'chopped off with an ax.' Rather it is maintained that in physical reality, 'all is in all' (ἐν παντὶ πάντα), that there is no strictly separated particular existence, but that everything has a share in everything else (οὐδὲ χωρὶς ἔστιν εἶναι ἀλλὰ πάντα παντὸς μοῖραν μετέχει). Anaxagoras, who made this notion the basic principle of natural explanation, thought of being in strictly qualitative and substantial terms. His assertion of this universal sharing therefore had to be grasped in the conception that in every apparently separate existence all the elements of being are really contained and are present in a sense as pieces. Every so-called particular thing is a *panspermia,* it contains in itself all the seeds out of which the universe as a whole is made up, so that flesh is not only flesh, but at the same time blood and bone, and blood and bone are simultaneously also flesh. This interfusion of all the elements of being had to give way to another conception when science liberated itself from the naive schema of things and substance. The postulate of 'all is in all' could no longer be satisfied by an intermixing of either things or original qualities, but only by way of an intermixing of laws. The statements of laws now provide the single and only allowable way of joining the particular to the whole and of uniting the whole with the particular, and thus of establishing that 'harmony' between them which is the real aim of all scientific knowledge. Since the Renaissance, since Kepler and Galileo, Descartes and Leibniz, scientific and philosophic thought has devised in the mathematical concept of function the ideal means of satisfying this demand. . . . Once thrown into this form, the phenomena of nature have been made fast in 'enduring thoughts,' in thoughts whose duration often extends far beyond the occasion which first instigated the setting up of the particular forms." [134] But Aristotle observed in his *On The Heavens* III.8: ". . . the shape of all the simple bodies is observed to be determined by the place in which they are contained. . . . therefore the shapes of the elements are not defined. . . . Here, as in everything else, the underlying matter must be devoid of form or shape, for so, as is said in the *Timaeus,* the 'receiver of all' will be best able to submit to modification. It is like this that we must conceive of the elements, as the matter of their compounds, and this is why they can change into each other, and lose their qualitative differences." [135] By this juxtaposition of Aristotle's thought with that of Anaxagoras, with that of Cassirer, it is readily seen that what ancient natural science shares with modern science, over against Aristotle, is an understanding of matter's intelligibility *in se.* Whereas for Aristotle matter's form, its potential intelligibility, is a product of an essentially determined preexistent natural order, while its actual intelligibility, presupposing that natural order, prescinds therefrom to essential identity with particulars, *qua* particular. Nevertheless Anaxagoras shares with Aristotle a *qualitative* form, as Cassirer emphasizes; this indicates to us that he is ancient, that is, that he does *not* understand as Kant does that nature's sensible substrate is itself a

[134] E. Cassirer, *Determinism And Indeterminism In Modern Physics* (trans. O. T. Benfey, New Haven, 1956), 37–38.

[135] Aristotle, *On The Heavens* (trans. W. K. C. Guthrie, Cambridge, Mass., 1945), 319.

product of human intuition; whereas so to understand, as Kant does, is to be modern: It is to perceive that intuition is conceptualized only by virtue of pure reason's original synthetic unity of apperception, in the image of reason's purely functional concept of its very own identity. Thus, mathematics is modern science's ideal language simply because nature is a reflection of pure reason's internal order: infinite indifference to its own particularity, as a condition of its mastery over nature, in knowledge, over self, in action. In other words what has happened since the Renaissance happened before the Renaissance. Nor did it involve a rebirth of ancient thought or consist in using a mathematical method. It presupposes nothing other than an alteration in being or a radical reorientation of reason's self-conception, in which substantiality of self or of object known is reduced to being a matter of pure subjective judgment. In this consideration of pure reason's structure of infinite indifference there opens before us yet another perspective on this change of mind in history. It is this: Pure reason's transcendental unity is *pure potentiality*. In fact, if reason were not by definition a form one would be tempted to understand it as matter itself; or, what emerges out of the original synthetic unity of apperception is that its indifference is in itself an indifference to being matter or form. It is, indeed, as a pure abstraction an identification of these two, but since, in Kant, this identity is still an abstraction pure reason is, as a matter of fact, conditioned by matter, or divided. In Kant therefore is most clearly seen that absolute potentiality or state of pure abstraction that reason is in essence's absence. It is understood that this state is prepared for in sacred doctrine. Thomas Aquinas writes in *Summa Theologica* I,86,1: "Our intellect cannot know the singular in material things directly and primarily. The reason of this is that the principle of singularity in material things is individual matter, whereas our intellect . . . understands by abstracting the intelligible species from such matter. Now what is abstracted from individual matter is the universal. Hence our intellect knows directly the universal only." [136] An object is

[136] *Summa Theologica,* op. cit., 440–441.

known as an actually existing other; therefore not, as in Aristotle, by intellect's identification with its essence, but rather, in Thomas, through a medium of universal being it is known as an object materially individuated, a form divided by matter.[137] Now while Thomas states very clearly that soul's intellectual power, *qua* intellectual, is not the act of a corporeal organ, it, nevertheless, as he states in *Summa Theologica* I,76,1 ad 1: ". . . exists in matter so far as the soul itself, to which this power belongs, is the form of the body, and the term of human generation." [138] It therefore follows that natural reason is, *as a matter of fact,* divided according to the number of bodies, that is, divided by matter insofar as it is a power of soul. Thus the constitution of the object known reflects its knower's constitution. Thomas then says in *Summa Theologica* I,76,2 ad 2: "Everything has unity in the same way that it has being; consequently we must judge of the multiplicity of a thing as we judge of its being. Now it is clear that the intellectual soul, by virtue of its very being, is united to the body as its form; yet, after the dissolution of the body, the intellectual soul retains its own being. In like manner the multiplicity of souls is in proportion to the multiplicity of bodies; yet, after the dissolution of the bodies, the souls retain their multiplied being." [139] Here, Thomas steps out of Aristotle's metaphysical order into a realm of transcendental indifference: As a matter of fact, reason's individual identity is eternally, coincidentally a matter of form, a form of matter. In Thomas this indifference is merely a matter of fact, not a matter of necessity, since difference belongs to individuality by virtue of its being essentially, that is, absolutely, in its very act of being, related to another, God, as its creator. Or, God's omnipotent creative act enables natural reason to exist as if it were simply, as Aristotle would see it, its own essence, thereby elevating matter itself to intelligibility as *materia signata.* As long as natural reason exists explicitly in revelation's light its self-subsistence as form, its priority to matter, is con-

[137] Cf. above, p. 70.
[138] *Summa Theologica,* op. cit., 372.
[139] Ibid., 374.

crete. Once, however, it becomes in Descartes a matter of its own thinking, it is in principle, as is explicit in Kant, a purely logical entity, pure reason, whose nature is, in God's absence, conditioned by matter as a matter of necessity. Pure reason stands under a law of its own necessity, or indifferently, under a law of its own freedom; but, divided from matter, apart in its freedom as pure form; but experiencing matter as its limiting condition. For matter has, since its creation, acquired a reality for thought that it lacked in Aristotle's metaphysics; this is clear in Thomas' epistemology as it relates to matter as a principle of individuality; in Descartes' establishing an existence for material entities on a basis of God's veracity; but finally in Kant's understanding that natural entities themselves appear only through pure sensible intuition's own formality. We come then to see that *pure reason's infinite indifference to particularity when directed externally is actually a reflection of its own internal state of objectivity, or pure potentiality, which is the obverse of matter's reality for thought.*

For Hegel, Kant's transcendental Ideal, although the latter did not see it so, is real. Listen to Hegel in his *Lectures On The History of Philosophy* III: "The reason why that true Idea should not be the truth is therefore that the empty abstractions of an understanding which keeps itself in the abstract universal, and of a sensuous material of individuality standing in opposition to the same, are presupposed as the truth. Kant no doubt expressly advances to the conception of an intuitive or perceiving understanding, which, while it gives universal laws, at the same time determines the particular. . . . For 'to knowledge there also belongs intuitive perception, and the possession of a perfect spontaneity of intuition would be a faculty of knowledge' specifically 'distinct from the sensuous, and quite independent thereof, and therefore it would be understanding in the most universal sense. Consequently it is possible to think of an intuitive understanding which does not pass from the universal to the particular . . .' . . . . But that this '*intellectus archetypus*' is the true Idea of the understanding, is a thought which does not strike Kant. Strange to say, he certainly has this idea of the intuitive; and he does not know why it should have no truth—

except because our understanding is otherwise constituted, namely such 'that it proceeds from the analytic universal to the particular.' But absolute Reason and Understanding in itself . . . in Kant's view . . . have no reality in themselves. . . . In spite of their directly and definitely expressed non-absoluteness, they are yet looked on as true knowledge; and intuitive Understanding, which holds Notion and sensuous perception in one unity, is looked on as a mere thought which we make for ourselves." [140] If with Descartes was discovered, as the content of a judgment of perception, the universal; if Kant introduced the particular into the universal, as the content of a judgment of experience; Hegel may be understood to arrive at the particular itself, as the content of a judgment of intuition. It is Hegel's thought that is a solution of that abstract division within pure reason by which is maintained, irresolutely, an infinite coincidence of freedom with necessity, form with matter. In this respect Hegel is to Kant as Aristotle was to Plato; as with Aristotle so with Hegel the solution lies in discovering *essence.* But Hegel is a modern intellect, his essence is Kant's ideal realized. It is a divine reason or intuitive judgment that determines particulars to exist through the synthetic universality of its thought; that is, it determines particulars to exist through an absolute disjunctive judgment as parts of a totality.[141] It is evident that this is not Aristotle's essence, nor Aristotle's God, for Aristotle's pure thought is neither a judgment nor a thought of a totality of existence. It might well be Thomas Aquinas' God, except that it is pure reason's essence. That is, *Hegel discovers human self-consciousness to be essentially divine reason.*[142] In Descartes the idea of God as creator is innate; in Kant it is reason's transcendental ideal; in Hegel it is the very essence of humanity, but, of humanity as universal or species. With the realization of the synthetic universal in Hegel pure reason's infinite indifference is superseded. That is, its radical duality is

[140] *Hegel's Lectures On The History of Philosophy* III, op. cit., 472–474. Hegel's citations here from Kant are to be found in *The Critique of Judgement* II, op. cit., 60ff.

[141] Cf. above, p. 102.

[142] Cf. above, pp. 66–67. Also pp. 85–86.

overcome in absolute reason's identity in difference, so that, to speak abstractly about a most concrete matter, absolute reason *identifies two as such.* Now, since our analysis has led us to see that modern thought's objectivity increases in proportion to matter's reality for thought we are prepared to find in Hegel's philosophy of absolute reason not only that perfect objectivity of reason's being itself divine Idea, but also matter's perfect reality; nor shall we be disappointed. Absolute reason in identifying *two as such,* identifies, in Platonic terms, the *duad,* or matter *per se:* It is identified as essence. If in Kant pure reason, on the one hand, resolved itself into transcendental indifference, while, on the other hand, reality itself disappeared into indifferent *noumena,* it is Hegel's thought that determines this coincidence, or indifference, in the totality of things as matter's self-relating essence. Hegel says in *The Logic:* "Matter, being the immediate unity of existence with itself, is also indifferent towards specific character. Hence the numerous diverse matters coalesce into the one Matter, or into existence under the reflective characteristic of identity. In contrast to this one Matter these distinct properties and their external relation which they have to one another in the thing, constitute the Form,—the reflective category of difference, but a difference which exists and is a totality. This one featureless Matter is also the same as the Thing-by-itself was: only the latter is intrinsically quite abstract, while the former essentially implies relation to something else, and in the first place to the Form. . . . Thus the Thing suffers a disruption into Matter and Form. Each of these is the totality of thinghood and subsists for itself. But Matter, which is meant to be the positive and indeterminate existence, contains, as an existence, reflection-on-another, every whit as much as it contains self-enclosed being. . . . But Form, being a complete whole of characteristics, *ipso facto* involves reflection-into-self; in other words, as self-relating Form it has the very function attributed to Matter. Both are at bottom the same. . . . The Thing, being this totality, is a contradiction. . . . Thus the thing is the essential existence, in such a way as to be an existence that suspends or absorbs itself in itself. In other words, the thing is an Appearance or Phenomenon. . . . The Essence must appear or shine forth." [143] Descartes' rational intuition was a presuppositionless phenomenon, but totally immaterial. Kant, understanding Descartes' perception as mere subjectivity, concretized it as an appearance, but presupposing pure reason's transcendental horizon of objectivity, also presupposing *noumena.* Hegel, once again, eliminates presuppositions by taking the appearance as the thing in itself, the totality of which is an internal mutual presupposing of matter with form. So appearance is an intelligible revelation of matter as essence. *The appearance is the thing, or essence as self-determining difference in existence.* The glass is absolutely black; no, the glass is taken away. Hegel doubts absolutely. Not only is the object determined as to knowledge as in Kant. But in Hegel it is absolutely determined in being determined as to knowledge; that is, it is determined to existence by absolute reason, since in Hegel is reestablished Descartes' principle, apparently suspended in Kant, of the identity of thought and being. Thus absolute reason creates particulars whose very existence, *qua* particular, that is, whose very essence, is determined by reference to a synthetic universal. In other words, absolute reason determines uniqueness itself to be related to its essential totality as self-determination. Descartes understood himself to be a *res cogitans,* a thinking thing, but for Hegel it is that there is a thinking thing for which this man's existence is a thought; this thing that thinks my being is absolute reason. That pure potentiality of Kant's reason is, by Hegel's expedient absolute doubt, whatever actually is, so that he is able to write in *The Logic:* "*Logic therefore coincides with Metaphysics, the science of things set and held in thoughts,*—thoughts accredited able to express the essential reality of things." [144] Or again, "The real nature of the object is brought to light in reflection; but it is no less true that this exertion of thought is *my* act. If this be so, the real nature is a *product* of *my* mind, in its character of thinking subject—generated by me in my simple universality, self-collected and

---

[143] *The Logic of Hegel* (trans. W. Wallace, 2nd ed., Oxford, 1892), 236–239.
[144] Ibid., 45.

removed from extraneous influences,—in one word, in my Freedom." [145] Essence or real nature is simply the content of absolute reason or of a mind that contradicts itself. Absolute reason takes up the position of Aristotle's intellect, as it were; that is, matter in itself is nothing for it apart from its own identity. In Hegel, Aristotle's self-not-evident-to-reason is absolutely self-evident reason or objectivity. Absolute reason is the divine *technē*, a synthetic universal, so that God, who for Aristotle was *like* man's essence (therefore identically different), is human reason's essence, indifferent to the distinction of matter. Again Hegel says in *The Logic:* ". . . it is an abstraction of the understanding which isolates matter into a certain natural formlessness. . . . no formless matter appears anywhere even in experience as existing. Still the conception of matter as original and pre-existent, and as naturally formless, is a very ancient one; it meets us even among the Greeks. . . . Such a conception must of necessity tend to make God not the Creator of the world, but a mere world-moulder or demiurge. A deeper insight into nature reveals God as creating the world out of nothing. And that teaches two things. On the one hand it enunciates that matter, as such, has no independent subsistence, and on the other that the form does not supervene upon matter from without, but as a totality involves the principle of matter in itself." [146] Thus God's creation of the world is not a fact, as in Thomas' *novitas mundi,* but the necessity of a deeper insight; when absolute reason comprehends creation it does so universally as the existent individuality (essence) of all things. Hegel looks back on Kant's abstract understanding as something disowned; Hegel has passed through into the Looking Glass, into the realm of absolute thought. He writes in his *Science of Logic:* "This realm is truth as it is without veil and in its own absolute nature. It can therefore be said that this content is the exposition of God as he is in his eternal essence before the creation of nature and a finite mind. . . . What we are dealing with in

[145] Ibid., 44.
[146] Ibid., 236–237.

logic is not a thinking *about* something . . . on the contrary, the necessary forms and self-determinations of thought are the content and the ultimate truth itself." [147]

The practical implications of absolute reason Hegel lays out in his introduction to *The Philosophy of History* where he writes: "The insight then to which—in opposition to these ideals [ideals of imagination]—philosophy should lead us is that the actual world is as it ought to be, that the truly good, the universal divine Reason is the power capable of actualizing itself. This good, this Reason, in its most concrete representation, is God. God governs the world. The actual working of His government, the carrying out of His plan is the history of the world. . . . Simply and abstractly, it is the activity of the subjects in whom Reason is present as their substantial essence in itself, but still obscure and concealed from them." [148] Or further, "World history . . . represents the phases in the development of the principle whose *content* is the consciousness of freedom. . . . All we have to indicate here is that Spirit begins with its infinite possibility, but *only* its possibility. As such it contains its absolute content within itself, as its aim and goal, which it attains only as result of its activity. Then and only then has Spirit attained its reality. Thus, in existence, progress appears as an advance from the imperfect to the more perfect. But the former must not only be taken in abstraction as the merely imperfect, but as that which contains at the same time its own opposite, the so-called perfect, as germ, as urge within itself. In the same way, at least in thought, possibility points to something which shall become real; more precisely, the Aristotelian *dynamis* is also *potentia,* force and power. The imperfect, thus, as the opposite of itself in itself, is its own antithesis, which on the one hand exists, but, on the other, is annulled and resolved. It is the urge, the impulse of spiritual life in itself, to break through the hull of nature, of sensuousness, of its own self-alienation, and to attain

[147] *Science of Logic* (trans. A. V. Miller, London, 1969), 50.

[148] G. W. F. Hegel, *Reason In History* (trans. R. S. Hartman, Indianapolis, 1953), 47–48.

the light of consciousness, namely, its own self." [149] Absolute reason is the idea of freedom actualizing itself, according to that dynamic structure of essence, self-contradiction, in historical process; it is a perfectly cognizable providence whose reality lies in its result. As totality, its conclusion is not in any particular person or nation, but lies beyond all particulars, although its essence is theirs, namely, consciousness of freedom. This notion of development is spoken of in *The Logic:* "The movement of the notion is as it were to be looked upon merely as play: the other which it sets up is in reality not an other. Or, as it is expressed in the teaching of Christianity: not merely has God created a world which confronts Him as an other; He has also from all eternity begotten a Son in whom He, a Spirit, is at home with Himself." [150] For Kant Christianity introduced a new content for philosophy by chance, as it were. But for Hegel Christianity is the manifestation of the truth of the idea of history, as he writes in the *Philosophy of History:* " 'When the fulness of the time was come, God sent his Son,' is the statement of the Bible. This means nothing else than that *self-consciousness* had reached the phases of development [Momente], whose resultant constitutes the Idea of Spirit, and had come to feel the necessity of comprehending those phases absolutely." [151] Or again, of Christ, ". . . in him the idea of eternal truth is recognized, the essence of man acknowledged to be Spirit . . ." [152] So for Hegel is Christianity the Absolute Religion of absolute reason. But the *novitas mundi,* the novelty of this universe, is nowhere in sight, it has been swallowed up in the 'feeling of necessity.'

---

[149] Ibid., 70–71.

[150] *The Logic of Hegel,* op. cit., 289.

[151] G. W. F. Hegel, *The Philosophy of History* (trans. J. Sibree, New York, 1956), 319.

[152] Ibid., 328.

Chapter 6

KIERKEGAARD AND LESSING: THE LEAP OF FAITH

Kierkegaard writes in his *Concluding Unscientific Postscript To The Philosophical Fragments:* "From the speculative standpoint, Christianity is viewed as an historical phenomenon. The problem of its truth therefore becomes the problem of so interpenetrating it with thought, that Christianity at last reveals itself as the eternal truth. The speculative approach to the problem is characterized by one excellent trait: it has no presuppositions. It proceeds from nothing, it assumes nothing as given, it begs no postulates. . . . And yet, something is after all assumed: Christianity is assumed as given. Alas and alack! philosophy is altogether too polite. How strange is the way of the world! Once it was at the risk of his life that a man dared to profess himself a Christian; now it is to make oneself suspect to venture to doubt that one is a Christian. Especially when this doubt does not mean that the individual launches a violent attack against Christianity with a view to abolishing it; for in that case it would perhaps be admitted that there was something in it. But if a man were to say quite simply and unassumingly, that he was concerned for himself, lest perhaps he had no right to call himself a Christian, he would indeed not suffer persecution or be put to death, but he would be smothered in angry glances, and people would say: 'How tiresome to make such a fuss about nothing at all; why can't he behave like the rest of us, who are all Christians? It is just as it is with F.F., who refuses to wear a hat on his head like others, but insists on dressing differently.' And if he happened to be married, his wife would say to him: 'Dear husband of mine, how can you get such notions into your head? How can you doubt that you are a Christian? Are you not a Dane, and does not the geography say that the Lutheran form of the Christian religion is the ruling religion in Denmark? For you are surely not a Jew, nor are you a Mohammedan; what then can you be if not a Christian? It is a thousand years since paganism was driven out of Denmark, so I know you are not a pagan. Do you not perform your duties at the office like a conscientious civil servant; are you not a good citizen of a Christian nation, a Lutheran Christian state? So then of course you must be a Christian.' We have become so objective, it seems, that even the wife of a civil servant argues to the particular individual from the totality, from the state, from the community-idea, from the scientific standpoint of geography." [153] It is Hegel who carries Descartes' doubt through to absolute existence. Thus, in Hegel, *thought is existence itself;* Kant's dualism, in which Descartes' doubt was extended to the frontier of the noumenal, thereby securing that intervening space of experience for thought, is superseded by Hegel's absolute reason which identifies reality itself with appearance. That is, Hegel *doubts difference itself* or his absolute reason thinks what is different in things in themselves to be what appears. The difference in things themselves, their essence, appears as the thought of an absolute reason. In this way is realized Kant's transcendental ideal, namely, that there exists a being which determines everything to exist by having its being as a part of the totality of being, or, identically, of the totality of its universal thought. The particular, *qua* particular, is known to exist by virtue of its participation in that totality of absolute reason which, as divine Spirit, actualizes itself in history through a dynamic process of self-contradiction, which, in turn, is reason's realizing its own unconditional freedom. It is as instruments in this self-realization of Spirit in history that both individuals and states

[153] S. Kierkegaard, *Concluding Unscientific Postscript To The Philosophical Fragments* (trans. D. F. Swenson and W. Lowrie, Princeton, 1941), 49–50.

have their substantial existence. Hegel writes in the *Philosophy of Mind:* "This liberation of mind, in which it proceeds to come to itself and to realize its truth, and the business of so doing, is the supreme right, the absolute Law. The self-consciousness of a particular nation is a vehicle for the contemporary development of the collective spirit in its actual existence: it is the objective actuality in which the spirit for the time invests its will. Against this absolute will the other particular natural minds have no rights: *that* nation dominates the world: but yet the universal will steps onward over its property for the time being, as over a special grade, and then delivers it over to its chance and doom. To such an extent as this business of actuality appears as an action, and therefore as a work of *individuals,* these individuals, as regards the substantial issue of their labor, are *instruments,* and their subjectivity, which is what is peculiar to them, is the empty form of activity. What they personally have gained therefore through the individual share they took in the substantial business (prepared and appointed independently of them) is a formal universality or subjective mental idea—*Fame,* which is their reward." [154] Nor does Hegel deny that individuals possess in themselves an infinite value as ends in themselves.[155] As Kant understood abstractly in his *Groundwork Of The Metaphysic of Morals:* ". . . rational beings all stand under the *law* that each of them should treat himself and all others, *never merely as a means,* but always *at the same time as an end in himself*. But by so doing there arises a systematic union of rational beings under common objective laws—that is, a kingdom. Since these laws are directed precisely to the relation of such beings to one another as ends and means, this kingdom can be called a kingdom of ends (which is admittedly only an Ideal)." [156] But here as everywhere Hegel takes the Kantian *ideal* as *essence:* Kant's kingdom comes in Hegel to be the Absolute Law of the World Spirit which manifests itself in the *state,* which understood comprehensively is the total life of a people,

---

[154] G. W. F. Hegel, *Philosophy of Mind* (trans. W. Wallace, Oxford, 1971), 281.

[155] Cf. *Reason In History*, op. cit., 43–49.

[156] I. Kant, *Groundwork Of The Metaphysic of Morals* (trans. H. J. Paton, New York, 1964), 101.

a 'spiritual individual,' the 'divine Idea as it exists on earth. Thus Hegel understands, as over against Kant, that as a matter of necessity the duality of ends and means, as for Hegel every duality, is overcome in this spiritual life of the state, so that, while Hegel admits abstractly this distinction between private, subjective life on the one hand, and objective, public life on the other, essentially, that is, actually, self-contradiction prevails over Kant's horror of this same, and Hegel is able to say in the introduction to the *Philosophy of History:* "This spiritual content then constitutes the essence of the individual as well as that of the people. It is the holy bond that ties the men, the spirits together. It is one life in all, a grand object, a great purpose and content on which depend all individual happiness and all private decisions. The state does not exist for the citizens; on the contrary, one could say that the state is the end and they are its means. But the means-end relation is not fitting here. For the state is not the abstract confronting the citizens; they are parts of it, like members of an organic body, where no member is end and none is means. It is the realization of Freedom, of the absolute, final purpose, and exists for its own sake. All the value man has, all spiritual reality, he has only through the state. For his spiritual reality is the knowing presence to him of his own essence, of rationality, of its objective, immediate actuality present in and for him. Only thus is he truly a consciousness, only thus does he partake in morality, in the legal and moral life of the state. For the True is the unity of the universal and particular will." [157] So, for Hegel, the end which the individual is in himself is none other than the end of the state: Absolute reason *identifies the two as such,* this identity in difference is Hegel's divine essence of man, which, as Descartes had superalternated doubt to identity[158], absolutely internalizes contradiction so as to radically identify every other, *qua* other, with self-actualization through the medium of the synthetic universal. Only insofar as an individual is considered with respect to a corrupt or otherwise bankrupt state is it real to speak of his

[157] *Reason In History*, op. cit., 52–53.
[158] Cf. above, p. 80.

essence in isolation, but, as Hegel says in his introduction to the *Philosophy of History*, this individual is then thought of in opposition not to the *essence* of the state, but merely to one of its *forms*, which according to the limits of nature is transitory, or corruptible.[159] The realization of Kant's Ideal in Hegel is the positive philosophical expression of the totalitarian idea of the Spirit of Man. The Hegelian Idea or *essence* embraces, as it were, every *other* as its own in a substantial unity of purpose. This unity of purpose extending, as it does, to an individual's essence, indeed, constituting his essence, is substantially different from Aristotle's conception of political association, which, although it subordinated the good of the individual to that of the state, did not do so absolutely, but rather recognized a limit, namely, an essentially ordered preexistent nature, within which human nature was one among others, so that, as a matter of fact, the principle of identity, not being able to be God, as in Hegel, had to be each individual's essential priority to his own nature, to human nature.[160] Or, in terms of our previous analysis, Aristotle distinguished essence from nature so that he knew no self-subsisting entity *humanity; a fortiori* no universal essence or absolute reason in process of achieving its freedom. But humanity is a modern invention, attributable to Descartes, the will of which to freedom is related to particular individuals *abstractly* in Descartes, *ideally* in Kant, *actually* in Hegel. The subjective isolation of Descartes' *cogito ergo sum* is reflected in Kant's autonomy of the will. But in Kant humanity's moral autonomy is divided from nature by that abstract, purely formal, self-conception that could not know its unconditioned other without contradicting itself. In its abstract formulation by Kant it is, however, already explicit that pure reason is a law unto itself prior to nature as the latter's *ought*. It is left to Hegel to unite Reason's autonomy of will to natural necessity, that is, to perceive that *what is actually is what ought to be*. As a result, not only is science, as in Kant, ultimately subordinated to practical interest, to human interest, but, since, in Hegel, logic is become

[159] *Reason In History*, op. cit., 48.
[160] Cf. above, pp. 44–46.

metaphysics, that is, since Kant's Ideal is recognized as Absolute Reason's synthetic universality determining particulars, in determining them as to knowledge, to existence itself, then, manifestly, existence itself is ordered to practical, that is, to human interests, and the particular individual discovers his very being to be determined in this totalitarian circle of humanity.[161] This man stands over against the infinite practical reason; he finds himself in the situation of either/or, as Hegel writes in the *Philosophy of Right:* "Once self-consciousness has reduced all otherwise valid duties to emptiness and itself to the sheer inwardness of the will, it has become the potentiality of either making the absolutely universal its principle, or equally well of elevating above the universal the self-will of private particularity, taking that as its principle and realizing it through its actions, i.e., it has become potentially evil." [162] Hegel further adds, "To have a conscience, if conscience is only formal subjectivity, is simply to be on the verge of slipping into evil; in independent self-certainty, with its independence of knowledge and decision, both morality and evil have their common root." [163] To will one's own private particularity, or, for Hegel, immediate, natural existence, is evil; its opposite, the good, consists in willing one's rational essence, the universal. Beyond this either/or nothing exists.

It is this absolute objectivity that Kierkegaard attributes to the good Danish civil servant's wife as she argues to his being a Christian (it having occurred to him to doubt it) from the fact that he is a citizen of a Lutheran Christian State, from the fact that he conscientiously does his duty, from the fact that historically Christianity supplanted paganism in Denmark over a thousand years ago. The truth of the wife's analysis is not, as Kierkegaard sees it, that Christianity can be so determined, but that modern thought takes it to be so determined, that is, to be simply an historical phenomenon, therefore, as Kant understood it, a subject of possible knowledge, or simply a matter of

[161] Cf. above, pp. 130–131.
[162] G. W. F. Hegel, *Philosophy of Right* (trans. T. M. Knox, Oxford, 1952), 92.
[163] Ibid.

fact.[164] Kant, of course, presupposed the unknowableness of things in themselves. Therefore, he distinguished historical facts of Christianity from its dogmatic understanding of transcendent reality; truths of this latter kind Kant reinterpreted, where they were compatible with pure reason's autonomy, as archetypes of human freedom, or, where that procedure is not feasible, declared them inscrutable mysteries irrelevant for practical purposes, if not an actual danger to right thinking in moral matters. But that irresolute presupposition of knowledge's being limited to conditions of sensible intuition, that is, of knowing *phenomena* but not *noumena,* appeared to Hegel, without doubt, as an incomprehensible philistinism on Kant's part, especially with respect to Christianity wherein Hegel perceived not a symbolic truth but a revelation in history of man's essential identity with God, or the World Spirit actualizing itself in self-consciousness. Since Hegel understands that appearance is reality, that reason is intuitively cognizant of essence as an object's substance known, it follows that what Kant understood as transcendent, or unknowable for us, is, for Hegel, immanent, knowable for us; Christianity is that historical appearance in which self-consciousness manifests itself as Spirit; that is, it shows itself in its transcendence to nature, in its infinity. Listen to Hegel in the *Philosophy of History:* "Man himself . . . is comprehended in the Idea of God, and this comprehension may be thus expressed—that the unity of Man with God is posited in the Christian Religion. But this unity must not be superficially conceived, as if God were only Man, and Man, without further condition, were God. Man, on the contrary, is God only in so far as he annuls the merely Natural and Limited in his Spirit and elevates himself to God. That is to say, it is obligatory on him who is a partaker of the truth, and knows that he himself is a constituent [Moment] of the Divine Idea, to give up his merely natural being: for the Natural is the Unspiritual. . . . This implicit unity exists in the first place only for the thinking speculative consciousness; but it must also exist for the sensuous, representative consciousness—it must become an object

[164] Cf. above, p. 120.

for the World—it must *appear,* and that in the sensuous form appropriate to Spirit, which is the human. *Christ has appeared*— **a Man who is God—God who is Man . . . . in him the idea of** eternal truth is recognized, the essence of man acknowledged to be Spirit, and the fact proclaimed that only by stripping himself of his finiteness and surrendering himself to pure self-consciousness, does he attain the truth." [165] Here is to be seen that Christianity is for Hegel that historical phenomenon whereby is recognized the *eternal* truth of man's Spiritual essence, his infinity, as opposed to his finitude, that is, his immediate, natural being. But as we know, Hegel understands that this immediate, natural existence constitutes that private particularity which is *evil,* unspiritual, opposed to the universal. So Christianity is that objectivity so objective that for our good Dane to question his being a Christian, to place himself in opposition to his neighbors, to his wife, constitutes what evil is for objectivity, or, at least, constitutes a state of conscience on the question of this individual's being for himself a Christian, therefore, placing him on the verge of that self-will by which he might dissociate himself from that totality that is Christian Denmark. But Kierkegaard perceives, precisely in that nest of everydayness he portrays, that to be a Christian is, in reality, no little difficulty in a situation where, Christianity being assumed to be an eternal truth, a man is understood to be essentially, that is, *qua* man, a Christian. This difficulty was further compounded by Hegel's thought that Christianity's content is God's being essential humanity. It is not that Hegel would deny to Kierkegaard's civil servant his subjective freedom; indeed it is recognized to result precisely from Christianity's advent in history. Hegel says in the *Philosophy of Right:* "The right of the subject's particularity, his right to be satisfied, or in other words the right of subjective freedom, is the pivot and centre of the difference between antiquity and modern times. This right in its infinity is given expression in Christianity and it has become the universal effective principle of a new form of civilization. Amongst the primary shapes which this right assumes are love,

[165] *The Philosophy of History*, op. cit., 324, 328.

romanticism, the quest for the eternal salvation of the individual, &c.; next come moral convictions and conscience; and, finally the other forms. . . . Now this principle of particularity is, to be sure, one moment of the antithesis, and in the first place at least it is just as much identical with the universal as distinct from it. Abstract reflection, however, fixes this moment in its distinction from and opposition to the universal and so produces a view of morality as nothing but a bitter, unending, struggle against self-satisfaction . . ." [166] So for Hegel this principle of particularity is Christianity's product, but it is not absolute freedom. Absolute freedom is its necessary unity with humanity, that is, universal will. Particularity is finally related to universality; its right to private satisfaction is to absolute reason's concretion only an abstraction. So once again with Hegel this particular man is confronted with an either/or: Either surrender to universal humanity, or pursue a merely private desire: either good/or evil. But for Kierkegaard these alternatives deceitfully conceal a third possibility, that of *faith,* that category whereby an individual, *qua* individual, is related neither, on the one hand, to universal humanity, nor, on the other hand, to his own private desires, but absolutely to God. Kierkegaard explores this question of faith by examining, in *Fear and Trembling,* the story of Abraham's sacrifice of Isaac at God's command. In *Genesis* 22:1–2 is to be read: "It happened some time later that God put Abraham to the test. 'Abraham, Abraham' he called. 'Here I am' he replied. 'Take your son,' God said 'your only child Isaac, whom you love, and go to the land of Moriah. There you shall offer him as a burnt offering, on a mountain I will point out to you.' " [167] Now, for such a command, reason possesses no context (Kant, who is less absolute than Hegel, would take this situation as a case of demonic possession, or, if it were assumed that Abraham is after all dealing with God himself, Kant would introduce a doubt as to whether or not Abraham might have misunderstood what God required, especially since what is required is plainly an evil act; [168] Hegel is abso-

[166] *Philosophy of Right*, op. cit., 84.

[167] *The Jerusalem Bible: O.T.,* op. cit., 38.

[168] Cf. *Religion Within The Limits of Reason Alone,* op. cit., 81–82, and 175, for the basis of this estimate of Kant's reaction.

lutely silent). Even Abraham is silent; he simply obeys. Kierkegaard considers Abraham explicitly in light of Hegel's thought. He writes in *Fear and Trembling:* "The ethical as such is the universal, and as the universal it applies to everyone, which may be expressed from another point of view by saying that it applies every instant. . . . Conceived immediately as physical and psychical, the particular individual is the individual who has his *telos* in the universal, and his ethical task is to express himself constantly in it, to abolish his particularity in order to become the universal. . . . If this be the highest thing that can be said of man and of his existence, then the ethical has the same character as man's eternal blessedness, which to all eternity and at every instant is his *telos*. . . . If such be the case, then Hegel is right when in his chapter on 'The Good and the Conscience,' he characterizes man merely as the particular and regards this character as 'a moral form of evil' which is to be annulled in the teleology of the moral, so that the individual who remains in this stage is either sinning or subjected to temptation (*Anfechtung*). On the other hand, Hegel is wrong in talking of faith, wrong in not protesting loudly and clearly against the fact that Abraham enjoys honor and glory as the father of faith, whereas he ought to be prosecuted and convicted of murder. For faith is this paradox, that the particular is higher than the universal—yet in such a way, be it observed, that the movement repeats itself, and that consequently the individual, after having been in the universal, now as the particular isolates himself as higher than the universal. If this be not faith, then Abraham is lost, then faith has never existed in the world . . . because it has always existed." [169] For Kierkegaard, if Abraham is the father of faith, faith is manifestly other than modern thought's universal practical idea. Abraham's faith is absolute inwardness that is, as it were, *pure act of faith;* as opposed to that passivity of reason, absolute in Hegel, which has its other as a moment to be annulled in reason's universal self-realization, faith as pure act obeys another absolutely, that is, an other, absolutely another, God. This is demonstrated by Abraham's

[169] S. Kierkegaard, *Fear and Trembling and The Sickness Unto Death* (trans. W. Lowrie, Princeton, 1941), 64–65.

being commanded to sacrifice Isaac whom he loves: 'Take your son,' God said, 'your only child Isaac, whom you love, and go to the land of Moriah.' In his obedience to God he obeys an absolute other because in obedience he slays himself, both his desire for his son, his love, as well as his humanity, his universal self-consciousness; in his act of faith Abraham lost himself essentially; his very being for thought, which, as Hegel makes finally evident, is only what is real, is lost, his very substance. As Kierkegaard says: "In so far as the universal was present, it was indeed cryptically present in Isaac, hidden as it were in Isaac's loins, and must therefore cry out with Isaac's mouth, 'Do it not! Thou art bringing everything to naught!' "[170] Abraham's act is a transcendence to self (on the assumption that this act of his is faith); but not a transcendence of public to private self, or of good to evil, rather of a consecrated self, a holy will, to moral integrity (not that Abraham lacks moral integrity: as Kierkegaard stresses, he is first in the universal; Abraham loves Isaac). If in this act of faith there is no other self, a self existing absolutely in its particularity, prior to thought, not prior in natural immediacy, but prior to thought after thought in a new immediacy, Abraham is lost. If faith is, as it is for Kant, a moral disposition of pure practical reason, or, as with Hegel, a form of knowledge, the content of which is an eternal truth or eternal history of Spirit, if faith is a conception introduced by Christianity, but freely appropriated by Kant's pure reason as 'confidence in the promise of the moral law,'[171] or Christianity's comprehension of Hegel's absolute reason, then faith has always existed implicitly or explicitly, as a necessary corollary of reason's eternality. But then Abraham is not the father of faith, so that faith, because it has always existed, does not, for Kierkegaard, exist. It is clear then that faith involves a transcendence to self-consciousness, an absolute otherness, on the one hand, on God's part who communicates his will, and, on the other hand, on Abraham's part, an absolute particularity in existence, an infinite personal otherness (think of the good

[170] Ibid., 70.
[171] Cf. above, pp. 119–120.

Danish civil servant!) whereby God's will is able to be done in opposition to one's very own humanity, that is, to one's universality or participation in totality (in this connection Kierkegaard cites the New Testament: "In Luke 14:26, as everybody knows, there is a striking doctrine taught about the absolute duty toward God: 'If any man cometh unto me and hateth not his own father and mother and wife and children and brethren and sisters, yea, and his own life also, he cannot be my disciple.' This is a hard saying, who can bear to hear it? For this reason it is heard very seldom" [172]). Now faith is also for Kierkegaard something that must not always have existed, not even, we might imagine, as an innate idea of Descartes. But faith, if it exists, is unprepared for even in that totality which has internalized presuppositions in self-actualizing contradiction because that system, Hegel's synthetic universal, is an eternal Idea which 'feels faith's necessity'; but, for Kierkegaard, faith, if it exists, is a novelty, indeed, it is a freedom undetermined absolutely by reason, but absolutely obedient to God's word. If modern science following Descartes' *cogito ergo sum* exists absolutely within the horizon of its own thought, so, for Kierkegaard, the knight of faith exists absolutely within the horizon of his own belief, *credo ergo sum*. But, whereas modern human consciousness is potentially-actually determinate of the totality of being, whereas it is 'at home with itself,' keeping itself company, as it were, like the Triune God before creation, Kierkegaard's knight of faith is absolutely isolated, an other infinitely. Were he an Aristotelian essence he might be a happy *daimon* in his finite otherness, resting on himself; but as it is, the knight of faith is unnaturally, infinitely alone with God's will. If faith exists, Abraham is lost to universal understanding, but he has a new understanding of himself, by himself, in intimate communion with God with whom he is on speaking terms, as Kierkegaard says: "he becomes God's intimate acquaintance, the Lord's friend, and (to speak quite humanly) . . . he says 'Thou' to God in heaven, whereas even the tragic hero only addresses Him in the third person." [173] The tragic hero exists

[172] *Fear and Trembling*, op. cit., 82.
[173] Ibid., 88.

ultimately in the universal. The knight of faith lacks this medium; he exists paradoxically. We spoke of paradox in connection with Aristotle, but there paradox was in the way of knowing, but knowing was not simply existing: Aristotle knew no universal middle, but he knew particularity; the knight of faith who lacks the universal lacks his own essential particularity. Insofar as his form is the form of faith, paradox is here not in the way of knowing but in the way of existing; the knight of faith is more than he is, not in essence, but in existence; but not in existence determined with reference to totality, but in existence determined with reference to absolute existence: *If God exists, faith exists.*

Kierkegaard deals with the question of the existence of God in his *Philosophical Fragments* where the issue is that of Christianity's understanding of what Kierkegaard distinguishes as God's *factual* being; as he says in a footnote: ". . . the difficulty is to lay hold of God's factual being and to introduce God's ideal essence dialectically into the sphere of factual being."[174] Christianity is faith in God's factual existence in time. Thus, at a certain moment God entered into human existence, not that God is essentially or ideally present in human nature, but that God was this particular man. This is the Truth of Christianity, or, what amounts to the same thing, this is the God of Christianity. Kierkegaard characterizes this situation as the coming into existence of the Eternal; God, moved by love, resolves to reveal himself to man, who erroneously thinks that God exists in and with human existence. That is, God will reveal himself as absolutely other, but, says Kierkegaard, love will not constrain the beloved. It desires a free requital. Therefore, a direct manifestation of God's glory is out of the question, but, for the same reason, a transformation of man is not desirable. So Kierkegaard writes: "Since we found that the union could not be brought about by an elevation it must be attempted by a descent. Let the learner be *x* [Kierkegaard throughout treats the God (Christ) as the Teacher]. In this *x* we must include the

[174] S. Kierkegaard, *Philosophical Fragments* (trans. D. F. Swenson, rev. H. V. Hong, 2nd Ed., Princeton, 1962), 52.

lowliest; for if even Socrates refused to establish a false fellowship with the clever, how can we suppose that the God would make a distinction! In order that the union may be brought about, the God must therefore become the equal of such a one, and so he will appear in the likeness of the humblest. But the humblest is one who must serve others, and the God will therefore appear in the form of a *servant*. But this servant-form is no mere outer garment. . . . It is his true form and figure. For this is the unfathomable nature of love, that it desires equality with the beloved, not in jest merely, but in earnest and truth. And it is the omnipotence of the love which is so resolved that it is able to accomplish its purpose. . . . To sustain the heavens and the earth by the fiat of his omnipotent word, so that if this word were withdrawn for the fraction of a second the universe would be plunged into chaos—how light a task compared with bearing the burden that mankind may take offense, when one has been constrained by love to become its saviour! [But] . . . Every other form of revelation would be a deception from the standpoint of the divine love." [175] That God so exists as a matter of fact as this man is only to be believed. The learner, *qua* man, knows that God exists as the eternal counterpart of his self-consciousness, but for this very reason he lacks faith; he cannot possibly know that his eternal happiness or unhappiness depends upon this Moment in time. Therefore the God himself, as the Absolute Paradox, also provides that faith which is necessary, which then comes into existence itself for the first time. What is to be believed by the learner is an historical fact, but an absolute fact in time, with respect to which the eternal state of the learner's self-consciousness is, likewise in time, decided. With respect to this absolute historical fact, since it is an absolute contradiction of appearance to reality, no essential difference exists between one near or far in time from the event, since it is only to be recognized with faith's eyes. But since faith itself is a novelty in a man's existence, a condition which he at first lacks, indeed, essentially lacks, *qua* man, then faith, if it exists, is also that condition of this particular man, *qua*

[175] Ibid., 39–41.

particular. If this is faith, it is not to be assimilated to universal humanity, *qua* universal. That is, it is not able to be naturalized. Kierkegaard writes: "In the individual life the hypothesis of naturalization is expressed in the principle that the individual is born with faith; in the life of the race it must be expressed in the proposition that the human race, after the introduction of this fact, has become an entirely different race, though determined in continuity with the first. In that event the race ought to adopt a new name; for there is indeed nothing inhuman about faith as we have proposed to conceive it, as a birth within a birth (the new birth); but if it were as the proposed objection would conceive it, it would be a fabulous monstrosity." [176] Actually in the Moment when this man becomes a Christian, when faith comes into existence, this man becomes a sinner, not only that, but he sees that the race exists in sin. His fellow-feeling is extended not to men, *qua* men, but to those like himself related to the salvation brought by that absolute historical event; finally, there is ever-present to him that possibility of offense at the paradox by which he could fall from faith.[177] Christianity, so sharply defined by Kierkegaard in antithesis to that optimistic, collective self-love or spiritlessness of modern times, is not to be naturalized; it is, he says, essentially "discriminative, selective, and polemical." [178] Nor is Kierkegaard unaware that with such presuppositions as it has Christianity is a missionary religion. But he is aware that Christianity is offensive to humanity, since love of God radically divides humanity's totalitarian spirit.

Hegel, in his *Philosophy of Right,* says: "The history of mind is its own act. Mind is only what it does, and its act is to make itself the object of its own consciousness. In history its act is to gain consciousness of itself as mind, to apprehend itself in its interpretation of itself to itself. This apprehension is its being and its principle, and the completion of apprehension at one stage is at the same time the rejection of that stage and its transition to a higher. . . . The question of the perfectibility and *Education of the Human Race* arises here. Those who have maintained this

[176] Ibid., 122.
[177] Cf. *Concluding Unscientific Postscript*, op. cit., 516–519.
[178] Ibid., 516.

perfectibility have divined something of the nature of mind, something of the fact that it is its nature to have *γνῶθι σεαυτόν* as the law of its being, and, since it apprehends that which it is, to have a form higher than that which constituted its mere being."[179] Here come together three ideas: Hegel's self-actualizing World Spirit, Gotthold Lessing's longing for the ultimate illumination of the human species, and Apollo's watchword, wonderfully exemplified in Socrates' life, 'Know Thyself.' Lessing, in his work referred to by Hegel with approval, *The Education of the Human Race,* includes, among others, these thoughts: "What education is to the individual man, revelation is to the whole human race. Education is revelation coming to the individual man; and revelation is education which has come, and is still coming, to the human race. . . . Education gives man nothing which he could not also get from within himself; it gives him that which he could get from within himself, only quicker and more easily. In the same way too, revelation gives nothing to the human race which human reason could not arrive at on its own; only it has given, and still gives to it, the most important of these things sooner. . . . Christ was the first *reliable, practical* teacher of the immortality of the soul. The first *reliable* teacher. Reliable, by reason of the prophecies which were fulfilled in him; reliable by reason of the miracles which he achieved; reliable by reason of his own revival after a death by which he had put the seal to his teaching. Whether we can still *prove* this revival, these miracles, I put aside, as I leave on one side *who* the person of Christ was. All *that* may have been at the time of great importance for the first acceptance of his teaching, but it is now no longer of the same importance for the recognition of the *truth* of his teaching. The first *practical* teacher. For it is one thing to conjecture, to wish, and to believe in the immortality of the soul, as a philosophic speculation: quite another thing to direct one's inner and outer actions in accordance with it. And this at least Christ was the first to teach."[180] If we pause here a moment it is quite clear that not

---

[179] *Philosophy of Right*, op. cit., 216–217.

[180] *Lessing's Theological Writings* (trans. H. Chadwick, Stanford, 1957), 82–92.

only is Kierkegaard's polemic against Hegel, but equally, specifically against Lessing. For Kierkegaard Christ's teachings certainly fall within humanity's potentiality, although great strength is needed to exist in that perfection Jesus enunciated in his Sermon on the Mount, which he himself existed as a man. But for Kierkegaard this perfect teaching merely emphasizes that it is not Jesus' teachings that require faith, but the Teacher himself is what is to be learned, his Person is the message. In this it is precisely Socrates who exemplifies for Kierkegaard the limits of human teachers who wisely refer their pupils to their own, that is, their pupils' own, humanity for truth. Socrates' teaching, if he could be said to have any, is merely an occasion for a learner; most certainly, Socrates' person is not what is requisite. But for Kierkegaard it is only Christ's Person, Divine Love, that saves. It is quite obvious also that in countering Lessing, Kierkegaard counters Kant's moral interpretation of Christianity. But Lessing shares not only Kant's moral view, but is, as it were, a bridge to Hegel's Idea in History; he writes further in *The Education of the Human Race:* "It is absolutely necessary for [human understanding] to be exercised on spiritual objects, if it is to attain its perfect illumination, and bring out that purity of heart which makes us capable of loving virtue for its own sake alone. Or is the human species never to arrive at this highest step of illumination and purity?—Never? Never?—Let me not think this blasphemy, All Merciful! Education has its goal, in the race, no less than in the individual. . . . It will assuredly come! the time of a new eternal gospel, which is promised us in the primers of the New Covenant itself! . . . Go thine inscrutable way, Eternal Providence! . . . And what if it were as good as proved that the great, slow wheel, which brings mankind nearer to its perfection, is only set in motion by smaller, faster wheels, each of which contributes its own individual part to the whole? It is so! Must every individual man—one sooner, another later—have travelled along the very same path by which the race reaches its perfection? Have travelled along it in one and the same life? Can he have been, in one and the self-same life, a sensual Jew and a spiritual Christian? . . . Surely not that! But why should not every individual man have been present more than once in this world? Is this hypothesis so laughable merely because it is the oldest? Because human understanding, before the sophistries of the Schools had dissipated and weakened it, lighted upon it at once? . . . Or is it a reason against the hypothesis that so much time would have been lost to me? Lost?—And what then have I to lose?—Is not the whole of eternity mine?" [181] No doubt Kierkegaard perceives in Lessing, on a level of subjective imagination, a face of that 'fabulous monstrosity' that a new species would be; in any event it is evident that Lessing, as he himself says, has nothing to lose; eternity is his. To compensate the individual for being but a small wheel in a big machine (we may understand that spiritual machine that humanity is) Lessing suggests reviving Plato's hypothesis of reincarnation. To understand what is at stake here, it is to be recalled that for Kierkegaard an eternal decision is to be made in time; eternity is *before* one in Christianity, not *behind*. One enters eternity taking one's departure on a decision made in time, that decision itself relating one to an absolute historical fact, or to a Moment in time when Eternity itself comes into existence. Time, *this man's temporal existence, becomes infinitely intelligible to God as that time in which an absolute difference is made for eternity,* in which something new comes into being. The task then, for Kierkegaard, is an infinite labor in most precious time for an eternity with God, not by any stretch of Lessing's imagination an eternal self-assurance essentially indifferent to one or more hypothetical lifetimes. Kierkegaard shudders at this sinful waste of self, which for Kierkegaard *is* sin, is despair. Kierkegaard writes in *The Sickness Unto Death:* "This then is the formula which describes the condition of the self when despair is completely eradicated: by relating itself to its own self and by willing to be itself the self is grounded transparently in the Power which posited it." [182] We may understand that to be transparently grounded in the Power of God is to be this infinite particularity, this absolute other, one's self; the intelligibility of being so abso-

---

[181] Ibid., 96–98.
[182] *The Sickness Unto Death*, op. cit., 147.

lutely distinct, that is, transparent, exists for no rational system, no totality, but immediately for God's own infinite existent self. At this point, which is Christianity's point, Kierkegaard touches Thomas Aquinas' idea of 'participated existence.' But for Kierkegaard this *act of being* is *act of faith,* this most intimate interiority prior to every idea, every world, every theology; it is the point of contact of man with God made in time, or else, despair, eternal loss of self (loss of God). The forms of despair vary, but despair is sin, it is universal sickness, it is universal. Kierkegaard writes: "At any rate there has lived no one and there lives no one outside of Christendom who is not in despair, and no one in Christendom, unless he be a true Christian, and if he is not quite that, he is somewhat in despair after all." [183] Kierkegaard says further: "Every human existence which is not conscious of itself as spirit, or conscious of itself before God as spirit, every human existence which is not thus grounded transparently in God but obscurely reposes or terminates in some abstract universality (state, nation, etc.), or in obscurity about itself takes its faculties merely as active powers, without in a deeper sense being conscious whence it has them, which regards itself as an inexplicable something which is to be understood from without—every such existence, whatever it accomplishes, though it be the most amazing exploit, whatever it explains, though it were the whole of existence, however intensely it enjoys life aesthetically—every such existence is after all despair. It was this the old theologians meant when they talked about the virtues of the pagans being splendid vices. They meant that the most inward experience of the pagan was despair, that the pagan was not conscious of himself before God as spirit. . . . Nevertheless there is and remains a distinction, and a qualitative one, between paganism in the narrowest sense, and paganism within Christendom. The distinction . . . is this, that paganism, though to be sure it lacks spirit; is definitely oriented in the direction of spirit, whereas paganism within Christendom lacks spirit with a direction away from it, or by apostasy, and hence in the strictest sense is spiritlessness." [184]

---

[183] Ibid., 155.

[184] Ibid., 179–180.

For Kierkegaard everything Christian points to inwardness. Unless a man goes through that intimate union with God which constitutes his new self in faith, he cannot possibly, lost himself, see God. Therefore in Lessing's case Kierkegaard detects a certain deception. He says in the *Concluding Unscientific Postscript* concerning Lessing's inability to overcome historical limitations with respect to Christianity: "Everything that becomes historical is accidental or contingent; it is precisely through coming into being, and thus becoming historical, that it has its moment of contingency, for contingency is precisely one factor in all becoming. Here again we have the root of the incommensurability that subsists between an historical truth and an eternal decision. Understood in this manner, the transition by which something historical and the relationship to it becomes decisive for an eternal happiness, is μετάβασις εἰς ἄλλο γένος, a leap, both for a contemporary and for a member of some later generation. Lessing even says, 'If this is not what it is, then I do not understand what Aristotle has meant by it.' It is a leap, and this is also the word that Lessing has used about it. . . . His words are as follows: '*Das, das ist der garstige breite Graben, über den ich nicht kommen kann, so oft und ernstlich ich auch den Sprung versucht habe.*' ['That then is the ugly, broad ditch which I cannot get across, however often and however earnestly I have tried to make the leap.' [185]] It is possible that the word *Sprung* [Leap] is merely a stylistic phrase, and that perhaps it is for this reason that the metaphor is expanded for the imagination by adding the predicate *breit* [broad]; as if the least leap did not have the characteristic of making the chasm infinitely wide; as if it were not equally difficult for one who absolutely cannot leap whether the chasm is wide or narrow; as if it were not the passionate dialectical abhorrence for the leap which makes the chasm so infinitely wide. . . . Possibly also it is a bit of cunning on Lessing's part to make use of the word *ernstlich* [earnestly]; for in connection with a leap, especially when the metaphor is developed for the imagination, the reference to earnestness is droll enough, because it stands in no relation, or in a comic relation, to the leap; for it is not externally the width of the chasm which pre-

[185] Translation from *Lessing's Theological Writings*, op. cit., 55.

vents the leap, but internally the dialectical passion which makes the chasm infinitely wide. To have been very near doing something has in itself a comic aspect; but to have been very near making the leap is absolutely nothing, because the leap is the category of decision. And now to have tried with utmost earnestness to make the leap—aye, that man Lessing is indeed a wag; for it is no doubt rather with the utmost earnestness that he has endeavored to make the chasm wide: does it not seem as if he were making fun of people?" [186]

Kierkegaard says later on in the *Postscript:* "Not even God, then, enters into a direct relationship with derivative spirits. And this is the miracle of creation, not the creation of something which is nothing over against the Creator, but the creation of something which is something, and which in true worship of God can use this something in order by its true self to become nothing before God. . . . Nature, the totality of created things, is the work of God. And yet God is not there; but within the individual man there is a potentiality (man is potentially spirit) which is awakened in inwardness to become a God-relationship, and then it becomes possible to see God everywhere." [187] So, in the autopsy of faith, in the inwardness of the God-relationship, arises the perception of creation itself; the *novitas mundi* is not a doctrine without, but a revelation to faith.

---

[186] *Concluding Unscientific Postscript*, op. cit., 90–91.
[187] Ibid., 220–221.

Part II

THE PROSPECT: INTRODUCTORY PRESENTATIONS IN THE ESSENTIAL HISTORY OF THOUGHT

Chapter 7

AUGUSTINE: THE KNOWLEDGE OF EXISTENCE

If it is understood that metaphysics has truth solely on the condition that human reason exists essentially in itself, but that this identity in existence is prior to reason's thinking of either itself or others, prior as that which makes reason's thinking to be in the first place, prior as an identity in pure, disinterested thought, then, it may strictly be said, since Aristotle there has been in truth no metaphysics.[188] In the light of metaphysics, actual knowledge is identically its object,[189] but this object of knowledge, in virtue of that very identity, wherein it is indistinguishably pure thought, is, on that account, different in existence than reason. That is, the object actually exists apart from human thought as potentially intelligible. Indeed, this object's actual difference in existence from the mind that comes to know it depends on its essential identity with that mind as its object.[190] In the way in which Aristotle knows existence he expresses wonderfully Greece's classic good sense, the beautiful paradox of the Parthenon: that existence reposes in essence. In essence the movements of existence come to rest: Difference is for the sake of identity. This 'for the sake of' is of the essence of reality in Aristotle. Therefore, strictly speaking, there is in Aristotle's universe no coexistence; the object exists in itself, apart from reason, on condition that in coming to be it is in itself

[188] Cf. above, pp. 52–53.
[189] Cf. above, p. 50.
[190] Cf. above, pp. 54–56.

subordinate to reason's idea of it as capable of being known. The elevation of existence to essence in actual knowledge does not raise what is subordinate in existence to coexistence with reason in a single moment of time, but, rather, to an identity with reason's essence transcending time. Knowledge itself, in Aristotle, is transcendent, but it is not knowledge of what is transcendent. It is eternally incapable of overcoming the difference in the order of existence between itself and its object, which existential difference is, for Aristotle, quite particular, since it is separated from its object by an infinitely divisible time. Indeed, this self-conscious distinction of its object from itself is not knowledge, but potential knowledge (reason), itself subject to conditions of time together with its object. But, as far as the conditions of time reach, reason has no basis for knowing this object; that is, in itself, conditioned by time, reason finds its object existing outside of itself: In this being forever outside in existence that belongs to the object, *qua* object, reason finds that the object is never coincident with reason, not even with reason's ideal form, since, insofar as this coincidence is able to be imagined, it resides within reason's temporality, an abstraction, as it were, of an object potentially, but not actually, known.

Knowledge is incompatible with time, since time is infinitely divisible in Aristotle's universe; the infinite divisibility of time is that condition of the material universe which makes it impossible for any two things to coexist. Aristotle says in his *Physics* VI.8: ". . . there is no smallest or irreducible first component of time or of dimension or of anything that is continuous, for all such are divisible without limit. And since whatever is in motion moves in a period of time and changes from one position to another, it is impossible that the mobile should in its entirety be exactly over against any definite (stationary) thing [κατά τι] during the period occupied by its motion—occupied, that is to say, in the proper sense, not in the sense that the motion falls within some part of the period in question. . . . For since time is divisible without limit and during *every* part of the proper time of the motion the mobile must be in motion, if it could be stationary in *any* part of it, however small, it could be stationary in such parts successively and so in the whole period. If the assertion [that the mobile is coincident with a stationary object] refers not to an interval between two 'nows' but to one single 'now', then the moving thing will not be so situated during any period of time at all but only at a limit of such a period. Now it is true that at any particular instant the moving thing is always situated over against some stationary thing, but it is not 'at rest,' for in the indivisible instant there is neither rest nor motion. Rather, while it is true to say of the moving thing that *at the indivisible instant* [ἐν τῷ νῦν] it 'does not move' and is over against some definite thing, it cannot *during any period of time* [ἐν χρόνῳ] be over against something that is at rest; for if it were it would be both moving and resting." [191] The indivisible instant, the Now, in Aristotle, is not to be confused with time itself; the Now is a measure of time, a limit of the unlimited dimensionality of time. The Now itself is not a dimension. Aristotle speaks as follows in *Physics* IV.11: "It is evident . . . that neither would time be if there were no 'now', nor would 'now' be if there were no time; for they belong to each other as the moving thing and the motion do, so that whatever ticks off the position of the one ticks off the other. For time is the dimension proper to motion, and the 'now' corresponds to the moving object as the numerical monad." [192] Further, in *Physics* IV.13: ". . . it is through the 'now' that time is continuous, for it holds time past and future time together; and in its general character of 'limit' it is at once the beginning of time to come and the end of time past. . . . the 'now' . . . is a divider in mental potentiality, but a continuing unifier as the coincident end-term and beginning-term of past and future time . . ." [193] Finally, in *Physics* IV.14: ". . . if nothing can count except consciousness, and consciousness only as intellect (not as sensation merely), it is impossible that time should exist if consciousness did not; unless as the 'objective thing' which is subjectively time to us, if we may suppose that movement could thus objectively exist without there being any consciousness. For 'before' and 'after' are

[191] *Physics* II, op. cit., 173–175.
[192] *Physics* I, op. cit., 391.
[193] Ibid., 409, 411.

objectively involved in motion, and these, *qua* capable of numeration, constitute time." [194] The Now, for Aristotle, is to time as the moving object is to motion: that by which consciousness counts the time of motion (or rest); nothing moves through, or rests in, this Now. The coincidence of two times in one Now is actually no time at all; in actuality there is no coincidence in time. If two things, or two times, exist together, this is in virtue of their being between two Nows; this is to say that their coincidence in time is not an actuality, but a potentiality, since the identity of the period of time between the two Nows exists only in consciousness. In the time-consciousness of reason this period of time is constituted as the *same* by reason's reckoning with Nows. We may conclude, then, that not only is it impossible for a mobile to be coincident with a stationary object (as Aristotle demonstrates in *Physics* VI.8), but, also, in Aristotle's universe, it is impossible for two stationary objects to coexist in time, except potentially, that is, by a purely formal determination of reason, which, as such, prescinds from the actual infinite divisibility of the dimension. By this brief excursus in time we are returned to our starting point, namely, that in Aristotle's universe an object exists in itself, apart from reason, only in subordination to reason's idea of its potential intelligibility. But, now, we see more clearly that, with respect to objects in time, this potential intelligibility is coincident with potential coexistence with other objects in time. Thus, when the potential intelligibility of the object is reduced to actual identity in knowledge, thereby transcending reason's purely formal determination of its coexistence with other objects in time, we see more clearly still that what is actually intelligible is this object, *qua* particular. Its actual existence in time is incommensurable with reason, *qua* reason, on the one hand, but, *a fortiori,* on the other hand, with the universe: Its actual existence is commensurable with its essence alone. Again, by this short tour in time, we are enabled to see that if the Now, Aristotle's indivisible instant, were, as a matter of fact, divided, thereby becoming its own dimension, capable of containing within itself both motion and rest, then,

[194] Ibid.,419, 421.

this extended Now would be a logical basis for an actual coexistence: a coexistence of reason with objects themselves coexisting in time. Within this hypothetical Now reality would become a matter for reason's formal determination whereby things would be understood actually to exist simply by virtue of satisfying conditions of potential existence, that is, conditions of reason. But, within the hypothetical site of this dimensional Now, judgments of existence would lack essential insight, and, resting on grounds of reason rather than grounds of being (to adapt Schopenhauer's neat distinction [195]), would constitute not cognition, but, in place thereof, *conviction*.

By anticipating what the Now is not yet in Aristotle, we recognize that Stoicism is that thinking in which the Aristotelian Now, that 'numerical monad' by which time, the measure of motion, is itself measured, undergoes a dissipation in time. As a consequence, time itself is dissipated in motion, so that, time, in Aristotle itself a dimension actual in consciousness, is reduced to being a dimension of motion. Diogenes Laertius, in his *Lives of Eminent Philosophers* VII.141, speaks of the Stoic conception of time in this way: "Time . . . is incorporeal, being the measure [διάστημα = interval = dimension] of the world's motion. And time past and time future are infinite, but time present is finite [πεπερασμένον]. They hold that the world must come to an end, inasmuch as it had a beginning, on the analogy of those things which are understood by the senses. And that of which the parts are perishable is perishable as a whole. Now the parts of the world are perishable, seeing that they are transformed one into the other. Therefore the world itself is doomed to perish." [196] The motion of the whole is taken into a dissipated dimension of time; the latter embraces the former within a limitation which was merely potential in Aristotle, but actual in Stoicism: perishability. But this actual perishability of the whole

[195] A. Schopenhauer, *The Fourfold Root of the Principle of Sufficient Reason* (trans. E. F. J. Payne, Illinois, 1974), 200. Perhaps it is not being too careful to stress here that this distinction is adapted to a context very different in reality than Schopenhauer's.

[196] Diogenes Laertius, *Lives of Eminent Philosophers* II (trans. R. D. Hicks, Cambridge, Mass., 1925), 245.

motion of the world is, nevertheless, not absolute in Stoicism, for the reason that it is made possible by the dissipation of the Now, not its destruction. The universe suffers that destruction as a whole, which, in Aristotle, is reserved to particulars in time, *qua* temporal, since existence is prior in essence to motion or time, to the Now or to consciousness. But the Stoic universe suffers its destruction within the limits of a consciousness itself dissipated in matter. Diogenes Laertius, in his *Lives* VII. 135ff., speaks of this dissipated consciousness as follows: "God is one and the same with Reason, Fate, and Zeus; he is also called by many other names. In the beginning he was by himself; he transformed the whole of substance through air into water, and just as in animal generation the seed has a moist vehicle, so in cosmic moisture God, who is the seminal reason of the universe, remains behind in the moisture as such an agent, adapting matter to himself with a view to the next stage of creation. . . . The term universe or cosmos is used [by the Stoics] in three senses: (1) of God himself, the individual being whose quality is derived from the whole of substance . . . who at stated periods of time absorbs into himself the whole of substance and again creates it from himself. (2) Again, they give the name of cosmos to the orderly arrangement of the heavenly bodies in itself as such; and (3) in the third place to that whole of which these two are parts . . . . a system made up of heaven and earth and the natures in them, or, again, . . . a system constituted by gods and men and all things created for their sake." [197] Consciousness, following time and its Now to the limits of the universe, becomes itself a dimension; existing without essence, it exists abstractly; its infinite divisibility guarantees periodically its ability to regenerate that world whose radical unintelligibility, that is, whose primarily material nature foredooms it; it will spring up from its own ashes again. The time of Stoicism is time extended as a definite present to the limits of motion. From this abstract limitation is, *ex hypothesi*, excluded both past and future as different times, or, as times definite in themselves: There is only the one Now from the

[197] Ibid., 241, 243.

limits of which consciousness perpetually returns upon itself in what is for it the sole meaningful concept of past or future, *a present perpetually reiterated*.

If Stoicism is reticent concerning time, it is less so concerning knowledge. In this area the consequences of the dissipation of the Aristotelian Now are, perhaps, more tractable. In the absence of insight into reality, Stoicism discovers the criterion for knowing. Cicero, in his *Academica* I.11, recounts the Stoic Zeno's teaching on this subject: ". . . he made some new pronouncements about sensation itself, which he held to be a combination of a sort of impact offered from outside (which he called *phantasia* and we may call a presentation . . .),—well, to these presentations received by the senses he joins the act of mental assent which he makes out to reside within us and to be a voluntary act. He held that not all presentations are trustworthy but only those which have a 'manifestation' [*declarationem*], peculiar to themselves, of the objects presented; and a trustworthy presentation, being perceived as such by its own intrinsic nature, he termed 'graspable' . . . . after it had been received and accepted as true, he termed it a 'grasp,' resembling objects gripped in the hand. . . . Well, a thing grasped by sensation he called a sensation itself, and a sensation so firmly grasped as to be irremovable by reasoning he termed knowledge, but a sensation not so grasped he termed ignorance. . . . between knowledge and ignorance he placed that 'grasp' of which I have spoken . . . and he . . . said that it was only 'credible.' . . . he held that a grasp achieved by the senses was both true and trustworthy, not because it grasped all the properties of the thing, but because it let go nothing that was capable of being its object, and because nature had bestowed as it were a 'measuring rod' of knowledge [*normam scientiae*] and a first principle of itself [*principium sui*] from which subsequently notions of things could be impressed upon the mind, out of which not first principles only but certain broader roads to the discovery of reasoned truth were opened up." [198] We have seen that

[198] Cicero, *De Natura Deorum, Academica* (trans. H. Rackham, Cambridge, Mass.,1933), 449, 451.

objects in the Stoic universe coexist as parts of a systematic whole, so thoroughly permeated by the abstract dimension of time that from the destructibility of these parts it is possible to conclude to that of the whole, while both whole as well as parts are reconstituted by that cosmic, conscious Zeus with whom everything coexists as with that original determination of reason in agreement with whom it possesses its proper nature. When reason's essence, in Aristotle, prior to reason as its unknowable identity, was by reason utterly forgotten in Stoicism during its dissipation in matter, it substituted for that forgotten identity, as a matter of conscience, a promise made to itself guaranteeing itself an identity. The periodic reiteration of this world-time is an expression of this self-guaranteeing identity. Happily, it excludes, in principle, verification! In Stoicism, reason not only promises itself this identity, but, Stoicism is that firm resolution to trust itself alone in this matter, as in all matters. In its epistemology Stoicism trusts reason within to decide for itself concerning the credibility of extrinsic presentations in their claim to be actually those natures that they appear to be. A sensation, then, as a presentation assented to, trusted, is taken as a criterion of knowledge. Knowledge proper is a trust whose ground is unshakeable by reason. Knowledge, then, rests upon grounds of reason, not on grounds of being: It is *conviction*, irrefutable or not. To bring vividly to mind the Aristotelian world of essence, here completely forgotten, it is not necessary to rehearse yet again 'knowledge's identity with its object,' but it is sufficient to turn to Aristotle's understanding of sensation, itself never transformed by conviction into knowledge. Aristotle says in the *De Anima* III.3, in the course of distinguishing imagination (φαντασία) from sensation (αἴσθησις), "Sensation is either potential or actual, *e.g.*, either sight or seeing, but imagination occurs when neither of these is present, as when objects are seen in dreams. . . . sensation is always present but imagination is not . . . . Again, all sensations are true, but most imaginations are false. Nor do we say 'I imagine that it is a man' when our sense is functioning accurately with regard to its object, but only when we do not perceive distinctly. And, as we have said before, visions are seen by men even with their eyes shut." [199] To look back in this way to Aristotle is to be refreshed by a crisp clarity of perception, the ground of which is in being, not in reason; which perception is in no sense a voluntary act of a free agent, for, in Aristotle's world, reason itself is not an agent. One cannot help but reflect that Aristotle's yet intact Now is that condition of the measure of the motion of the physical world which makes possible sensation's inerrancy; whereas Stoicism's dissipated Now so disturbs sensation with an entire universe of motion that men see visions with their eyes *open*. Thus, consciousness must retire into its resolution to inspect credentials, confident, nevertheless, that nature is ultimately in accord with reason in the knower. But, at the same time, it is aware that this work of reason is night work. Nothing is so striking as Epictetus' epitome of this epistemological situation; he says in the *Discourses* III.12: ". . . just as Socrates used to tell us not to live a life unsubjected to examination, so we ought not to accept a sense-impression unsubjected to examination, but should say, 'Wait, allow me to see who you are and whence you come' (just as the night-watch say, 'Show me your tokens'). 'Do you have your token from nature, the one which every sense-impression which is to be accepted must have?' " [200] In this dark night of a reiterative time, the Stoic stands guard at the door to himself.

What lies within? A meretricious identity. Diogenes Laertius, in his *Lives* VII.86ff., speaks of this identity in this way: ". . . Nature's rule is to follow the direction of impulse. But when reason by way of a more perfect leadership has been bestowed on the beings we call rational, for them life according to reason rightly becomes the natural life. For reason supervenes to shape impulse scientifically. This is why Zeno was the first . . . to designate as the end 'life in agreement with nature' . . . , which is the same as a virtuous life, virtue being the goal towards which nature guides us. . . . Again, living virtuously is equivalent to living in accordance with experience of the

[199] *On the Soul,* op. cit., 159.

[200] Epictetus, *The Discourses As Reported by Arrian, The Manual, and Fragments* I (trans. W. A. Oldfather, Cambridge, Mass.,1928), 85, 87.

actual course of nature, as Chrysippus says . . . ; for our individual natures are parts of the nature of the whole universe [μέρη γάρ εἰσιν . . . τοῦ ὅλου]. And this is why the end may be defined as life . . . in accordance with our own human nature as well as that of the universe, a life in which we refrain from every action forbidden by the law common to all things [ὁ νόμος ὁ κοινός], that is to say, the right reason which pervades all things, and is identical with this Zeus, lord and ruler of all that is. And this very thing constitutes the virtue of the happy man . . . when all actions promote the harmony of the spirit dwelling in the individual man with the will of him who orders the universe." [201] The Stoic has his very being in his participation in this natural order; there is here no essentially individual existence (as Diogenes Laertius points out in the same place,[202] "Cleanthes takes the nature of the universe alone as that which should be followed, without adding the nature of the individual"). The priority of the individual man, *qua* individual, to his ethical, political nature, that is, to his specifically human nature, which is a face of the essentially paradoxical thought of Aristotle,[203] by which the individual in his identity prescinds from nature, both human and universal, has long since been dissipated. By an inversion, as it were, individual identity is, in Stoicism, a work, a deed of reason, not of the individual, but, rather, the selfless self-guaranteeing of reason itself. The virtue of the Stoic is what it is only if it is 'in front of' others; it is a leadership following the leadership of Zeus, a leadership of reason self-authenticating grounds of knowledge: it is not for virtue's sake that knowledge is sought, but, virtue, the deed of reason, establishes truth. It is, as it were, a retrospective declaration of the ultimate trustworthiness of sensation, together with that body of knowledge dependent upon it. Cicero, in his *Academica* II.8, makes this point on behalf of the Stoic: "The greatest proof . . . of our capacity to perceive and grasp many things is afforded by the study of Ethics. Our percepts alone we actually pronounce to form the basis of knowledge . . . , and

[201] *Lives of Eminent Philosophers* II, op. cit., 195, 197.
[202] Ibid.
[203] Cf. above, chapter on Aristotle, especially pp. 43ff.

likewise of wisdom, the science of living, which is its own source of consistency. But if this consistency had nothing that it grasped and knew, whence, I ask, or how would it be engendered? consider also the ideal good man, who has resolved to endure all torments and to be mangled by intolerable pain rather than betray either his duty or his promise—why, I ask, has he saddled himself with such burdensome rules as this when he had no grasp or perception or knowledge or certainty of any fact that furnished a reason why it was his duty to do so? It is therefore absolutely impossible . . . unless he has given assent to things that cannot possibly be false." [204] The dogmatic self-assertion of the Stoic, his immovable decisiveness, his consistency derive their firm foundation from his refusal to betray those previous decisions, scientific as well as wise, which continue in their dependence into this ratifying present moment: At this present time, into this extended Now, there culminates a totality of virtue, or else, in this moment, which goes at a bound to the limits of the universe, there is no virtue, in this dimensional instant which now extends to the miniature of presentations, out of the magnitude of cosmic motion. To see Stoicism in this structure of the dissipated Now is to see the reason within, upon which everything, otherwise arbitrary in Stoicism, depends, indeed, to see it whole, because it is to see reason entire within time. There is no 'moral improvement' as in Aristotle, not because there is no time for it, but because there is nothing but time, this present eternally reiterated. Stoic identity is meretricious not only because it is only by participation in the whole, nor, only because it merits or earns itself by constancy, but also because it is not sincere, which is not to say that it is insincere in a psychological sense, but that its insincerity is constitutional. That is, it pours itself out in world-saving activity, but it remains behind as the moist residue of Zeus, in the cup of its consciousness: It will die to the world, but not to itself. In this way reason's self continues to be as a shadow of the 'luminous opacity' of the Aristotelian essence.[205]

[204] *Academica*, op. cit., 497, 499.
[205] Cf. above, p. 59.

This shadowlike existence of Stoicism's self-in-nature is plunged by Skepticism into obscurity. Cicero, in his *Academica* I.12, sets forth the position of the Academic skeptic, Arcesilaus, as follows: ". . . truth (in Democritus' phrase) is sunk in an abyss [ἐν βυθῷ ἡ ἀλήθεια], opinion and custom are all-prevailing, no place is left for truth, all things successively are wrapped in darkness [*deinceps omnia tenebris circumfusa*]. Accordingly Arcesilas said that there is nothing that can be known, not even that residuum of knowledge that Socrates had left himself—the truth of this very dictum: so hidden in obscurity did he believe that everything lies, nor is there anything that can be perceived or understood, and for these reasons, he said, no one must make any positive statement or affirmation or give the approval of his assent to any proposition, and a man must always restrain his rashness and hold it back from every slip, as it would be glaring rashness to give assent either to a falsehood or to something not certainly known, and nothing is more disgraceful than for assent and approval to outstrip knowledge and perception." [206] If in Stoicism there is a meretricious identity, that is, if reason preserves itself in the systematic order of reasons which guarantees to it its self above and beyond that system's perishability as nature embraced in the dimensional Now, Skepticism experiences reason's totality in pure immediacy. That is, each successive event occupies a Now of its own: Each successive thing exists in its own dimensionality. Thus, Skepticism is reason lost in its potentiality, without nature, without self. The moon is dark: Reason is indistinguishable from matter. To state it most abstractly, but, most accurately: Reason in Skepticism is absolute coexistence or coincidence—with what? Not with its self, but with its self as a possibility. Or, absolute coexistence is absolutely possible, or unverifiable, by virtue of the fact that the abyss overflows each and every ground of reason. It is not merely unverifiable in the manner of the Stoic's periodic regeneration of the world (the latter is a matter of principle laid down by reason), but in Skepticism the unverifiability of truth is a matter of necessity arising out of the

[206] *Academica,* op. cit., 453.

infinite depth of its own reluctance. Because Skepticism is reason without self, that is, reason as pure parasite, looking upon everything as mere occasion, it is instructive to notice the limits of its reluctance on the occasion offered it by Stoicism; to wit, Stoicism provided Skepticism, for its 'refusal of assent,' only grounds of reason, but never grounds of being. While Skepticism, therefore, remains unconvinced of the truth of knowledge, it is totally oblivious of the question of essence, so that it is able to enshroud all things in perfect darkness while allowing their continuance in existence, this not without a certain inconsequence. Bearing this in mind we will not confuse it with modern forms of scepticism, which, beginning with Descartes, concern existence itself, for reasons, therefore, which have to do with the history of being, rather than with the history of thought, the latter running off from time to time into its own dead ends. Be that as it may, the limitation of Skepticism here taken note of is not to be controverted by reference to the distinction between 'appearance' and 'underlying reality' attributed to the Skeptic Aenesidemus, set forth as follows by Sextus Empiricus, in his *Against The Logicians* II.367ff.: "But, say they [Dogmatists], one ought not to ask for proof of everything, but accept some things by assumption, since the argument will not be able to go forward unless it be granted that there is something which is of itself trustworthy. But we shall reply . . . that . . . as apparent things merely establish the fact that they appear, and are not capable also of showing that they subsist, let us assume also that the premises of the proof appear, and the conclusion likewise. But even so the matter in question will not be deduced, nor will the truth be introduced, so long as we abide by our bare assertion and our own affection. And the attempt to establish that apparent things not merely appear but also subsist is the act of men who are not satisfied with what is necessary for practical purposes but are eager also to assume hastily what is possible."[207] Aenesidemus refuses to accept a purely logical law of contradiction, which law, as we can see

[207] Sextus Empiricus, *Against The Logicians* (trans. R. G. Bury, Cambridge, Mass., 1935), 433, 435.

retrospectively, is taken to operate without regard to the limitations of time imposed on reason by essence in an Aristotelian universe. Therefore, Aenesidemus' refusal to admit the possibility of knowing what appears to subsist is but a form of the pure infinity of reason without self. It is not, as with Kant, a denial of this possibility of knowing a thing-in-itself, set out as a foundation of a secure science on the basis of the principle of noncontradiction, where there is knowledge but not existence. In Aenesidemus there is no 'underlying reality' only insofar as there is no knowledge. It is clear then that there is no ground in reality for alleging a continuity between ancient and modern so-called 'phenomenism.' [208] In Skepticism, then, the absolute coincidence of appearance and reality is merely unverifiable; the reality 'too hastily assumed.' The concession that reality is the 'possible' is not, as in Kant, purely formal, but it is substantive in Skepticism, where, in the midst of the indefinite multiplication of Nows, reason-without-self retains its possible self in the form of the object, whose purely material being is never denied. The argument of Carneades against the criterion of truth manifestly shows this. It is presented by Sextus Empiricus in his *Against The Logicians* I.159ff.: ". . . his first argument, aimed at all alike, is that by which he establishes that there is absolutely no criterion of truth—neither reason, nor sense, nor presentation, nor anything else that exists; for these things, one and all, play us false. Second comes the argument by which he shows that even if a criterion exists, it does not subsist apart from the affection produced by the evidence of sense. . . . But when the sense is unmoved and unaffected and undisturbed, neither is it sense nor perceptive of anything; but when it is disturbed and somehow affected owing to the impact of things evident, then it indicates the objects. . . . Hence we must say that the presentation is an affection of the living creature capable of presenting both itself and the other object. . . . But

[208] Cf. C. J. De Vogel, *Greek Philosophy: A Collection of Texts* III (Leiden, 1964), 223, where this continuity is alleged. But the *inability* to know *what is transcendent* presupposes a *change in being*, and *consequently* in mind, of which Skepticism could know nothing: cf. above, chapter on Aquinas, especially pp. 57–58. On Kant and contradiction, cf. above, pp. 100–101.

since it does not always indicate the true object, but often deceives and, like bad messengers, misreports those who dispatched it, it has necessarily resulted that we cannot admit every presentation as a criterion of truth . . . . since there is no true presentation of such a kind that it cannot be false, but a false presentation is found to exist exactly resembling every apparently true presentation, the criterion will consist of a presentation which contains the true and the false alike. But the presentation which contains them both is not apprehensive, and not being apprehensive, it will not be a criterion . . . . neither will reason be a criterion; for it is derived from presentation. And naturally so; for that which is judged must first be presented, and nothing can be presented without sense which is irrational. Therefore neither irrational sense nor reason is the criterion." [209] In the case of Stoicism we have spoken of reason's self-guaranteeing activity whereby it establishes as trustworthy a claim of a presentation to be what it appears to be; but in the extremity of its temporality, which Skepticism is, reason discovers that what formerly was its own formal determination of coincidence or coexistence of object with its self has, now, become a determination of matter to be, *qua* appearance, the coincidence of 'the true and the false alike.' Reason has lost its true self in the abyss; sense presents itself as the irrational: a bad messenger; reason is left to make its way on the basis of probable appearances, about which not even conviction is possible, let alone insight.

Augustine is a man claimed by God. The spirit of Augustine is prophetic. All things wrapped in dark of night; each alone in his infinite dimension, equal only in their insignificance; those best among them signifying their own emptiness. In *Isaiah* 21:11–12, the prophet writes: "Oracle on Edom: Someone shouts to me from Seir, 'Watchman, what time of night? Watchman, what time of night?' The watchman answers, 'Morning is coming, then night again. If you want to, why not ask, turn round, come back?' " [210] What shall we understand Isaiah to mean? Is conversion an answer to the question, 'What

[209] *Against The Logicians,* op. cit., 87, 89, 91.
[210] *The Jerusalem Bible: O.T.,* op. cit., 1174.

time of night?' Is the repetition, 'Morning is coming, then night again,' a statement of the greater power of night, that night prevails in the question's being asked? Does Isaiah perceive that this question, under a cloak of night, quietly solicits an invitation, such as, 'If you want to, why not ask, turn round, come back?' Shall conversion come over the power of night, bringing with it a morning unanticipated? Is the claim of God upon man so ready a response as Isaiah's? Is time to come to an end in conversion? Augustine says, in *The City of God* V.10–11, "No man sins unless it is his choice to sin; and his choice not to sin, that, too, God foresaw. This supreme and true God—with His Word and Holy Spirit which are one with Him—this one omnipotent God is the creator and maker of every soul and of every body. All who find their joy in truth and not in mere shadows derive their happiness from Him. He made man a rational animal, composed of soul and body. He permitted man to sin—but not with impunity—and He pursued him with His mercy." [211] The answer to the prophet's question—the invitation—lies in man's freedom: a freedom, in the first instance, a gift of creation, but, secondly, after sin, a work of God's mercy in the person of Jesus Christ.

Augustine writes concerning creation, in his *Confessions* 13.2: "Your creation subsists out of the fullness of your goodness, to the end that a good that would profit you nothing, and that was not of your substance and thus equal to you, would nevertheless not be non-existent, since it could be made by you. What claim on you had heaven and earth, which you made in the beginning? And those spiritual and corporeal natures which you made in your Wisdom, that on it might depend even things inchoate and formless . . . let them tell me what claim they had on you unless they were recalled by that same Word to your unity and given form and, from you, the one and supreme good, would all be very good. What claim did they have on you even to be formless? . . . What claim did corporeal matter have upon you, merely to be invisible and without form, since it

[211] Augustine, *The City of God* (An Abridged Version, trans. G. G. Walsh *et al.*, Garden City, 1958), 110–111.

would not even be such except because you made it? Hence, since it did not exist, it could have no claim on you to exist. What claim did inchoate spiritual creation have on you, even to float and flow about, darksome like the deep, but all unlike you, unless it were converted by that same Word to the same Word by whom it was made, and were enlightened by him and made into light, although not equal to the form equal to you, yet conformed to it? . . . for a spiritual creature, to live is not the same as to live wisely, otherwise it would be immutably wise. But it is good always for it to adhere to you, lest by aversion from you it lose the light gained by conversion, and fall back into a life similar to the darksome deep. For we also, who are spiritual as to the soul, being turned away from you, our light, were sometimes darkness in this life. Still do we labor amid the remains of our obscurity, until in your Only-begotten we may be your justice, as the mountains of God. For we have been your judgments, which are like a great deep." [212] For Augustine creation is coalescent with conversion. But in this coalescence of being with mercy that radical discontinuity is open to view by which creation stands before God in its own identity, so that, by no ultimate divisibility is it capable of being reduced to an original continuity with God; so that, even in its aversion from God it does not lose itself, but becomes similar to the abyss, for even this abyss has an identity of its own by virtue of the fact that it, like matter itself, depends upon God's creation for the fact that it exists. The fact of existence is creation; the fact of dependence is conversion. But, for Augustine, in their growing together in his mind, the form of creation is conversion; the form of creation is dependence: In its dependence upon God creation has its form, its potential intelligibility. So then, that world which lies in the abyss, overcome in that perfect darkness of the infinite reluctance of reason wherein each thing is divided into nothing in its own private dimension, is to be restored not by a retrospective reconstruction of time, and the Now, and the Self, moving backwards, as it were, through Skepticism, Stoicism, to

[212] Augustine, *The Confessions* (trans. J. K. Ryan, Garden City, 1960), 336–337.

a source of identity in Aristotle, which source, in the progress of reconstruction, will have been transmuted into an infinite principle of continuity, unknown to Aristotle (or, for that matter, to Plato), a transmutation necessary to effect the 'flight of the alone to the alone' wherein dimensionality is ultimately overcome spiritually on the ground of an absolute ecstatic withdrawal. That world, wrapped in darkness, is to be enlightened not by Neoplatonism's reconstruction of metaphysics, if that were possible, which redeems this world by denying it, but, rather, by the fact of the conversion of the world, in the person of the Word, which converts by invitation only. To say that conversion is by invitation only is to say, first, that man is not in origin a member of God's household, that he has no claim in nature or reason, as Stoicism held, to share with God a common-wealth, but then God's invitation to man is not an expression of self-love or affinity; it is not *οἰκείωσις*, a manner of God's appropriation of man to his own advantage as if he had need of a creature upon whom to lavish his love. Second, to say that the Word converts the world by invitation only is to say that God does not compel man, rather challenges man; in the words of Isaiah, 'If you want to, why not ask, turn round, come back?' Indeed, in conversion by invitation only, God challenges that Skepticism which understands itself to be, precisely, a want of self, under the domination of which, as of necessity, it lies cut off in night. God challenges man by refusing to acknowledge man's impotence, his destiny; the very necessity of the question, 'What time of night?,' is challenged by the invitation, which reminds him that he himself remains apart in this night by his own volition. But, third, the fact that conversion is by invitation only points to the most original fact of all, namely, that God speaks. It, therefore, means to say that without an invitation, without hearing God speak, one may not enter the light of day; but to stand before God so as to hear him speak is precisely obedience, to respond to the invitation that places one under obligation to come as invited, not, indeed, an obligation to come, but, to come as called. 'If you want to, why not ask, turn round, come back?' The invitation puts man under an obligation to ask, to acknowledge his unworthiness, therefore, to assume freely whatever conditions accompany the invitation. In *Matthew* 22:1–14 Jesus tells the parable of the wedding feast to which those originally invited would not come: " 'The wedding is ready; but as those who were invited proved to be unworthy, go to the crossroads in the town and invite everyone you can find to the wedding.' So these servants went out onto the roads and collected together everyone they could find, bad and good alike; and the wedding hall was filled with guests. When the king came in to look at the guests he noticed one man who was not wearing a wedding garment, and said to him, 'How did you get in here, my friend, without a wedding garment?' And the man was silent. Then the king said to the attendants, 'Bind him hand and foot and throw him out into the dark, where there will be weeping and grinding of teeth'." [213] A conversion, then, for whom one has no one to thank but one's self, or one's good fortune or fate, which is not mindful that another has called, or, a vocation to which one has called oneself, is not clothed in obedience. But without obedience, by which one acknowledges that it is God who speaks, who invites, a man's being welcome at the wedding is only his surmise. Consequently, a conversion that lacks the fact of existence, or, creation out of nothing, is, if it is not simply a matter of feeling, a matter of metaphysics, and, as such, an imposition on God of one's self.

In his *Confessions* 13.22, Augustine writes: "For behold, O Lord our God, our creator, when our affections have been restrained from love of this world, in which affections we were dying by living evilly, and when by living well a living soul has begun to exist, and your Word, by which you spoke to us through your apostle, has been fulfilled in us, namely, 'Do not be conformed to this world,' there follows what you immediately adjoined, and said, 'But be reformed in the newness of your mind.' No longer is this 'after one's kind,' as though imitating our neighbor who goes on before us, or living according to the example of some better man. You did not say, 'Let man be made according to his kind,' but 'Let us make man to our image and likeness,' so that we may prove what is your will

[213] *The Jerusalem Bible: N.T.*, op. cit., 50.

. . . . since he is renewed in mind and perceives your truth that he has understood, he does not need a man to point the way so that he may imitate his own kind. By your direction, he himself establishes what is your will, what is the good, and the acceptable, and the perfect thing. Since he is now capable of receiving it, you teach him to see the Trinity of unity and the unity of the Trinity. . . . Thus man 'is renewed unto knowledge of God, according to the image of him who created him,' and being made spiritual, he 'judges all things,' that is, all things that are to be judged, but 'he himself is judged by no man.' "[214]

If the content of conversion is the fact of creation, then by conversion a man is raised to a life above human nature, above reason, not, as in Aristotle,[215] prescinding from nature to his own divine essence within the limits of his nature, not, as in Plotinus, to an ecstatic transcendence of God himself,[216] but, to a new identity in the Truth or Word of God: This is the identity of a person in union with that Son of God upon whom depends everything that exists. To be converted is to enter into that One in which all things exist in an intelligibility transcendent to reason. Augustine writes, in *Of True Religion* 72ff., "Do not go abroad. Return within yourself. In the inward man dwells truth. If you find that you are by nature mutable, transcend yourself. But remember in doing so that you must also transcend yourself even as a reasoning soul. Make for the place where the light of reason is kindled. What does every good reasoner attain but truth? And yet truth is not reached by reasoning, but is itself the goal of all who reason. There is an agreeableness than which there can be no greater. Agree, then, with it. Confess that you are not as it is. It has to do no seeking, but you reach it by seeking, not in space, but by a disposition of mind, so that the inward man may agree with the indwelling truth in a pleasure that is not low and carnal but supremely spiritual. If you do not grasp what I say and doubt whether it is

---

[214] *The Confessions,* op. cit., 355–356.

[215] Cf. above, chapter on Aristotle, especially pp. 51ff.

[216] Cf. Plotinus, *The Enneads* (trans. S. MacKenna, 4th ed., London, 1969), VI.9.

true, at least make up your mind whether you have any doubt about your doubts. If it is certain that you do indeed have doubts, inquire whence comes that certainty. . . . Everyone who knows that he has doubts knows with certainty something that is true, namely, that he doubts. He is certain, therefore, about *a* truth. Therefore everyone who doubts whether there be such a thing as *the* truth has at least *a* truth to set a limit to his doubt; and nothing can be true except truth be in it. Accordingly, no one ought to have doubts about the existence of *the* truth, even if doubts arise for him from every possible quarter. Wherever this is seen, there is light that transcends space and time and all phantasms that spring from spatial and temporal things. . . . Reasoning does not create truth but discovers it. Before it is discovered it abides in itself; and when it is discovered it renews us." [217] Descartes' *cogito ergo sum* meant that Descartes, who had methodically set out to discover an undoubtable truth, found it in his existence as a *res cogitans,* a thinking thing. But Augustine suggests that, immediately from your doubt, you possess, in your certitude that you doubt, in this thought, a *truth* on the basis of which one can know that *truth itself* exists, not that *you* exist. For Augustine, *you do not truly exist if you do not exist in the knowledge of the truth;* for Augustine, *Truth is the medium for existence,* transcending space and time and phantasms, that is, rational animality. For Augustine, in the act of knowledge of existence, the form of which is an act of acknowledgment of dependence, man transcends himself. Therefore, we may understand that in conversion to Truth a man *prehends* God, so that he may come to *apprehend* in truth. As he says in his *Confessions* 11.29: "But since 'your mercy is better than lives,' behold, my life is a distention, or distraction. But 'your right hand has upheld me' in my Lord, the Son of man, mediator between you, the One, and us, the many, who are dissipated in many ways upon many things; so that by him 'I may apprehend, in whom I have been apprehended,' and may be gathered together again from my former days, to follow the

[217] Augustine, *Of True Religion* (trans. J. H. S. Burleigh, Chicago, 1959), 69–71.

One; 'forgetting the things that are behind' and not distended but extended, not to things that shall be and shall pass away, but 'to those things which are before.' . . . But now . . . I am distracted amid times, whose order I do not know. . . ." [218] If we understand the sacred doctrine of Thomas Aquinas to be the *transcendental form of natural reason,* [219] in which is seen, for the first time, existence itself, then, we may understand that Augustine sets out the *transcendental essence of existence* in the form of conversion. Of this essence of existence, Augustine says, in *Of True Religion* 110, "All rational life obeys the voice of unchangeable truth speaking silently within the soul. If it does not so obey it is vicious. Rational life therefore does not owe its excellence to itself, but to the truth which it willingly obeys. The lowest man must worship the same God as is worshipped by the highest angel. In fact it is by refusing to worship him that human nature has been brought low. The source of wisdom and truth is the same for angel and man, namely the one unchangeable Wisdom and Truth." [220] As for Descartes: from a purely transcendental viewpoint, setting aside, if it is possible, theological considerations,[221] his mistake is that he includes within the essence of transcendental existence a notion of time, thereby confusing existence with humanity's eventuality, in such a way, that this existence becomes a *medium* for truth.[222] The 'mistake' is to be corrected by Husserl; to be sure, in his own way. In the meantime, it is clear that, in truth, Descartes' doubt is related to Augustine's by way of contradiction.

[218] *Confessions,* op. cit., 302.
[219] Cf. above, pp. 65ff.
[220] *Of True Religion,* op. cit., 103.
[221] Cf. above, chapter on Descartes.
[222] Cf. above, pp. 93ff.

Chapter 8

LEIBNIZ: THE IDEAL OF THE HISTORY OF BEING

History properly belongs to man alone. It is the story of what has occurred to him in the course of his being in the world. History is essentially not an account of man's deeds or accomplishments; for, what these are remains to be seen. The works of man can only be recounted from a perspective in the future; or, again, it is of the essence of works that they endure, so that, whatever has been done is now present, it is open to inspection. What man has wrought, even if it lies in the past, is a proper object of scientific investigation; whether this science is a 'natural' or so-called 'social' science is a matter of perfect indifference: It is not history. Nor will discovering or cataloging or theorizing about conditions, material or cultural, which shape or create human work, constitute history. History, therefore, is essentially not a record of what man has become, or, what has become of man in the course of time or of any period of time. It is not anthropology, in whatever form it may take, be it biology, or logic, or religion. Nor, of course, is history a matter of journalism. Neither is it anecdotage, nor does it have a cause: It is radically not information. Indeed, history is that form of knowledge in which the story of what has occurred is what has occurred; history, then, is the identity of the storyteller with what has occurred to him; it is the identity of the storyteller with the fact of his being in the world. History belongs properly only to man, then, because only in his case is the fact of being in the world identically that of the historian. The historian, *qua* historian, is what has occurred to him in the course of his being. History, therefore, if it exists, exists only as a matter of personal experience. But as soon as this is said it seems that history is declared to be an impossible ideal, situated, in principle, infinitely beyond precisely that factual experience which delimits any one historian or group of historians within their own time or place. That the essence of history is that it is a personal experience of man seems, then, to be merely a regulative principle, in the manner of Kant's ideas of pure reason, for the organization of a science of history within the spatiotemporal limits of understanding.[223] But by such an inversion we would be immediately returned to those perspectives that take man as a creature active in a natural environment, which views, since they lack the identity of the storyteller with the story of what has occurred to him, stand back in their freedom from the thought of the unconditioned, concerned essentially with man's future, but not with his history. This is to say that modernity, for all its profuse modes of investigation, understands that *essentially* nothing has occurred to man in the course of his worldly being, for the contrary is self-contradictory.[224] If history actually exists, to the contrary, it is a matter of personal experience of an occurrence to the essence of man; it is necessarily freed of conditions of Kant's sensuous intuition. But, it is instructive to compare this thought of the unconditioned existing as history with what Kant writes in his *The End of All Things,* where he speaks about what he understands as the "*mystic* (supernatural) end in the order of efficient causes of which we comprehend nothing," where he says, "In the Apocalypse ([i.e., Revelations] 10:5, 6), 'An angel lifted up his hand to heaven, And sware by him that liveth forever and ever, who created heaven, etc.: that there should be time no longer.' If one does not assume that this angel 'with his voice of seven thunders' . . . desired to cry out nonsense, then he must have meant with these words that henceforth there should be no *change;* for if there were still change in the world, time, too, would be there,

[223] Cf. above, chapter on Kant, especially pp. 108ff.
[224] Cf. above, pp. 100–101.

because change can only take place in time and is completely unthinkable without the presupposition of time. Here now is represented an end of all things as objects of the senses whereof we can formulate absolutely no concept, because inevitably we entangle ourselves in contradictions if we choose to take one single step out of the sensible world into the intelligible. This happens through the fact that the moment which determines the end of the sensible world is also supposed to be the beginning of the intelligible world; therefore the latter is brought into one and the same temporal series with the former, and this is self-contradictory. But we also say that we conceive a duration as infinite (as eternity), not because we have any ascertainable concept at all of its enormity, for that is impossible since eternity lacks time altogether as a measure of itself; but rather, that concept is a purely negative one of the eternal duration, because where there is no time, also *no end* is possible. . . . The rule for the practical use of reason according to this Idea, therefore, intends to express nothing more than that we must take our maxims as if, in all its changes from good to better which proceed into the infinite, our moral state, with respect to its disposition (the *homo noumenon,* 'whose change takes place in heaven') would not be subjected at all to temporal change. But that some time a moment will make its appearance when all change—and with it time itself—will cease is a notion that revolts our imagination."[225] To deny, as Kant does, the end of change, here the only meaningful interpretation of the end of time, since change presupposes time for Kant, is to deny that history is possible as the identity of the historian with what has occurred to him in the course of his being in the world, since that identity is nothing other than an end of a change, an end or identity of person eternally established, but which occurred to a man during the course of his being in time. It is fairly obvious that it is Kant's presupposition of the forms of imagination as the conditions of any knowledge which renders it impossible for him to conceive of any end to time, and, therefore, given this same presupposition, of any end to change, since he

[225] I. Kant, *On History* (trans. L. W. Beck *et al.,* Indianapolis, 1963), 76–78.

must conceive the end of time as the *beginning* of eternity, that is, as a final Now in the same sense as every other Now, namely, as a divider which actually unites time past with time future.[226] So profound is this commitment to the sensibility of its own understanding on the part of pure reason that, in the first instance, Kant's objection is the purely formal one, namely, that two different worlds would as a matter of fact be continuous in that moment announced by the angel, so that *no end* of change would occur. Then, the intelligible world, or eternity, would be for Kant, if experienced, an infinite duration. But, in the second instance, this concept is understood by Kant to be purely negative in its import; that is, what lacks time can have *no end.* It may be recalled that even where Kant thinks of 'personal identity' as a regulative principle for a 'transcendental psychology of the soul' this *as if* identity is not construed as necessarily lasting beyond this life, in its origin being nothing other than an extension of inner sense or time.[227] It is evident that this purely provisional concept of personal identity is not able to be a foundation of history in its essential sense, but, rather, with its inherent limitation, is subordinated to mankind's progress in time toward a point infinitely distant when, fulfilling Nature's destiny, there will be established a universal political order. Kant writes, in his *Idea For A Universal History,* "It is strange and apparently silly to wish to write a history in accordance with an Idea of how the course of the world must be if it is to lead to certain rational ends. It seems that with such an Idea only a romance could be written. Nevertheless, if one may assume that Nature, even in the play of human freedom, works not without plan or purpose, this Idea could still be of use. Even if we are too blind to see the secret mechanism of its workings, this Idea may still serve as a guiding thread for presenting as a system, at least in broad outlines, what would otherwise be a planless conglomeration of human actions." [228] In the absence of an actual

[226] Cf. above, p. 159, for this fundamental Aristotelian concept, here at work within the limits of Kant's understanding.

[227] Cf. above, pp. 108ff.

[228] *On History,* op. cit., 24.

identity of man with what has occurred to him in the course of his being in the world, that is, in the absence of the essence of personal identity, there exists no thought of essential history, so that, Kant is able to conclude his *Idea For A Universal History* as follows: "Otherwise the notorious complexity of a history of our time must naturally lead to serious doubt as to how our descendants will begin to grasp the burden of the history we shall leave to them after a few centuries. They will naturally value the history of earlier times, from which the documents may long since have disappeared, only from the point of view of what interests them, i.e., in answer to the question of what the various nations and governments have contributed to the goal of world citizenship, and what they have done to damage it." [229] Here is the ultimate indifference of the pure potentiality of reason to what as a matter of fact has occurred to it; the Aristotelian Now, otherwise intact (along with the principle of noncontradiction), is (along with contradiction) confined to a purely analytic function within the limits of Kant's pure understanding. It is not sufficient to organize the synthetic manifold of time as the condition of change from one state to another; its purely logical utility is subordinated to the *Moment* which functions as the cause of change, insofar as it is considered an aspect or phase of that time which, in turn, is subject to the ultimate legislative freedom of pure reason itself in the light of its transcendental ideal of the systematic intelligibility of an infinity of instants.[230] It is just such a Moment that Goethe bestows on Faust at his life's end:

A swamp there by the mountain lies,
Infecting everything attained;
If that foul pool could once be drained,
The feat would outstrip every prize.
For many millions I shall open spaces
Where they, not safe but active-free, have dwelling places.

---

[229] Ibid., 25–26.
[230] Cf. Kant's exclusion of time itself from the principle of contradiction, *Critique of Pure Reason,* op. cit., 189ff. Also, Kant on the Moment, ibid., 230ff.

Verdant the fields and fruitful; man and beast
Alike upon that newest earth well pleased,
Shall settle soon the mighty strength of hill
Raised by a bold and busy people's will,
And here inside, a land like Paradise.
Then let the outer flood to dike's rim rise,
And as it eats and seeks to crush by force,
The common will will rush to stem its course.
To this opinion I am given wholly
And this is wisdom's final say:
Freedom and life belong to that man solely
Who must reconquer them each day.
Thus child and man and old man will live here
Beset by peril year on busy year.
Such in their multitudes I hope to see
On free soil standing with a people free.
Then to that moment I could say:
Linger on, you are so fair!
Nor can the traces of my earthly day
In many aeons pass away.—
Foresensing all the rapture of that dream,
This present moment gives me joy supreme.[231]

The Now, in Aristotle, was time's measure, that indivisible monad with which consciousness measured time, itself the measure of motion.[232] Reason, then, understood its true self to be an existent individuality, or essence, prior to its thinking, therefore, prior to its temporality, eternally thinking as an Active Intellect. To this essence of itself, to which reason, *qua* reason, could never think of transcending, neither time nor the Now could be properly applied.[233] Actual knowledge, in Aristotle, is a transcendent, timeless, identity with its object; [234] it is never

[231] J. W. von Goethe, *Faust I and II* (trans. C. E. Passage, Indianapolis, 1965), 392–393.
[232] Cf. above, p. 159.
[233] Cf. above, pp. 157–161.
[234] Cf. above, p. 50.

Now, but, indeed, is a prior being of reason (precisely not to be understood in any temporal sense, which was not so with Plato's doctrine of recollection, consequently, with his idea of immortality [235]). This identity Aristotle explicitly tells us 'we do not remember.' [236] This immemorial identity of the Aristotelian individual is that by which he is constituted a person, that is, one who with respect to others is essentially independent in existence. This same individual essence is immortal, but, this person, *qua* person, is not known to be immortal by Aristotle.[237] The Now, then, known to this person by virtue of its existence as a rational animal in this universe, is not able to be transferred, merely by virtue of this personhood, to the realm of essence. In fact, Aristotle states quite plainly, in *Physics* IV.11: "It is evident . . . that neither would time be if there were no 'now,' nor would 'now' be if there were no time . . ." [238] But if we turn our attention to Thomas Aquinas' *Commentary On Aristotle's Physics* IV, Lecture 18, we read the following: "Moreover, from these considerations an understanding of eternity can be easily had. For the 'now', insofar as it corresponds to a mobile object differently related, distinguishes the before and after in time. And by its flux it produces time [*et suo fluxu tempus facit*], as a point produces a line. Therefore, when different dispositions are removed from a mobile object, there remains a substance which is always the same. Hence the 'now' is understood as always stationary [*ut semper stans*], and not as flowing or as having a before and after. Therefore, as the 'now' of time is understood as the number of a mobile object, the 'now' of eternity is understood as the number, or rather the unity, of a thing which is always the same [*unitas rei semper eodem modo se habentis*]." [239] Here Thomas detaches the Aristotelian Now from its necessary dependence upon time. The Now is no longer essentially tem-

---

[235] Cf. above, pp. 49–52.
[236] Ibid.
[237] Cf. above, pp. 48ff.
[238] Cf. above, p. 159, note 192.
[239] Thomas Aquinas, *Commentary On Aristotle's Physics* (trans. R. J. Blackwell *et al.*, London, 1963), 261–262.

poral, but it is equally, indeed, easily, conceivable as the Now of eternity. But if the Now is that by which reason reckons time, if it is reason's measure, but not essentially so, then the way is open for understanding the Now of time to be deficient by virtue of its being determined by the before and after of time. Time itself is understood to be the product of the Now in precisely that aspect which for Aristotle is its potentiality *in mente;* that is, the Now is understood to correspond to time essentially when the former is a divider and the latter is actual. But, when the Now is considered in its actuality as a unifier, then, time is merely potentially its matter, while eternity is that immutable, not merely whose number, but whose unity it is. In the *Summa Theologica* I,10,5, Thomas Aquinas discusses the question: "Whether This Is a Good Definition of Eternity, 'The Simultaneously-Whole and Perfect Possession of Interminable Life'?" In the course of his discussion he says (ad 5): "Two things are to be considered in time: time itself, which is successive; and the *now* of time, which is imperfect. Hence the expression *simultaneously-whole* is used to remove the idea of time, and the word *perfect* is used to exclude the *now* of time." [240] In the form of transcendental reason that Thomas' sacred doctrine is,[241] the Now of actual time is never itself actual, but time's continuation, which is the actuality of the Now and of time in an Aristotelian universe, is in Thomas merely potential. This is a situation radically different from that fate that overtook Aristotle's Now in Stoicism, where, suffering an internal division, the Now was dissipated, together with reason, in matter, in which state it accounts for the destruction and rebirth of the entire universe in a periodically reiterated present time. Neither is it Skepticism's infinite multiplication of the dimensional Now of the Stoics in which all things successively lie wrapped in the darkness of their own inaccessible grounds. Nor is the state of time and the Now in Thomas what it was in Neoplatonism's backward-looking reconstruction of meta-

[240] *Summa Theologica,* op. cit., 40–41.
[241] Cf. above, chapter on Aquinas.

physics, where, by virtue of its boundless desire, the world is grounded in an absolute principle of continuity, the One, and Time is understood as the omnipresent concomitant of the eternally formed Soul of the universe.[242] If, then, sacred doctrine understands time's continuation to be merely potential, if the measure of this potentiality is time's own imperfect Now, that is, a Now which does not as a matter of fact unite past with future time, it is further clear that one Now is not united with another Now through that time span of which in Aristotle's universe these two Nows would be the intelligible limits. But rather we see that the Nows in Thomas' created universe, as known in the transcendental form of sacred doctrine, while still internally indivisible, nevertheless, exist in isolation; these isolated Nows of sacred doctrine, indivisible instants as in Aristotle, point to *a rift in time itself;* while, in themselves, these isolated Nows possess no actual existence. This radically contingent state of the isolated Nows of sacred doctrine endures on the ground of the transcendental essence of existence, this transcendental essence of existence being existence itself without respect to time, or, existence *indifferently* that of the Now of time *or* of the Now of eternity, that is, existence not simply transcendent to time or to reason, as in Aristotle's universe, but transcendent to eternity or intellectual essence itself. In the fact of the transcendental essence of existence the universe of reason is, in itself, nothing; the universe of essence is, in itself, possible. In Augustine, this transcendental essence of existence is known *generically,* without respect to essential limits of intellect: Knowledge of this existence is immediately a pure will's obedience to God.[243] But, in Thomas Aquinas existence's transcendental essence is known *specifically,* that is, with respect to the limitations of human science. It is therefore known in its distinction from man's intellectual power as a content of faith, where faith is distinguished from knowledge as God's knowledge

---

[242] Cf. above, chapter on Augustine, especially pp. 161–171. Also, further on Time in Neoplatonism: *The Enneads,* op. cit., III.7.
[243] Cf. above, p. 178.

revealed to man, upon which, in turn, is founded that human science most secure in its principles, sacred doctrine.[244] When, therefore, we speak of sacred doctrine as the transcendental form of natural reason, it is to be understood that this transcendental form self-consciously distinguishes itself from its merely natural state. For the first time, it sees transparently the radical contingency of all things, itself included; through existence itself, it comes to know its object as known to be other than itself.[245] Its knowledge is knowledge of what is transcendent with respect to human science; the latter operates in the defective light of natural reason. But defective or not, natural reason is reformed as transcendental by sacred doctrine's very existence; to put it simply, reason lost its essence, or its reason for being, not as in ancient times by virtue of the dissipation of the Now in time, which accompanies a loss of original self-identity, leading to the abyss and to Neoplatonic attempts at reconstruction.[246] Rather, reason lost its essence, its being its own reason for being, by virtue of the isolation of the Nows, therefore, by virtue of the rift in time itself created by the appearance of the transcendental essence of existence. So that, after this appearance of existence itself in the course of time, reason comes to see its own contingency reflected in that of the merely potential succession of isolated Nows. Now, then, when natural reason, in the person of Descartes, resolved to take for itself this universe of contingencies, at the very moment that it had begun to exist within its own thought, existing as if it itself were, in the thought of its existence, existence itself or the transcendental essence, at that very moment, it bore witness to that rift in time, to the isolated Nows, which, indeed, made possible this resolution, if they did not make it necessary. Listen to Descartes in *Meditations On The First Philosophy* III: "But though I assume that perhaps I have always existed just as I am at present, neither can I escape the force of this reasoning, and imagine that the conclusion to be drawn from this is, that I need not

[244] Cf. above, pp. 56ff.
[245] Ibid.
[246] Cf. above, chapter on Augustine.

seek for any author of my existence. For all the course of my life may be divided into an infinite number of parts, none of which is in any way dependent on the other; and thus from the fact that I was in existence a short time ago it does not follow that I must be in existence now, unless some cause at this instant, so to speak, produces me anew, that is to say, conserves me. It is as a matter of fact perfectly clear and evident to all those who consider with attention the nature of time, that, in order to be conserved in each moment in which it endures, a substance has need of the same power and action as would be necessary to produce and create it anew, supposing it did not yet exist, so that the light of nature shows us clearly that the distinction between creation and conservation is solely a distinction of the reason." [247] This last remark of Descartes shows that just as he acknowledges the rift, he is busy covering it over with his own thought. For Descartes, God and creation are innate ideas of natural reason, which, we can see, serve to obscure the fact of the rift in time, namely, that something has occurred to man in the course of his being in the world, or, that there is an essential history of thought, or, that there is a history of being. Descartes provides himself with a cause sufficient to unite the isolated Nows: God. Or, in anticipation of Leibniz, Descartes understands God, as the God of reason, to be the sufficient cause of the existence of the radically contingent. Because this contingency is taken for granted, conservation is not distinguishable from creation. But to take contingency for granted is to remain under the interdict of sacred doctrine. That is, it is to be cut off from access to the transcendental essence of existence, cut off precisely as natural reason, *qua* natural, from the fact of existence, confined to the thought of existence, a confinement in which free reason abstracts from the simplicity of creation *ex nihilo* to the absoluteness of its own thought, from the isolation of the Now to the solitude of thought in which the existence of the totality of things is implicit. We are now in a position to see that Descartes' solitary thought of existence is the original Moment corresponding to the Moments of Kant,

[247] *The Philosophical Works of Descartes,* op. cit., 168. Cf. also, above, pp. 90ff.

which make possible the experience of time by generating uniformly an alteration from one state to another state within the limits of two instants, Nows themselves incorporated as parts of the total alteration or change, by virtue of what becomes in Kant the original synthetic unity of consciousness, itself the ultimate origin, through sensuous intuition, of time itself, as well as of the laws by which pure understanding conceptualizes the world of sensible experience.[248] The Moment, then, is pure reason generating out of its own resources a world of appearances; as the poetry of Faust's last moment reveals, it is by its very nature pregnant of the *future.* As we see in Descartes, it is in its very inception a denial of history because it takes for granted contingency, which, in Kant, reduced to sensible conditions of knowledge, *is* the world of appearances, so that, finally, the Moment is the actual time of Reason in which it establishes, according to its own laws, nature itself, or the contingency of things.[249] The interdict of sacred doctrine, under which reason lies cut off from the fact of existence, restricted to the thought of existence, is, ironically, that by which it is condemned to confusing itself for practical purposes with the transcendental essence. Since the matter of the confusion is time, reason's demonic insistence on its freedom to create, or better, to conserve what from a pure transcendental viewpoint is, in itself, nothing, is given over to an insatiable appetite for a future. The revulsion that Kant tells us that imagination experiences at the thought expressed by the angel of the Apocalypse that change itself will come to an end has its root in reason's confusion of the transcendental essence of existence with the conditions of its own understanding. It, therefore, appears to it that a last moment, by which time would be united to eternity, is formally impossible. We may understand that reason's original resolution to be for this sensible universe its transcendental existence sees in an announcement of an actual realm of intelligible existents an end of that reason for existence that it took itself to be. Now, while it is undoubtably true that, in the event of an actual

[248] Cf. above, pp. 182–183, note 230.
[249] Cf. above, p. 116, note 114.

transition from a realm of what is nothing in itself to a realm of what is in itself possible, pure reason would be destroyed, what is, perhaps, most remarkable is that reason is, by sacred doctrine's interdict, precluded altogether from even the faintest perception of what would be the true source of its downfall. This darkness in which it is bound is immediately shown by its assumption that the necessary condition of change is time, that therefore the end of time is the end of change, or simply eternity.[250] Eternity, in the words of Thomas Aquinas, is, however, 'the simultaneously-whole and perfect possession of interminable life,' [251] thereby indicating not the end of change, but the end of the succession of isolated Nows in which change occurs only potentially; indicating thereby that eternal Now at which all isolated Nows are determined to their potentiality in time, as well as, at which every actual change is determined to be without or within worldly being. While pure reason is able to conceive of an intelligible realm as distinct from the world of appearances, the world of *noumena* as opposed to *phenomena*, reason, in its commitment to the conditions of sensibility, is forever precluded from conceiving this intelligible realm itself in its contingency, that is, as merely possible. It is left no choice but to conceive of the *noumena* as things-in-themselves, since, as contingent, an intelligible realm would be, for pure reason, indistinguishable from its own proper realm of appearances. In this way, pure natural reason cooperates in the interdiction by which it is to know nothing of the transcendental essence of existence whatsoever: When it approaches toward it, it perceives the transcendental thought of existence, but, that, in Kant, it conceives merely as the transcendental ideal of reason itself.[252] In this perspective we may catch a glimpse of that towering grandeur that is, in Augustine, the transcendental essence of existence. He writes in his *Confessions* 12.2: "The lowliness of my tongue confesses to your highness that you have made heaven and earth, this heaven which I see, this earth on

[250] Cf. above, pp. 180–181.
[251] Cf. above, pp. 186–187.
[252] Cf. above, pp. 108–110, 114–115.

which I tread and from which comes this earth I bear about with myself. You have made them. But where is that heaven of heaven, O Lord, which we hear of in the words of the psalm: 'The heaven of heaven is the Lord's, but the earth he has given to the children of men?' Where is the heaven that we do not see, before which all this which we see is earth? For this corporeal whole, since it is not everywhere whole, has in such wise received form and beauty in its least parts, of which the very lowest is our earth, but to that heaven of heaven even our earth's heaven is but earth. Not unreasonably are both these great bodies but earth before that indescribable heaven which is the Lord's and not the sons' of men." [253]

We are now in position to notice that when pure reason, in Kant, denies the possibility of knowing, as anything more than a regulative principle, the existence of personal identity,[254] that existence is denied in the form of the purely intelligible identity of person, such as we understand Aristotle to have known.[255] To put it directly: pure reason denies that *essential immortality* by which one is constituted a person (which, by the way, is the manifestly essential presupposition of Aristotle's entire scientific enterprise), but, it does not deny *personal immortality,* that is, immortality of the person, *qua* person, since, by the interdict of sacred doctrine, it is blind to the issue of the transcendental essence of existence or to its appearance in the course of human existence. But, then, we begin to perceive how, under the interdict, pure reason is, as it were, retrospectively stripped of the possibility of any substance whatsoever, including that of its own identity, which has become, in the form of the original synthetic unity of apperception, a merely logical entity. This it has become in company with Aristotle's law of contradiction (which has been denuded of its synthetic function by eliminating its reference to time),[256] and with, as well, the Aristotelian Now (which, intact, has yet been incorporated within the Mo-

[253] *The Confessions,* op. cit., 305–306.
[254] Cf. above, 181–182.
[255] Cf. above, 184–185.
[256] Ibid., note 230.

ment as part of the totality of change grounded in reason's freedom—this, ironically, in imitation of the isolation of the Now by the appearance of the transcendental essence of existence, of which pure reason possesses no knowledge, but, in lieu thereof, is possessed of its own thought of existence, or, of the thought of God),[257] and, finally, with the possibility of a true metaphysics: reason's identity as essence.[258] It is the peculiar fate of pure reason that in its denial of the possibility of history, or, that story which is identically what has occurred to the storyteller, it is limited, by its perspective, to denying the ground of history in the form of the essential identity of the man of Aristotle. But, while this essential immortality was for Aristotle ground to nature as well as science, it was not in Aristotle a ground of history, since, however we construe this essence, it is certain to Aristotle that nothing has occurred to it during the course of its being in the world. Therefore, the most that can be attributed to pure reason's denial of history on the ground of the purely intelligible identity of persons is that it denies the ultimate significance of the past *as* past for determining, or limiting, the future of man. Its denial in this peculiar form, which is of a piece with its perspective under sacred doctrine's interdict, is its understanding of man's freedom from nature, at least as its ideal, or the ideal of its pure practical reason. *It would be an error of the first order to mistake pure reason's confinement to statements conditioned by its own sensibility as a form of neutrality on the issue of the transcendental essence of existence, when, as a matter of fact, this confinement is nothing but its absolute incompatibility therewith, grounded, as it is, in its own will.* However, if there is such a thing as history, its possibility is not the essential immortality of Aristotle, but the personal immortality of Augustine, the meaning of which for history is that in the course of its being in the world of time, of space, indeed, as a necessary condition thereof, there has occurred to the intelligible essence of man the transcendental essence of existence, bringing into being, by a conversion of spatiotemporal man to intelligible

[257] Cf. above, p. 190.
[258] Cf. above, pp. 52ff.

man, an immortal person whose essence is to be a manifestation of the knowledge itself of existence. The essence of this person being the manifestation of the glory of God, although not of God, but of his glory. If there is history, then, it cannot be grounded on the rejection of the Aristotelian essence, as with modern thought, in which emerges not history, but a blueprint for engineering human destiny by rearranging mental habits or material conditions, in either event, both. Nor is history able to be grounded on the acceptance of the Aristotelian essence, since, in itself, it prescinds from the necessary spatiotemporal conditions of history; indeed, when it is forgotten in favor of the sensible conditions of human existence, its priority is felt not in the emergence of history, but of fatality or doubt, as Stoicism or Skepticism will testify. No, if there is to be history, it must be grounded precisely in the *clarification* of the Aristotelian essence: the clarification of the luminous opacity of the essence of the human person is the historical occurrence *par excellence.* That occurrence took place when the knowledge itself of existence became the essence of a human person in such a way that, by not ceasing to be itself, that is, the glory of God itself, it displayed by its simple identity with a human person, by this perfect ease of access (unqualified by the limits of human intelligibility) of the Word of God to man, the merely conditional opacity of the human essence. The latter, since the Word became flesh, could not be attributed to its nature, but, henceforth, to a decision of that essence itself, upon which, as Aristotle knew, a person's existence depended. At this same time in which the transcendental essence of existence appears, it announces its availability to each and every human essence, not, indeed, that it might be simply identified with another human person, but, rather, that through that personal relationship established by its appearance in time, the latter person might be called to that clarification of his own essence in which he exists transparently in the knowledge itself of existence, or as himself a manifestation of the glory of God, in his very existence. In the human person with whom the transcendental essence of existence was simply identical, God received his own glory; by this perfect glorification of God, the man's essence was likewise clear; by this perfect clarification of essence, the man was raised from death; by this resurrection of the man in whom was God's glory, the transcendental essence of existence called attention to its appearance in the course of time, creating a rift in time. The Now was isolated; time's continuity came to an end. This was the beginning of history. It began, as required if it is to be history, with the identity of the storyteller with the story of what occurred to him in the course of his being in the world. This beginning of human history is, in that instant, the history of being, first, in that the transcendental essence of existence enters worldly being; second, in that the clarification of human essence is metaphysically a change of being itself; third, in that the rift in time displays the historical structure of the transcendental essence of existence itself as in itself an eternity of relationships in the Trinity. Because of history's precise significance, it may be understood that being is properly said to have a history in each of these three respects, though in none of them is its essence exhausted by history.

According to Leibniz this world is the best of all possible worlds. *The best of all possible worlds exists* is reason's essential determination of the fact of existence, that is, of creation *ex nihilo.* It is the discovery of the sufficient reason for the existence of this world, the contingency of which it takes for granted—as in Descartes, so in Leibniz. This contingency is experienced by Descartes within the solitude of the *cogito ergo sum,* within the Moment of free reason that knows its contingency immediately as its merely potential continuance in *time,* but which upon reflection discovers as the essential thought of its existence God's conservation of that existence in time (thereby anticipating, in this respect, Kant's Moments, by which transcendental self-consciousness [of which God will be explicitly symbolic] organizes originally its own experience). Leibniz, in contrast, realizes God as creator of *this world;* thereby falling short of Kant's critical philosophy in which *self, world,* and *God* are mere ideas of pure reason. But he excels Descartes, insofar as he, Leibniz, immediately distributes that Cartesian thought of contingent existence to an infinite totality of existing things, or things-in-themselves, whose coordinate existence is that of a perfectly intelligible universe. What is real in Leibniz is exactly that which is directly denied in Kant, namely, the contingency of the intelligible realm, its being known to reason. The contingency of Kant is purely phenomenal. We have seen how a necessary or actual essence is indistinguishable in Kant from what it would be in Aristotle, except that, in Kant, its existence is merely logical. We may add that this existence *in principle* of intelligible entities, while ultimately the result of sacred doctrine's interdict, is proximately traceable to Leibniz' perpetuation of the separation of spirit from matter effected by Descartes. It is, in any event, not directly related to Aristotle, since his essence never existed as a matter of principle, but the intelligible realm in Kant is Aristotelian for practical purposes, and only so. Now, in Leibniz, this essentially determined universe of intelligible entities is, as such, a *specific* arrangement of what is possible (in this way Leibniz is to Descartes' *generic* thought of existence, as Thomas Aquinas' specificity is to Augustine's generic knowledge of the transcendental essence of existence). The determination of the specific arrangement of what is possible is that of the principle of sufficient reason, or, of God himself. Leibniz writes, in his *On The Ultimate Origination of The Universe,* ". . . the actually existing world is necessary only physically or hypothetically, but not absolutely or metaphysically. Suppose the world to be indeed in a determinate state now; then other determinate states will be necessarily engendered by it. Since therefore the ultimate root of the world must be in something which exists of metaphysical necessity, and since furthermore the reason of any existent can be only another existent, it follows that a unique entity must exist of metaphysical necessity, that is, that there is a being whose essence implies existence. Hence there exists a being whose existence is different from the plurality of beings, that is, from the world. . . . But let us explain somewhat more distinctly how, from the eternal or essential—that is, metaphysical—truths, the temporal or contingent—that is, physical—truths are derived. First we must recognize this: from the fact that something rather than nothing exists, it follows that in possible things, or in their possibility or essence itself, there is a certain demand or (so to speak) a claim for existence; in short, that essence tends by itself toward existence. From this it follows, furthermore, that everything possible, that is, all that expresses a possible essence or reality, tends with equal right toward existence, the degree of this tendency being proportionate to the quantity of essence or reality, that is, to the degree of perfection of the possible involved. . . . From this it must be concluded that, out of the infinite number of combinations and series of possibles, one exists through which a maximum of essence of possibles is produced into existence." [259] In the sacred doctrine of Thomas Aquinas (wherein is subordinated a true metaphysics, Aristotle's, to an explication of reality in revelation's light [260]) the transcendental essence of existence, known to Augustine in its immediacy, is known to the transcendental form of reason as the participated existence (*esse participatum*).[261] This, together with what otherwise would be an essence independent in existence, is, as a matter of fact in the context of the isolated Now of the history of being, the actuality of the essence with respect to which the essence in itself is, henceforth, understood to be merely potential. Leibniz, through the medium of Descartes' thought of existence, in itself absolute as a matter of principle,[262] translates this potential essence of the transcendental form of natural reason, where it exists as a matter of fact, into a possible essence of natural reason itself, where it is understood to exist as a matter of right. This natural right of the essence to exist by virtue of its possibility is recognized by God himself, in Leibniz; indeed, God is, as the sufficient reason of the existence of the universe itself, the Supreme Reason who is essentially distinguished from other essences by reason of the fact that his possibility, his essence, is immediately his existence, while the existence of possible essences outside of God is mediated by a

[259] G. W. von Leibniz, *Monadology and Other Philosophical Essays* (trans. P. and A. M. Schrecker, Indianapolis, 1965), 85–86.

[260] Cf. above, chapter on Aquinas.

[261] Cf. above, p. 63.

[262] Cf. above, pp. 188–189. Also, cf. above, p. 35, where 'thought as an absolute principle,' in introducing Aristotle, is clearly not the *thought of existence*.

requirement, in consideration of the claim of all possibles to exist, *to produce existence to the greatest extent possible absolutely*. It is not by virtue of their *participated existence* that creatures are in the image of God, as in Thomas Aquinas; but, rather, in Leibniz' continuation of the translation of the fact of the history of being into terms of reason's own comprehension, that is, of the *fact* into the *ideal* of the history of being, creatures are in the image of God in Leibniz by virtue of their essential likeness. Hence, their claim to exist is recognized by God in bringing them to existence within the limits of the ideal, or within the limits of the best possible world. In furtherance, therefore, of the contradiction that the doubt of Descartes is, in truth, to that of Augustine,[263] Leibniz contradicts that coalescence of being with mercy which, in Augustine, is God's creating of the world by converting it *de nihilo* into existence from a nothingness in which, as Augustine says, since it existed in no way whatsoever, it could possess no claim upon God to exist.[264] (Thomas Aquinas expressed this as the *novitas mundi*, or, simple novelty of this universe before the first Now of which it existed not at all.[265]) Leibniz understands creation to be a matter, not of mercy, but of justice; it is not Augustine's City of God, in which the citizens are constituted in *vera iustitia*, or true justice, by their love of God, but, for Leibniz, a Republic of Being, in which the interests of the citizens are satisfied within the limits of the common good. He writes, in his *On The Ultimate Origination of The Universe:* ". . . it must be realized that in the best constituted republic care is taken to grant everyone the greatest possible good, and that, analogously, the universe would not be sufficiently perfect unless the interest of everyone were taken into consideration, without prejudice, of course, to the harmony. There is no better verification of this than the very law of justice which ordains that each one should have his share in the perfection of the universe, and that his happiness should be proportionate to his virtue and to his voluntary contribution to

[263] Cf. above, p. 178.
[264] Cf. above, pp. 172–173.
[265] Cf. above, pp. 63ff, also, pp. 73ff.

the common good; this is what we call charity and love of God, and in this alone consists also the essence and power of the Christian religion, according to the judgments of learned theologians. Nor should one be surprised that in the universe spirits are the objects of so much solicitude, since they most faithfully reflect the image of the Supreme Creator, and since their relation to him is not so much that of a machine to its artificer (as is the case with all other created things), as it is that of a citizen to his prince. Moreover, the spirits will last as long as the universe itself, and in a certain way they express the whole and concentrate it in themselves, so that we may well call them total parts."[266] This republic of being, then, is the power of the essence of reason itself, as the sufficient reason of its own existence. What modern thought sees in the manifestation of the transcendental essence of existence itself, what it sees by virtue of its retrospective anticipation of revelation's matter as due it in the original possibility of its own essence, or, what it sees in the Incarnation of the Word by virtue of its translation of charity into love of self or creation, what it sees in Christianity, is the power of the essence of reason itself.[267] Leibniz writes, in his *What Is Nature?* ". . . one must infer that in a corporeal substance there must be located a *first entelechy,* a first capacity . . . for action . . . the primitive moving force: this has to be added to extension . . . and mass. And it acts constantly, though variously modified by the concurrence of other bodies. . . . This same substantial principle is what is called *soul* in living beings and *substantial form* in other corporeal objects. In so far as it constitutes, joined to matter, a truly unified substance, that is, a unit by itself (*unum per se*), it is what I call a Monad. Eliminate these genuine and real units, and you will have only beings which are but aggregates and, consequently, bodies will not be real beings at all. For although there are atoms of substance, namely our Monads which lack parts, there are no atoms of mass or of minimum extension as the ultimate elements, since the continuum is not composed of points. Neither . . . is there

---

[266] *Monadology etc.,* op. cit., 92–93.
[267] Cf. above, p. 71. Also, cf. *The City of God* XIV, op. cit.

a being endowed with a maximum mass or infinite extension, though there are always some beings larger than others. There exists, however, one being which is the greatest by degree of perfection, that is, a being of infinite power." [268] Of this infinitely perfect expression of the power of the universe as the ideal of the history of being Leibniz writes in *A Vindication of God's Justice:* "The strongest reason for the choice of the best series of events (namely, our world) was Jesus Christ, God become Man, who as a creature represents the highest degree of perfection. He had, therefore, to be contained in that series, noblest among all, as a part, indeed the head, of the created universe." [269] Here, with startling clarity, we see that retrospective bridge built over the rift in time, here in Leibniz' physics, which is to be seen in Descartes' logic, in which the idea of creation is an innate idea, and, in Kant's ethics, where the transcendental essence of existence that appears in history is but the 'archetype of the moral disposition in all its purity'.[270] The bridge is the principle of sufficient reason, superalternated to principles of identity and contradiction. It is expressed in the law of continuity: ". . . when the essential determinations of one being approximate those of another, as a consequence, all the properties of the former should also gradually approximate those of the latter." [271] Stoicism embraced this universe in that reiterative present of the dimensional Now, by which, forgetful of Aristotle's essence, it self-guaranteed to itself knowledge of the world's nature. In Leibniz, however, this universe is the greatest possible extension of the thought of existence, that is, of substance or possible essence, surpassing Descartes' subjectivity in the direction of an infinitely increasing appropriation of the power of the universe by reason itself, so that, as Leibniz says: ". . . everything is interconnected in the universe by virtue of metaphysical reasons so that *the present is always pregnant with the future,* and no given state is explicable naturally without

---

[268] *Monadology, etc.,* op. cit., 106.
[269] Ibid., 124.
[270] Cf. above, p. 119, note 129.
[271] *Leibniz: Selections* (ed. P. P. Wiener, New York, 1951), 187.

reference to its immediately preceding state."[272] The finitude of the Moment of Descartes' *cogito ergo sum,* its consciousness of the merely potential succession of the Nows, is seen to be not of the essence of the thought of existence. Leibniz writes in the *Letters to Samuel Clarke:* "It cannot be said, that (a certain) *duration* is eternal; but that *things,* which continue always, are eternal (by gaining always new duration). Whatever exists of time and of duration (being successive) perishes continually: and how can a thing exist eternally, which (to speak exactly) does never exist at all? For how can a thing exist, whereof no part does ever exist? Nothing of time does ever exist, but instants; and an instant is not even itself a part of time. Whoever considers these observations, will easily apprehend that time can only be an ideal thing. And the analogy between time and space, will easily make it appear, that the one is as merely ideal as the other."[273] The God of Leibniz is the sufficient reason of the infinite extension of power in the universe; that dimension of power is infinitely divisible into intelligible entities essentially intelligible, in themselves and to God. This principle of intelligibility by which God relates directly to every state of the universe reduces time, together with space, to a merely ideal coordination of a universe now understood to be of its very essence contingent. Time, *a fortiori* the rift in time, in the dimension of the power of the essence of reason itself suffers total abstraction.

[272] Ibid., 185.
[273] Ibid., 255.

Chapter 9

HEGEL: THE ABSOLUTE TRUTH

In Leibniz the history of being is translated into the continuity of being. The rift in time created by the appearance of the transcendental essence of existence itself, makes time's very own continuance a matter of potentiality. Time's constitution is displayed as a discontinuous succession of yet indivisible, or, Aristotelian, Nows. The rift, in Thomas Aquinas, leads to the detachment of the Now in its essentiality from time in such a way that, as the Now of a time itself contingent, it is compared as the imperfect to the perfect Now of eternity, where it is the number or unity of a thing possessing itself always in the same mode.[274] This rift in time, in Leibniz, is retrospectively anticipated in the form of the infinite power of the essence of reason itself which, together with this purely intelligible universe of innumerable substances whose existence it expresses in its own existence with the greatest possible perfection, is brought into being by God in recognition of its essential right to exist together with others of like substance in the best of all possible worlds. Since we are, here, engaged with the essential history of thought, possible only if history exists, further, only if, before all else, there is a history of being, the manifestation of which is the appearance of the transcendental essence of existence, and, if that appearance is *knowable,* in the generic mode of Augus-

[274] Cf. above, p. 185.

tine, or, *credible,* in the specific mode of Thomas Aquinas,[275] *the potential exists, in this light, to discern the essential interrelationships of all systems of thought in the history of thought, by relating them, together with the history of thought itself, to the transcendental essence itself, in relation to which, precisely, their existence as a matter of historical fact is.* We have seen that Leibniz understood that the contingency in time of the thought of existence upon which Descartes would found modern science was not of the essence of that thought, if, as is implicit in that thought of existence (explicit, of course, in Descartes' thought *about* the thought of existence),[276] reason is to be a universal instrument for all contingencies. Reason, for Leibniz, must not look outside of itself for the reason of its existence. But, time, as a condition of the sensible universe, is outside of reason. Moreover, just such a discontinuous temporality as Descartes' requires an abstract reason for its continuance, namely, God, who, as Descartes presents him, lacks a sufficient reason for continuing existence in time. Indeed, Leibniz understands that time's mere potentiality (the actuality of the Aristotelian Now, which, in itself, accounted for time's continuity,[277] but from access to which modern thought is cut off by the rift in time, no longer operative) in fact precludes the discovery of the sufficient reason in any but the barest of ideas: a causality *without motive,* attributed to God. But to discover a sufficient reason for existence in its concreteness, that is, to know, in principle, God's motive, is to have entered into that interior ground of things which reason itself is. It is, considering reason in its absolute form, to have entered, in principle, into God's interior. It is to discover that whatever exists for God exists because he was moved to bring it into existence because of an incentive in its very essence, namely, that in a superior manner it contributes to the harmony of the universe. In the case of God himself, Leibniz tells us that his mere possibility or essence necessitates his existence. The incentive is absolute; God sets the tune to the harmony of the universe. But in crea-

---

[275] Cf. above, p. 187.
[276] Cf. above, pp. 79ff; 89ff; 93ff.
[277] Cf. above, p. 159, note 193.

tures, the incentive is relative to the degree of perfection of their essence. Now, in the sacred doctrine of Thomas Aquinas existence, in the order of reality, precedes essence as the participated existence (*esse participatum*),[278] related, therefore, to the natural essence, in the transcendental form of reason, as actual to potential being. If we look closely, we see that this structure, namely, actual existence prior to potential essence is reminiscent of Aristotle's mode of thought in only one respect, formally, in that this order of reality reverses that temporal sequence in which potential being always precedes actual being in Aristotle, except insofar as one being in time is understood actually to precede another of the same kind that, at the time, is potential.[279] Materially, there is no agreement in reality between Thomas and Aristotle, since Aristotle's essence is not to be confused with the nature of which it itself, identically its own existence, is the cause.[280] Indeed, it is precisely the occurrence of the transcendental essence of existence to essence which clarifies it as, in itself, together with nature to which it is simultaneously assimilated, if not nothing, as in Augustine, at best potential, as in Thomas Aquinas. It is important to note, then, that creation is a matter of participated existence. That is, creatures, *qua* creatures, are not constituted by virtue of the limitations of their essence in the natural order. Simply, creatures *are*, by virtue of being raised from nothing to existence; whatever potentiality is attributed to an existing essence is after the fact of existence or creation. Leibniz, in contrast, attributes possibility to the existing essence after the thought of existence.[281] As a result, it is thought to contain within itself the reason for its own existence. This is a claim upon God to exist, or, an incentive by which God is moved to create it, to bring it to actuality. It is immediately evident that potentiality is what, precisely, is eliminated in modern thought. *Potentiality* is that form of possibility that *in the order of reality* always *follows* actuality, the significance

[278] Cf. above, p. 63.
[279] *Metaphysics*, op. cit., 191ff.
[280] Cf. above, chapters on Aristotle and Aquinas.
[281] Cf. above, p. 197.

of this statement being that *whatever exists potentially is ordered to being something particular,* or to exist according to its own form or essence prior to entering into ordered relationships with other entities. (It was, by the way, exactly this potentiality which Plato discovered in the Idea, which made possible the philosophy of Aristotle, but which, in Plato, remained confused with actuality. But this concreteness of the potential is Plato's immortality.) *Possibility* is, however, *what is in its very essence ordered to being something individual,* that is, to exist *contingently,* or *to be what it is in determination with other beings.* In the system of thought grounded on the thought of existence, in the absolute thought of existence (in Leibniz, still absolute in principle only, for between Leibniz and Hegel lies Kant) in which the potentiality of the transcendental form of natural reason is translated into the possibility of reason itself as the self-sufficient ground of the world, this possibility is to be realized as actuality *through the necessity of reason itself.* There is, then, the antithesis possibility-actuality to be mediated by necessity. If it is understood that this trinity of modern reason is the denial of the fact of the history of being, then one is able to see right at this place, in advance, how absolutely restricted in essence is Kierkegaard's worthy attempt to see the appearance of the transcendental essence of existence, as in his *Philosophical Fragments,*[282] where, however, he is freely determined to employ that very logic of modernity forged under the interdict of sacred doctrine. Indeed, he does so in explicit opposition to the metaphysics of Aristotle,[283] in clear indication that he, too, falls under the ban. So, for example, for Kierkegaard the Incarnation is the moment of the entrance of *the Eternal* into time. But more of this below. Leibniz not only translates *potentiality* into *possibility,* but reverses the paradoxical order of both Aristotle and Aquinas so that *actuality follows possibility* precisely in the order of reality. That is, in the order of the essences which constitute the real or intelligible substance of the universe as opposed to its merely apparent or sensible forms, the spatiotemporal order. The latter, for Leib-

[282] *Philosophical Fragments,* op. cit.; also, above, pp. 146ff.
[283] Ibid., op. cit., 92.

niz, is not real but ideal, a mental construction by which the mind locates the things themselves distributed through an infinite continuum of power or substantially formed matter. But as I think of Leibniz' universe in which it is of the essence of all things that *actuality follows possibility,* I begin to perceive that while, in essential correction and improvement of Descartes, the continua of time and space have been relegated to a merely ideal existence, nevertheless, I remain *formally* in the presence of time as long as actuality follows possibility in reality. I begin to see that while Leibniz assigns time (in particular that weak time which for historical reasons is radically potentiated, so as no longer to be self-supporting as in Aristotle) to a merely ideal being, his continuum of substantial being is shaped itself *on the analogy of time* as it appears in the transcendental form of natural reason as a matter of fact. Corresponding to the indefinite succession of isolated Nows is the infinite succession of individual Monads, no one of which is, however, itself, sufficient to account for its actuality without reference to God's essence as the sole necessary existent. But, since this Supreme Reason, through the preestablishment of harmony between extended mass and substantial form, has so provided for the coincidence of soul with body as to make actual the coexistence of essences, I see further that there is in this continuum of being *an analogy to the space* of the transcendental form of natural reason, where the transcendental reason in the transparency of its clarified essence knows its object *as known* to be *actually existing as an other.*[284] As a result of this coincidence of the analogies to the time and space of the transcendental form of reason, of sacred doctrine, I see clearly that the succession of the monads from the least to the greatest is simultaneously, intensity of individuality/extension of power. As the *numerical* monad or Now of the transcendental form of reason measures a merely potential time on the ground of the transcendental essence of existence, as it simply appears, so the *substantial* monad of scientific reason measures the essential possibility of power on the

[284] Cf. above, chapter on Aquinas, especially pp. 56ff; 65ff; 70ff.

ground of the essence of reason itself, as it absolutely exists of its own necessity. These analogies would be perfect except for the fact that Leibniz' God exists out of the necessity of its *possibility;* yet another indication that what elements of actuality exist in natural reason refer to nothing beyond the *potentiality* in the original or transcendental form of natural reason. In no way whatsoever, despite the fact that the only adequate analogy for natural reason is, as a matter of historical fact, the transcendental form, is reason itself able to recall that original actuality or transcendental essence of existence itself, or participated existence. Whereas the latter, the transcendental essence of existence, is a ground for the discontinuity of individuals as such, that is, for their particularity beyond continuity, the absolute thought of existence thinks through, in principle, that particularity to the law of the special arrangement of this world. Leibniz writes in a *Letter to Remond de Montmort:* "As to metempsychosis, I believe that the order of things makes it inadmissible, for that order requires everything to be distinctly explicable and nothing done by leaps. But the passage of the soul from one body into another would be a strange and inexplicable leap. There is always a presentiment in animals of what is to happen because the body is in a continual stream of change, and what we call birth or death is only a greater and more sudden change than usual, like the drop of a river or of a waterfall. But these leaps are not absolute and not the sort I disapprove of; e.g., I do not admit a body going from one place to another without passing through a medium. Such leaps are not only precluded in motions, but also in the whole order of things or of truths. . . . Now as in a geometric line there are certain special points called maxima, points of singularity, etc., and as there are lines which have an infinite number of such points, we must in like manner conceive in the animal's or person's life periods of extraordinary changes which are not outside general law, just as the special points on a curve may be determined by its general nature or its equation. . . . There are doubtless a thousand irregularities, a thousand disorders, in particulars. But it is impossible that there should be any in the whole, or even in each Monad, because each Monad is a living mirror of the Universe, according to its point of view." [285] The law governing this best possible arrangement of things is known in its infinity only to God, who, outside this infinite series, therefore, alone, perceives, not ideally, but in the reality of his creative operation, that identity of even the least of the infinite Monads, that is, that *unique essential possibility* in virtue of which Leibniz is able to maintain, together with the infinite order outside of which no Monad would be actual, an individuality of the Monad, not determined by the order, *qua* order, which is not *numerical* or ideal, but a real, though indiscernible, *unity of the thing*. So that, finally, I behold in Leibniz' Monad not only those blind analogies to the space and time of the transcendental form of natural reason as they illuminate the structure of the Monad in its function as a measure of the continuity of being, as a divider in mental potentiality (on the further analogy to Aristotle's Now), but, also, I see that the Monad, as a living mirror of the Universe in actuality, is, of its essence, illuminated by the analogy it is in substance to the Now of eternity in the sacred doctrine of Thomas Aquinas, which is the 'unity of a thing possessing itself always in the same way.' Here, in the total abstraction of reason from the rift in time, those disparate realms of the transcendental form of natural reason, namely, time and eternity, coalesce in the unity of substance and function in the Monad. There are no leaps, then, necessarily because whatever is already is all that it possibly might be. That is, in its very principle, or essence, and, all that it might possibly be is the infinite extension of being in the universe, concentrated in its individuality. Perhaps, nowhere better than in the philosophy of Leibniz are we able to perceive that the issue of the transcendental essence of existence, or the issue of history, or of a history of thought, is, essentially, the issue of *the history of being,* which Leibniz perfectly denies by translating *existence itself* into *an eternal region of essential possibilities,* so that what actually exists owes its existence to its pre-existing as *a* being, or, as the thought of *a* being, God. In this very special respect, then, in a

[285] *Leibniz: Selections,* op. cit., 188–189.

purely formal way, Leibniz' science is reminiscent of Aristotle's. For both, in radical distinction from Thomas Aquinas,[286] God is a substance. For Leibniz, God is the incorporeal substance or absolute species of the genus of possible essences; whereas, for Aristotle, God is that substance whose essence itself is absolute, having nothing in common with all other essences whose existence involves potentiality. We have previously noted how this merely abstract confusion of substance, as is Leibniz' with that of Aristotle, functions in Kant for all practical purposes without hindrance, since, indeed, it is grounded in a common remoteness, in different directions it would seem, from the fact of history.[287] Once again, the thought of Kierkegaard intrudes upon our consideration of the essential history of thought at this juncture of Leibniz with Aristotle, with Kant; for, it was Kierkegaard who, after all, distinguished ancient paganism from that paganism flourishing in modern Christendom, by characterizing the latter as spiritlessness in the direction away from 'spirit,' while the former, not able to be, we may take it, by the mere fact of history, apostasy, is understood by Kierkegaard to be spiritlessness in the direction *toward* 'spirit.'[288] But we must say that it only *seems* this way to Kierkegaard, since his entire logic, indeed his sensibility, is that of a modern man. Kierkegaard is the poetic individuality without a knowledge of the essential history of thought. His faith is neither the *knowledge* of Augustine, nor the *belief* of Aquinas, but it is *commitment:* the *pathos* of a man cut off in the continuity of being itself from the transcendental essence of existence itself, *qua* essence, and from the transcendental form of reason which organizes its knowledge secure on the foundation of its belief in existence itself, but, yet, encountering that essence of existence itself in his *transcendent passion to exist.* While, therefore, Kierkegaard's 'existentialism' points to the perfect simplicity of the transcendental essence of existence itself in its being that by which all things whatsoever are reconciled to existence itself, as Paul

[286] *Summa Theologica* I,3,5, ad 2, op. cit., 18.
[287] Cf. above, pp. 192ff; 196.
[288] Cf. above, pp. 151–152.

writes in his *Letter To The Colossians* 1:20, ". . . everything in heaven and everything on earth, when he made peace by his death on the cross,"[289] still, the simplicity of the transcendental essence itself is such that nothing of the narrowness of the conditions of its reception can be subsequently attributed to it as if they constituted its destiny or fate. Indeed, if there is history, then those forms by which it is to be apprehended at any subsequent time cannot be essentially determined in this Now. Therefore, to say that ancient paganism is spiritlessness directed toward spirit is strictly a retrospective thought grounded in the *pathos* or commitment of the individual. It may even be, as in Paul's speech to the Areopagus at Athens, in which he refers to the statue of the Unknown God, as well as to Greek poetic statements about God's being,[290] that there is essential justification for statements of this kind to be found in passion or *pathos* itself, in essence an unclear power, but it is only *the fact itself of history* that effects this clarification of natural impulse. Without the knowledge or belief in the transcendental essence of existence as it appears, *pathos* itself can only be directed to ends proposed to it by its own unclarified substance. That is, it is lost within its own finitude, as with the ancients, or, within its own infinitude, as with modern man. Indeed, as we see in our examination of the essential history of thought, neither ancient nor modern *pathos* is historical. The former, as with Stoicism and Skepticism, is given over to fate and doubt, the latter to its own magnification and mastery in the dimension of power. In fact, it may be said that the only medium for the transmission of the fact of history is the immortal person, but it belongs to the person not only to exist, but to believe, and to know.

To our surprise, this recent detour by which, again, we have been led to the thought of Kierkegaard, arising, as it did, at the moment when we were about to make the transition from Leibniz to Kant, turns out to our advantage, since it focuses our attention on the fact that in Leibniz the foundation of the indi-

---

[289] *The Jerusalem Bible: N.T.*, op. cit., 345.
[290] Cf. above, p. 118, note 122.

vidual substance is precisely in the *pathos* of the infinite divisibility of the universe which, in principle, it is. That is, the *pathos* in which the individuality itself, upon which, in the case of man, the person depends as a moral entity, is grounded in an infinite number of perceptions below apperceptive consciousness, but, ultimately determining it as to its content. Let us listen to Leibniz in the preface to his *New Essays On The Human Understanding:* ". . . there are a thousand indications which lead us to think that there are at every moment numberless *perceptions* in us, but without apperception and without reflection; that is to say, changes in the soul itself of which we are not conscious. . . . These *minute* (*petites*) *perceptions* are then of greater influence because of their consequences than is thought. It is they which form I know not what, these tastes, these images of the sensible qualities, clear in the mass but confused in the parts, these impressions which surrounding bodies make upon us, which embrace the infinite, this connection which each being has with all the rest of the universe. It may even be said that in consequence of these minute perceptions the present is big with the future and laden with the past, that all things conspire . . . ; and that in the least of substances eyes as piercing as those of God could read the whole course of the things in the universe. . . . These insensible perceptions indicate also and constitute the identity of the individual, who is characterized by the traces or expressions which they preserve of the preceding states of this individual, in making the connection with his present state; and these can be known by a superior mind, even if this individual himself should not be aware of them, that is to say, when a definite recollection of them will no longer be in him. But they (these perceptions, I say) furnish the means of recovering this recollection at need, by the periodic developments which may some day happen. It is for this reason that death can be but a sleep, and cannot indeed continue, the perceptions merely ceasing to be sufficiently distinguished and being, in animals, reduced to a state of confusion which suspends apperceptive consciousness, but which could not last always; not to speak here of man who must have in this respect great privileges in order to preserve his personality. It is also through the insensible perceptions that the admirable pre-established harmony of the soul and the body, and indeed of all monads or simple substances is to be explained; which supplies the place of the untenable influence of the one upon the others. . . ."[291] Thus the radical passivity of modern thought,[292] displayed in its inability to maintain itself transparently face to face with the transcendental essence of existence, is seen, in Leibniz, in the fact that the identity of the substance whose essence is possibility recedes into the infinite depth of the indiscernible, at the infinite dimensionality of which remove, the soul, in this intelligibly dynamic, progressive, neo-Stoicism, is, by pre-established harmony (the Supreme Creator's function), coincident with the body and all other monads, which, we may understand, constitute the *infinite material basis of this individual.* While this matter constituting the individual is itself infinitely real, or intelligible, its primacy in the form of the universe in Leibniz, to which he has distributed in all parts the Cartesian thought of existence, displaces time and space to a merely ideal realm. Thereby, Leibniz eliminates those dimensions that we may understand are just those of particularity. Leibniz moved from the identity of the individual and the universal in the abstract thought of Descartes to that same identity in the concrete of matter. The innate idea of God became the Sufficient Reason of the contingency of this intelligible world, which, at the point of infinity, unites the substantial form with the infinite universe through its pre-established harmony with its own body. This concrete intelligibility of the material universe embodies the Cartesian *extension* of matter in a *mass* with a *force* of its own. Nevertheless (for reasons founded upon its understanding that its intended appropriation of the contingency of the world could not go forward so long as time's reality remained to suggest that *as a matter of fact* that contingency was related to some *other* than reason itself), time, together with space, was now considered as ideal (as opposed to the reality of the things themselves). These material indi-

[291] *Leibniz: Selections,* op. cit., 374–377.
[292] Cf. above, pp. 115ff.

vidualities could be known not at all with respect to their apparent operations in time and space, since, as we saw above, each entity, on the analogy to the eternal Now of the transcendental form of natural reason as it appears in sacred doctrine, originally possesses its identity in its possibility, that is, prior to its actuality in the continuity of the best order.[293] What is lacking in Leibniz' matter, then, are precisely those *specific differences* by which the operations of matter in time and space might be able to be understood. Although Leibniz advanced from the generic thought of existence, which is the *cogito ergo sum* of Descartes, to the concept of the specific arrangement of the universe on the basis of the principle of sufficient reason, still, this universe consisted of what, from the point of view of the understanding, were entirely generic individualities. Or, we may understand that Leibniz' *actualizing of the infinite potentiality* of Descartes' thought *in matter,* yet was *not the actual intelligibility* of that matter which, as a matter of fact, is *experienced* in the world. But the world is the experience of objects arranged in time and space. It is the experience of particular objects existing *outside* of each other, so that, if there is to be any knowledge of these particulars, it must be a knowledge of their *relationships* to each other. If these relationships stand as matter for knowledge they must represent a real contingency, dependence of one on another, within this world of appearances, *qua* appearances, since the intelligible reality of Leibniz consists in simple generic individuals, who, as such, exist infinitely beyond relationships within the universe as it appears. It is, of course, Kant who marks out this region of appearances as a region of actual knowledge of particulars coexisting within a community of interactions. By placing the indifferent substantiality of matter outside possible knowledge, Kant makes room for the emergence of *difference* within time and space. These last appear to Kant as *forms* of sensible intuition within which the matter of possible knowledge, or sensations, must *first* be organized if they are to be able to be conceptualized by the understanding in actual knowledge. But this organization of the

[293] Cf. above, p. 204.

forms of time and space, within which matter is arranged according to specific differences, undertaken by pure natural reason itself, within the limits of its own understanding of experience, based as it is on sensible conditions, is, on Kant's part, a revocation of that distribution of the thought of existence to things-in-themselves effected by Leibniz. In this revocation of the thought of existence, while Descartes' thought, absolute phenomenon,[294] returns to itself, it returns without existence. Absolute thought is the thought of appearances. But in terms of modern thought's inevitable progress, reason possesses in the *noumena* of Kant's system *the concept of the intelligibility of matter itself,* which, while *a matter of indifference to Kant,* is, *in itself, indistinguishable from existence itself,* as conceived by reason itself. Indeed, it is not so much a matter of indifference to Kant that he will dispense with the *noumena.* On the contrary, this intelligible realm exists as the region of pure reason's ideals.[295] The revocation of thought from the continuity of being, while it sharply distinguishes appearance from reality, continues, as its presupposition, that continuity of being itself erected by Leibniz to provide reason itself with a defense against the appearance of the history of being. With the continuity of being itself secure in the *noumena,* Kant is free to introduce, once again, the discontinuity of spatiotemporal experience as a real appearance, the truth of which is pure reason's freedom to understand the manifold of nature as nothing other than the 'existence of things determined according to universal laws.' As a result, the actual continuity of appearances is in laws of relationship, whose actual source is the original synthetic unity of apperception, but which the understanding conceives to be the determinations of the object itself. Kant writes in his *Critique of Pure Reason:* "In an object of the pure understanding that only is inward which has no relation whatsoever . . . to anything different from itself. It is quite otherwise with a *substantia phaenomenon* in space; its inner determinations are nothing but relations, and it itself is entirely made up of mere relations. We

---

[294] Cf. above, chapter on Descartes, especially pp. 89ff.
[295] Cf. above, pp. 100–101.

are acquainted with substance in space only through forces which are active in this and that space, either bringing other objects to it (attraction), or preventing them penetrating into it (repulsion and impenetrability). We are not acquainted with any other properties constituting the concept of the substance which appears in space and which we call matter. As object of pure understanding, on the other hand, every substance must have inner determinations and powers which pertain to its inner reality. But what inner accidents can I entertain in thought, save only those which my inner sense presents to me? They must be something which is either itself a *thinking* or analogous to thinking. For this reason Leibniz, regarding substances as noumena, took away from them, by the manner in which he conceived them, whatever might signify outer relation, including also, therefore, *composition*. . . ." [296] Kant contrasts force, as the outer determination of objects in space, with thought, which everywhere in modern thought is absolute, as the inner determination of Leibniz' monads. In Kant's revocation of thought from matter, the forces inherent in the latter come to be seen as, in themselves, immediately related to a community of interaction in time and space. They are considered to be extrinsic determinations of matter, depending upon pure reason's synthetic apperception for their intelligibility. This is how it happens that in Kant, in distinction from Leibniz, form precedes matter, not as, for example in Aristotle, actual is prior to potential knowledge. Rather, on the analogy to Leibniz' possible essences which are prior to actual existence, Kant sets up the forms of sensible intuition, time and space, together with those specific differences inherent in the appearance of matter, such as weight as a constant of mass determined by gravitation. Kant takes these as the possible forms of an actual knowledge; these forms function for the construction of matter, indeed, for the composition of matter in anticipation of that conceptualization by understanding wherein it will become an actual object. The essential possibility of this knowledge is nothing else than pure reason's concept of its own capacity to

[296] *Critique of Pure Reason,* op. cit., 279.

think absolutely, that is, to think about the world in perfect indifference to its own particularity.[297] In Kant, pure reason itself withdraws (from the fact of the history of being) behind a world of appearance in which its own identity must appear to it to be a matter of its own construction, but, for the Moment, it thinks not of this. In the Moment, in the duration of which it constructs this universe of change, in which it itself is included, it does not think that it is in the process of the pure potentiality of self-construction; in its freedom, it knows itself subject to only one law, namely, that it so act that its maxim might become universal law.[298] Its morality, too, then, in that inner domain in which pure reason is intelligible to itself, is a Moment of action essentially directed to a self-creating legislation for the universal freedom of mankind. The essential possibility for this infinite indifference to its own particularity is nothing other than its own pure potentiality to itself be whatever it constructs in the way of the world.

It is Hegel who perceives this essential possibility in the form of Kant's pure reason. In order that I might in one simple illustration show clearly the essential movement of modern thought, think of *weight.* Weight exists in an order of actuality as the weight of matter. But, weight exists in an order of thought as the weight of reason (specifically as the weight of sufficient reason, as Leibniz understood it). *Step One*: In the order of thought, Descartes, in his thought of existence, lacked the weight of a sufficient reason for God's creating him; in the order of actuality, Descartes' matter lacked weight. *Step Two*: In the order of thought, Leibniz, in his distribution of the thought of existence to matter, provided a Supreme Creator who brought this universe into existence as a specific arrangement of the best possible world because he was moved by the weight of a sufficient reason whose specifics only he knows; in the order of actuality, Leibniz' matter, in the unity of extension and mass with substantial form, possessed weight, which, however, in the undifferentiated unity of substance and function, lacked

[297] Cf. above, pp. 122–125.
[298] Cf. above, pp. 110ff.

specificity. *Step Three*: In the order of thought, Kant, in his revocation of the thought of existence to the thought of appearances, held, within the horizon of understanding subject to conditions of sensible intuition, the weight of a sufficient reason for a confidence in real scientific knowledge, indeed progress; in the order of actuality, Kant's matter, constructed in a differentiated field of space and time, possessed specific weight. *Step Four*: In the order of thought, Hegel, for whom the essential possibility of pure reason's infinite indifference to its own particularity necessitates its actuality as appearance, knows that reason is its own weight, that it is Absolute Reason realizing itself in time and space; in the order of actuality, Hegel's matter is, in its very essence, *weight*. In Hegel, then, we discover the absolute intensification, or centering, of the weight of modern thought in its own essence, out of which center it need not go in order to be at the very center of matter itself. Here, in the absolute freedom of its own essence, it dwells as builder of the dimensions of its own existence. In Hegel, the appropriation of the contingency of this universe, which is of the essence of the thought of existence on which Descartes founds, not without a certain potentiality, modern science, this appropriation of the world by means of humanity's absolute self-utilization, this instrumentalizing of man's essence, whereby, under the interdict of sacred doctrine, reason is, in its own dynamism, confined to a perspective in which substance is increasingly dimensional,[299] is, in Hegel, essentially complete. In Hegel, this appropriation of the very being of the world by the human essence is absolute appropriation; it is the knowledge of the identity of the two, man and world, in essence. This essence of man and world is called God by Hegel. If we turn our attention back a moment to Leibniz, we may read the following in his *New Essays On The Human Understanding:* ". . . (however paradoxical it may seem) it is impossible for us to know individuals directly and to find a means of *determining* exactly the individuality of any thing except only to keep it itself; for all the circumstances may reappear; the smallest differences are imperceptible to us; space and

[299] Cf. above, p. 192.

time far from determining themselves need to be determined by the things they contain. Most important of all is the fact that *individuality* involves the infinite, and only he who understands the latter can have first-hand knowledge of the principle of individuation of this or that thing; this arises from the influence (conceived rightly) of all things in the universe on one another. It is true that it would not be so if the atoms of Democritus existed; but then it would also be true that there would be no *difference* between two *different* individuals of the same shape and size." [300] In Leibniz, then, we may understand that God's being outside the universe, he alone knowing the weight of sufficient reason for the actual existence of that universe, consequently for the existence of each Monad which itself mirrors the whole, is equivalent to understanding that *the world itself is the principle of individuality*. It is then the universe itself which comes between man and God; *it is the specific difference in a genus of possible essences*. Therefore, when Kant revoked the thought of existence from the universe itself, recalling thought from the *substantia noumenon*, the intelligible substance, to the *substantia phenomenon*, the substance as appearance, he, in effect, made the specific difference that the universe is in Leibniz a matter of pure reason's own construction. Thereby, the difference between God and pure reason became a matter of the latter's construction, indeed, a *symbol*, on the analogy to which reason understood its own relationship to its understanding, the latter being the transcendental horizon of possible knowledge of the world.[301] In this shift of Kant's from noumenal to phenomenal, from intelligible to phenomenal substance, God, by means of the real symbol, represents to reason, in the perfect similarity of analogous relations, its own generic identity with the Supreme Reason of the universe. God and man, considered without reference to those specific limitations that arise in the construction of the symbolic reality, considered without reference to their respective matters, are neither one different from the other. If Kant's improvement on Leibniz is the differentiation of func-

[300] *Leibniz: Selections,* op. cit., 450–451.
[301] Cf. above, pp. 109–110.

tion from substance, the former constituting the specific difference, then, as Hegel sees it, it is precisely this self-symbolizing function of pure reason that prevents it from realizing what, to Hegel, is most obvious, namely, that reason actually *exists* in this function, that it *is* what it represents to itself. In this way, Leibniz' dictum, 'space and time far from determining themselves need to be determined by the things they contain,' is most unexpectedly realized: Hegel introduces into the things themselves Kant's original synthetic function of pure reason. Thus, the community of interaction of things in time and space is the self-determining of absolute reason in those things. Hegel creates the synthesis of Kant with Leibniz out of which emerges *the differentiated unity of substance with function*. There remains, then, no matter without world-constructing thought by which the essential possiblity of God might be distinguished from the essential possibility of man. There is no longer a merely inner determination of particularity, as in Leibniz, the ground of which was the *pathos* of an infinite continuity of being. Nor, as in Kant, is there an outer determination of particularity, the ground of which was the *pathos* of reason's infinite indifference to its own particularity. But, in Hegel, *particularity is the immediate weight of absolute essence; it is absolute pathos*.

Hegel, in his *Philosophy of Nature* I.2, discussing the impact of bodies, writes as follows: "It is in the *contact* produced by the collision and pressure of these masses, between which there is no empty space, that the general ideality of matter begins. It is important to see how this internality of matter arises, for it is always important to see how the Notion arrives at existence. Masses create contact by being one for the other, merely because there are two material points or atoms in a single moment of identity, the being-for-self of which is *not* being-for-self. No matter how hard and inflexible one imagines matters to be, it is always possible to postulate an interstice between; as long as they touch one another, they have positedness within a unit, no matter how small one imagines this point to be. This is the higher continuity existing in matter, which is not external and merely spatial, but real. Similarly, the point of time is the unity of the past and the future; for here there are two in one, and in that they are in one, they are also not within it. The precise nature of motion consists in being in one place while at the same time being in another, and yet not being in another, but only in this place."[302] In order to better appreciate Hegel's thought here, let us compare it with Kant's thought. Concerning the coexistence of objects in space Kant writes in the *Critique of Pure Reason:* "Things are coexistent so far as they exist in one and the same time. But how do we know that they are in one and the same time? We do so when the order in the synthesis of apprehension of the manifold is a matter of indifference, that is, whether it be from A through B, C, D, to E, or reversewise from E to A. For if they were in succession to one another in time, in the order, say, which begins with A and ends in E, it is impossible that we should begin the apprehension in the perception of E and proceed backwards to A, since A belongs to past time and can no longer be an object of apprehension."[303] Further, on the concept of time in relation to motion, Kant writes in the *Critique:* ". . . the concept of alteration, and with it the concept of motion, as alteration of place, is possible only through and in the representation of time; and . . . if this representation were not an *a priori* (inner) intuition, no concept, no matter what it might be, could render comprehensible the possibility of an alteration, that is, of a combination of contradictorily opposed predicates in one and the same object, for instance the being and the not-being of one and the same thing in one and the same place. Only in time can two contradictorily opposed predicates meet in one and the same object, *one after the other*. Thus our concept of time explains the possibility of that body of *a priori* synthetic knowledge which is exhibited in the general doctrine of motion. . . ."[304] In Kant, then, motion is the contradiction to the coexistence of objects in space at one and the same time. This contradiction to coexistence is possible only in successive times. Time is the measure of coexistence; it is

---

[302] G. W. F. Hegel, *Philosophy of Nature* I (trans. M. J. Petry, London, 1970), 247.

[303] *The Critique of Pure Reason*, op. cit., 234.

[304] Ibid., 76.

the measure of *the absence* of motion, insofar as its Now is, on the analogy to the Now of eternity in sacred doctrine,[305] employed by reason to mark, in its pure abstractness, a continuing presence of a number of objects. Motion itself, as the alteration of place, requires time as its formal condition merely, since actual motion is the synthesis of the manifold of space in the Moment of apperception.[306] Time and motion actually exclude one another. Matter, in Kant, is formally determined by a relativity of forces outside of itself, as the condition for its comprehension by reason. But for Hegel the *interstice*, infinitely small though it may be, posited by bodies in contact, like the Now of time, is the opening, the breach in this externality of forces through which space is measured immediately; this *interstice* is the real continuity of matter. By the nature of motion two bodies occupy the same space in the same time, so that, Hegel understands that matter itself, by virtue of its own essence or weight generates space. The actuality of space is time; the truth of time is motion. The truth of motion is the self-determination of the Notion in matter, whose moments constitute the contradiction of being-for-self together with *not* being-for-self. Of contradiction Hegel says in his *Science of Logic:* ". . . it is one of the fundamental prejudices of logic as hitherto understood and of ordinary thinking, that contradiction is not so characteristically essential and immanent a determination as identity; but in fact . . . contradiction would have to be taken as the profounder determination and more characteristic of essence. For as against contradiction, identity is merely the determination of the simple immediate, of dead being; but contradiction is the root of all movement and vitality; it is only in so far as something has a contradiction within it that it moves, has an urge and activity. . . . External, sensuous motion itself is contradiction's immediate existence. . . . Similarly, internal self-movement proper, *instinctive urge* in general, (the appetite or *nisus* of the monad, the entelechy of absolutely simple essence), is nothing else but the fact that something is, in

[305] Cf. above, pp. 184ff.
[306] Cf. above, p. 183, note 230.

one and the same respect, *self-contained* and deficient, *the negative of itself*. . . . Something is therefore alive only in so far as it contains contradiction within it, and moreover is this power to hold and endure the contradiction within it. . . . *Speculative thinking* consists solely in the fact that thought holds fast contradiction. . . ." [307] If time in Kant is merely a measure of extrinsic relationships, if motion is merely a rearrangement of those relationships in successive instants of a purely hypothetical Moment; then Hegel infuses that motion, time, and space with an absolute vitality of the self-striving essence of reason itself. As a result, in the absolute thought of essence itself, space, time, and motion converge, as it were, with the entire weight of essential possibility in the *interstice* between bodies (between times), out of which wells up the self-determining absolute life of the universe. This absolute life is the life of what Hegel calls the Spirit; he says of it, in his *Philosophy of Nature* I.1: "The Idea or spirit is above time, because it is itself the Notion of time; in and for itself it is eternal and unbreached by time, because it does not lose itself in its own side of the process. This is not the case with the individual as such, on one side of which is the genus; the finest life is that which completely unites its individuality and the universal into one form. The individual is not then the same as the universal however, and is therefore one side of the process, or mutability, in accordance with which mortal moment it falls within time. Achilles, the flower of Greek life, and the infinitely powerful personality of Alexander the Great, are no more, and only their deeds and influences remain through the world that they have brought into being. Mediocrity endures, and finally governs the world. Thought also displays this mediocrity, with which it pesters the world about it, and which survives by extinguishing spiritual liveliness and turning it into flat formality. It endures precisely because it rests in untruth, never acquires its right, fails to honour the Notion, and never realizes the process of truth within it." [308] The absolute *pathos* of the self-realizing power of the essence of reason

[307] *Science of Logic*, op. cit., 439–440.
[308] *Philosophy of Nature* I, op. cit., 232.

itself in this world is heroic existence, which, though mortal, unifies its individuality with the genus of an eternal world process, with the truth of its essence. In Hegel, life is thoroughly, finally, itself a pure form. Particularity and personal immortality are not of the essence of this form of becoming that time is. What history is, then, in Hegel is nothing other than an 'eternal history of the Spirit,' that is, a form existing essentially in eternity which is manifested sensuously in the pictorial representation of Christian faith. But the truth of this 'eternal history' is, for Hegel, nothing other than the identity of human and divine nature.

In his *Lectures On The Philosophy of Religion* III, Hegel says: "In the Church Christ has been called the God-Man. This is the extraordinary combination which directly contradicts the Understanding; but the unity of the divine and human natures has here been brought into human consciousness and has become a certainty for it, implying that the otherness, or, as it is also expressed, the finitude, the weakness, the frailty of human nature is not incompatible with this unity, just as in the eternal Idea otherness in no way detracts from the unity which God is. . . . It involves the truth that the divine and human natures are not implicitly different. God in human form. The truth is that there is only one reason, one Spirit, that Spirit as finite has no true existence." [309] We may understand this absolute thought about the implicit identity of human and divine essence as reason's way of withholding itself in its eternal essence apart from existence itself. In this perfect lack of clarity about itself, ultimately derivative of its resolution to be for the contingent universe its reason for being, absolute reason becomes absolutely the thought of the absolute; it experiences the whole of creation, in its createdness, as nothing at all. Hegel writes in the *Science of Logic:* "In ordinary inference, the *being* of the finite appears as ground of the absolute; because the finite is, therefore the absolute is. But the truth is that the absolute is, because the finite is the inherently self-contradictory opposi-

[309] G. W. F. Hegel, *Lectures On The Philosophy of Religion* III (trans. E. B. Speirs and J. B. Sanderson, New York, 1962), 76–77.

tion, because it is *not*. In the former meaning, the inference runs thus: the being of the finite is the being of the absolute; but in the latter, thus: the non-being of the finite is the being of the absolute."[310] Hegel has defined the conditions which make it possible to live wholly within this world—in absolute dependence on the truth of its appearances.

[310] *Science of Logic*, op. cit., 443.

Chapter 10

CLARIFICATION OF THE ABSOLUTE. I: KIERKEGAARD

Shall we concede that the state of man is essentially *pathetic? Pathos* in itself is a suffering of its occurrence. It is, in itself, an interminable suffering of its own actuality, beyond which there is no possibility for it of another existence. Modern thought, beginning with Descartes, sees the progressive self-determination of humanity to exist within the absolute of its own thought. In this progress Leibniz secures the continuity of intelligible being, while Kant withdraws to a citadel of pure reason, a prospect commanding whatever might appear within an horizon of transcendental subjectivity. It is Hegel who declares the absolute impossibility of retreat from this position taken up by reason. There is no place for humanity to go: It has built its bridge behind itself. What Hegel, then, discovers for humanity's suffering of its own actuality, when he looks into that *interstice* between the points of contact, or, the point of time between two Moments, what he sees there is, as he says, 'a higher continuity'; Hegel's contribution is precisely to notice that the *pathos* is *interminable,* that, in essence, there is no limit to the continuity of modern man's passivity. Indeed, those last *termini,* the ideals of Kant, including the symbol of God, those ideals of constructive imagination fall to the ground within the limits of time; they are the last vestiges of the individual's unwillingness to recognize that, in the eternal history of Spirit, there is no difference between man and God *in essence.* With respect to this essence the particular in itself is nothing, nor is there any possibility of a transition of this universe into eternity, since eternity is present to it as its absolute possibility.[311] In Leibniz, God, as Supreme Reason, is related to the Monad as to another of the genus of possible essences, the universe being the specific difference between them, indeed, constituting the *pathetic* individuality of the Monad. Hegel (Kant having reduced specific difference to particular spatiotemporal conditions for understanding) introduces specific difference immediately into a *pure* individuality. The latter is indistinguishably *an absolute genus,* which, by force of life's very own self-contradictory structure, constitutes whatever is within actuality as simultaneously determining what an individual might be in itself, thereby relativizing absolutely, that is, *reducing to finitude in and of itself, each moment of the thought of existence itself,* namely, the individual, *qua* individual, the species, *qua* species, the universe, *qua* universe—each brought to a recognition, in face of the absolute genus, of its own *radical* finitude or nonexistence. We have previously noted that, in its inevitable progress, the thought of existence comes in Hegel to be the thought of the absolute itself, in such a way that creatures, *qua* creatures, become nothing.[312] This eventuality is understood to be a culmination of modern thought's translation of the appearance of the transcendental essence of existence into the power of reason itself. More particularly, this eventuality is the translation, as in Leibniz, of what it means to be a 'creature' from being related through a *participated existence* to God, as in Thomas Aquinas, to being related through its *essential possibility* to the best possible order of things. It is a translation therefore in which the essential characteristic of a creature is understood to be natural limitation.[313] Indeed, it may be understood that it is of the very essence of the thought of existence, beginning with Descartes, that this thought in its essence prefers its own absoluteness to the contingency of the world that it appropriates, or, that it

[311] Cf. above, pp. 132ff. Also, pp. 222ff.
[312] Cf. above, p. 223.
[313] Cf. above, pp. 197ff., notes 260–262. Also, p. 204.

appropriates that world on the condition that it is conformed to its essence. As is ultimately clear in Hegel, absolute thought is a pure form of existence itself without content: The 'non-being of the finite is the being of the absolute,' the condition of absolute thought is the pure formality of the moments of thought or existence. For Hegel, these pure forms constitute the truth of appearances, insofar as, *in themselves,* neither individual, nor species, nor universe *exist.* This is the state of *absolute pathos* to which reason comes by virtue of the interdict of sacred doctrine under which, cut off from the fact of existence in its simplicity, reason is left to the interminable occurrence of itself.[314] It is clear that if history is understood to be the identity of the storyteller with the story of what has occurred to him in the course of his worldly being, essential history is not possible in Hegel, since to the essence of man, indistinguishable from the Spirit of the world, nothing is able to occur.[315] Rather, it itself is this interminable occurrence to which one might imagine that only the *reappearance* of the transcendental essence of existence itself could provide an end or terminus: but to so imagine is to forget that this unlimited occurrence is grounded absolutely in its own will to power. But, what is more germane to our investigations in the history of being is the fact that, *in the absence of a knowledge of the essential history of thought,* a knowledge, in turn, only *possible* in light of knowledge of, or belief in, the appearance of the transcendental essence of existence, *criticism of absolute reason's appropriation of this universe of contingencies is confined to pathetic, that is, absolutely derived forms.* In the absence of this potential knowledge of essential history, criticism cannot be rationally or intellectually radical, that is, originally on its own foundations with respect to absolute reason, but concedes perforce to *pathos* an initiative that it enjoys only by the default of immortal persons whose commitment is otherwise radical. In this connection it might be a source of illumination to recall the words of Paul, in his *First Letter To The Corinthians* 2:1–16: "As for me, brothers, when I came to you, it was not with any show

[314] Cf. above, p. 190.
[315] Cf. above, pp. 179ff.

of oratory or philosophy, but simply to tell you what God had guaranteed. During my stay with you, the only knowledge I claimed to have was about Jesus, and only about him as the crucified Christ. Far from relying on any power of my own, I came among you in great 'fear and trembling' and in my speeches and the sermons that I gave, there were none of the arguments that belong to philosophy; only a demonstration of the power of the Spirit. And I did this so that your faith should not depend on human philosophy but on the power of God. But still we have a wisdom to offer those who have reached maturity: not a philosophy of our age, it is true, still less of the masters of our age, which are coming to their end. The hidden wisdom of God which we teach in our mysteries is the wisdom that God predestined to be for our glory before the ages began. It is a wisdom that none of the masters of this age have ever known, or they would not have crucified the Lord of Glory; we teach what scripture calls: *the things that no eye has seen and no ear has heard, things beyond the mind of man, all that God has prepared for those who love him.* These are the very things that God has revealed to us through the Spirit, for the Spirit reaches the depths of everything, even the depths of God. After all, the depths of a man can only be known by his own spirit, not by any other man, and in the same way the depths of God can only be known by the Spirit of God. Now instead of the spirit of the world, we have received the Spirit that comes from God, to teach us to understand the gifts that he has given us. Therefore we teach, not in the way in which philosophy is taught, but in the way that the Spirit teaches us: we teach spiritual things spiritually. An unspiritual person is one who does not accept anything of the Spirit of God: he sees it all as nonsense; it is beyond his understanding because it can only be understood by means of the Spirit. A spiritual man, on the other hand, is able to judge the value of everything, and his own value is not to be judged by other men. As scripture says: *Who can know the mind of the Lord, so who can teach him?* But we are those who have the mind of Christ." [316] It would be a mistake, then, to separate God's power

[316] *The Jerusalem Bible: N.T.*, op. cit., 293–294.

from God's wisdom, for by the acknowledgment of the former in the crucified Christ, *psycho-somatic* man, or natural man, is made *pneumatic* man, that is, spiritual; but, precisely by virtue of this spirituality he has in his clarified essence as a manifestation of the glory of God, the *nous* or intellect or understanding of that very appearance of the transcendental essence of existence, by whose death on the cross everything is reconciled to existence itself. It is in the light of the Glory of God itself, transparently presented by immortal persons, that everything's value is able to be submitted to the judgment of one who is not only *committed,* but who in perfect personal integrity *believes,* likewise *knows.* Clearly in this light we see that there is a spiritual work of the mind which, while it supposes the conversion of psycho-pathetic individuality, extends, *qua* mind, to the universe of thought absolutely, *qua* spiritual, to the fact of history, that is, to the question of *identity* that appears with the transcendental essence, which is henceforth that to which is to be related the universe of the mind *to the essence of which* will have occurred, as a fact of its own history, this change or clarification. Now this occurrence of the transcendental essence of existence is itself that which brings to an end, terminates (for that individual for whom the fact of history coincides with the fact of his own history, with the fact that in time he suffers a new identity), but, within the horizon of modern thought, transcends, the interminable *pathos* of absolute reason. Such an individuality, then, as is liberated from bondage to its own essential darkness by its obedient response to God's invitation, is, nevertheless, subject to perceiving this new reality solely through a mental glass of its own age, or, in the absence of a knowledge of the essential history of thought, is *liable* to so grasp it.[317]

By way of introduction to Kierkegaard, let us first compare Augustine to Thomas Aquinas on the question of faith's relationship to knowledge. In his *On Free Choice Of The Will* II, Augustine says: "Unless believing is different from understand-

[317] Cf. above, pp. 205–206; 209–210. Preliminary remarks on Kierkegaard's modernity.

ing, and unless we first believe the great and divine thing that we desire to understand, the prophet has said in vain, 'Unless you believe, you shall not understand.' Our Lord Himself, by His words and deeds, first urged those whom He called to salvation to believe. Afterwards, when he spoke about the gift He was to give to those who believed, He did not say, 'This is life eternal so that they may believe.' Instead He said, ' This is life eternal that they may know Thee, the one true God and Him whom Thou didst send, Jesus Christ.' Then, to those who believed, He said, 'Seek and you shall find.' For what is believed without being known cannot be said to have been found, and no one can become fit for finding God unless he believes first what he shall know afterwards. Therefore, in obedience to the teachings of our Lord, let us seek earnestly. That which we seek at God's bidding we shall find when He Himself shows us—as far as it can be found in this life and by such men as we are. We must believe that these things are seen and grasped more clearly and fully by better men even while they dwell in this world, and surely by all good and devout men after this life. So we must hope and, disdaining worldly and human things, must love and desire divine things." [318] But if we turn to Thomas Aquinas, in his *Summa Contra Gentiles* IV.1, we read: ". . . out of a superabundant goodness . . . so that man might have a firmer knowledge of Him, God revealed certain things about Himself that transcend the human intellect. In this revelation, in harmony with man, a certain order is preserved, so that little by little he comes from the imperfect to the perfect—just as happens in the rest of changeable things. First, therefore, these things are so revealed to man as, for all that, not to be understood, but only to be believed as heard, for the human intellect in this state in which it is connected with things sensible cannot be elevated entirely to gaze upon things which exceed every proportion of sense. But, when it shall have been freed from the connection with sensibles, then it will be elevated to gaze upon the things which are revealed. There is, then, in man a

[318] Augustine, *On Free Choice of the Will* (trans. A. S. Benjamin and L. H. Hackstaff, Indianapolis, 1964), 39.

threefold knowledge of things divine. Of these, the first is that in which man, by the natural light of reason, ascends to a knowledge of God through creatures. The second is that by which the divine truth—exceeding the human intellect—descends on us in the manner of revelation, not, however, as something made clear to be seen, but as something spoken in words to be believed. The third is that by which the human mind will be elevated to gaze perfectly upon the things revealed." [319] In Augustine, then, there is a continuity by which belief is united to knowledge; this continuity is, for Augustine, the fact of the *indwelling* Truth, of whom he writes in his *Concerning The Teacher* XI: ". . . referring now to all things which we understand we consult, not the speaker who utters words, but the guardian truth within the mind itself, because we have perhaps been reminded by words to do so. Moreover, He who is consulted teaches; for He who is said to reside in the interior man is Christ, that is, the unchangeable excellence of God and His everlasting wisdom, which every rational soul does indeed consult. But there is revealed to each one as much as he can apprehend through his will according as it is more perfect or less perfect." [320] We notice in Augustine, then, a perfect interiority of Truth, which abides in itself until it is discovered,[321] but, Christ himself, is immediately available to human nature. Its Self-revelation is directly proportionate to a man's purity of will, whereby, in obedience to God, he withdraws himself from his own self, distended, as it is, in this world, so as to be able, even in this life, to see what is revealed. In this interiority of Augustine man prescinds from 'worldly and human things' to contemplation of 'divine things'—thus the clarification of essence that is effected by the transcendental essence of existence itself, whose appearance in time, we may take it, constitutes, for all time, God's invitation to a man to be converted to him, is, once enjoyed, known immediately. But, in Thomas Aquinas,

[319] Thomas Aquinas, *Summa Contra Gentiles* IV (trans. C. J. O'Neil, Garden City, 1957), 36–37.

[320] Augustine, *Concerning the Teacher and On the Immortality of the Soul* (G. G. Leckie, New York, 1938), 47–48.

[321] Cf. above, p. 177, note 217.

while he concurs with Augustine that revelation transcends human intellect, there is *no possibility of self-transcendence* for the man in this life. Between faith and knowledge, in Thomas, God's exteriority to the human intellect lies as constituting a *specific* difference; in Thomas, it is not Christ himself, as intellect's interior master, who dwells within, but, rather, as he tells us in his *The Teacher,* ". . . the light of reason by which [self-evident] principles are evident to us is implanted in us by God as a kind of reflected likeness in us of the uncreated truth." [322] As a concomitant of this distance in Thomas between God and the human intellect, in which the former is present as the latter's created likeness to uncreated light, this intellect is, as a matter of fact, connected with sensibles in this life.[323] It is therefore incapable of knowing, or understanding, the truth of revelation which is neither sensible nor to be inferred from the sensible. Faith, in Aquinas, in its discontinuity with knowledge proper, acquires a specificity of its own as the *medium* for acquiring the truth of revelation; it is defined in the *Summa Theologica* II-II,4,1 as follows: ". . . faith is a habit of the mind, whereby eternal life is begun in us, making the intellect assent to what is non-apparent." [324] The act of faith that ensues upon this habit, as well as this state or disposition of the mind itself, has its inception in God's grace, since revealed Truth is an object outside man's natural capacity to know. Now, if we turn our attention to Kierkegaard, we read, in his *Concluding Unscientific Postscript,* the following: "The decision lies in the subject.The appropriation is the paradoxical inwardness which is specifically different from all other inwardness. The thing of being a Christian is not determined by the *what* of Christianity but by the *how* of the Christian. This *how* can only correspond with one thing, the absolute paradox. There is therefore no vague talk to the effect that being a Christian is to accept, and to accept, and to accept quite differently (all of them purely rhetorical and fictitious definitions); but *to believe* is specifically different from all

[322] Cf. above, pp. 64–65, note 34.
[323] Cf. above, pp. 125ff., notes 136ff.
[324] *Summa Theologica,* op. cit., 1190.

other appropriation and inwardness. Faith is the objective uncertainty due to the repulsion of the absurd held fast by the passion of inwardness, which in this instance is intensified to the utmost degree. This formula fits only the believer, no one else, not a lover, not an enthusiast, not a thinker, but simply and solely the believer who is related to the absolute paradox. Faith therefore cannot be any sort of provisional function. He who from the vantage point of a higher knowledge would know his faith as a factor resolved in a higher idea has *eo ipso* ceased to believe. Faith *must* not *rest content* with unintelligibility; for precisely the relation to or the repulsion from the unintelligible, the absurd, is the expression for the passion of faith. This definition of what it is to be a Christian prevents the erudite or anxious deliberation of approximation from enticing the individual into byways so that he becomes erudite instead of becoming a Christian, and in most cases a smatterer instead of becoming a Christian; for the decision lies in the subject. But inwardness has again found its specific mark whereby it is differentiated from all other inwardness and is not disposed of by the chatty category 'quite differently' which fits the case of every passion at the moment of passion." [325] Augustine, in intellectual sight, understood what first he believed. Thomas Aquinas, on the authority of God's Word, believed to be true what he heard, though the truth of what was revealed was not apparent. Kierkegaard, by virtue of the absurd, is committed in passionate inwardness to what, at the same time, repulses as absolute paradox. It was Hegel who, by his synthesis of Leibniz' undifferentiated unity of substance and function with Kant's original synthetic function of reason, discovered the *differentiated unity of substance and function* in things themselves, and, thereby, rendered absolute that *pathos* of modern thought in which each moment of the thought of existence is, itself, reduced to nonexistence. That is, individual, species, universe, each nothing itself; each a self-contradictory ground of Absolute Reason by virtue of its finitude.[326] If, in so doing, Hegel defined the

[325] *Concluding Unscientific Postscript,* op. cit., 540.
[326] Cf. above, pp. 216–224.

conditions of a life totally within the horizon of the truth of appearances, that this universe is immediately the Divine Idea manifesting itself in world process, then, it is Kierkegaard who, on behalf of *the individual,* delineates what it means for that individual to exist, not in contradiction to its self (as in Hegel), but, rather to exist in contradiction to its thought of itself, which is that universal function in terms of which it originally possesses *its existence* as, in itself, finite, or ultimately nothing at all outside of the universal.[327] This is Kierkegaard's *transcendent passion to exist.*[328] Indeed, we may understand that unlike Augustine, for whom the prehension of God is the apprehension of Truth, and unlike Thomas Aquinas, for whom the habit of faith is the beginning of eternal life, Kierkegaard's passionate faith begins in a perception of the nothingness of the individual coincidently before God and before Absolute Reason. The issue of faith is *immediately existence, immediate existence.* Kierkegaard writes in *The Book on Adler:* "To be entirely present to oneself is the highest thing and the highest task for the personal life, it is the power on account of which the Romans called the gods *presentes.* But this thing of being entirely present to oneself in self-concern is the highest in religion, for only thus can it be absolutely comprehended that one absolutely is in need of God every instant, so that everything belonging to time past or to time to come or generally to indefinite time, such as evasions, excuses, digressions, etc., grows pale and vanishes, as the other sort of jugglery, which also belongs to indefinite time of the gloaming and the twilight, retreats and vanishes before the bright light of day. When one is not present to oneself, then one is absent in the past or in the future time, then one's religiousness is recollection of an abstract purpose, then one dwells perhaps piously in the piety of an ancient and vanished age, or builds, religiously understood, the objective religiousness like the Tower of Babel—but this night shall thy soul be required of thee." [329] If, as Leibniz understood, it is of the essence of Des-

[327] Cf. above, chapter on Kierkegaard and Lessing.

[328] Cf. above, p. 209.

[329] S. Kierkegaard, *On Authority and Revelation* (trans. W. Lowrie, New York, 1966), 157.

cartes' thought of existence, of the Moment, that it appropriates for its self all contingencies, going out to this world through its idea of God, then Kierkegaard's passion is defined as *dwelling within precisely this modern Moment with immediate consciousness of one's own dependence upon God in every instant.* Thus, Kierkegaard stands in contradiction to the thought of existence within that thought. After Hegel, it is absolute pathos relating itself, in its finitude, to eternity, a relationship that cannot be thought. But it is striking to notice how absolutely circumscribed by the horizon of modernity is Kierkegaard's conception of this God-relationship in which the individuality dwells passionately attached to an absolute paradox behind appearances. If, in Hegel, absolute reason is understood to hold itself in its eternal essence apart from existence itself, then, in Kierkegaard, the individual, *qua* individual, withholds itself from its own eternal essence, indeed, refuses to be, in its finitude, a ground for absolute thought, withdraws its nonbeing as a ground for the being of thought itself. In Kierkegaard, the individual, in its willingness to forgo self-determination, sets itself up in opposition to thought, especially to the reality of thought. Kierkegaard speaks of this negativity of individuality in his *Concept of Dread:* ". . . 'self' signifies precisely the contradiction of positing the general as the particular (*Enkelte*). Only when the concept of the particular individual (*Enkelte*) is given can there be any question of the selfish. But although there have lived countless millions of such 'selves,' no science can state what the self is, without stating it in perfectly general terms. [It is worth while reflecting upon this, for precisely at this point it must become evident to what extent the new principle that thought and being are one is adequate, when we do not spoil it by misunderstandings which are inept and in part stupid, but on the other hand do not wish to have a highest principle which involves us in thoughtlessness. The general *is* only by the fact that it is thought or can be thought . . . and is *as* that which can be thought. The point in the particular is its negative, its repellent relationship to the general; but as soon as this is thought away, individuality is annulled, and as soon as it is thought it is transformed in such a way that either one does not think it but only imagines one is thinking it, or does think it and only imagines that it is included in the process of thought.] And this is the wonderful thing about life, that every man who gives heed to himself knows what no science knows, since he knows what he himself is; and this is the profundity of the Greek saying, γνῶθι σεαυτόν (know thyself), [The Latin saying, *unum noris omnes* (if you know one, you know all), expresses light-mindedly the same thing, and expresses it really if by *unum* one understands the thinker himself, and then does not inquisitively go scouting after the *omnes,* but seriously holds fast to this one, which really is all . . .] [know thyself] . . . so long has been understood in the German way as pure self-consciousness, the airiness of idealism. Surely it is high time to try to understand it in the Greek way, and then again in such a way as the Greeks would have understood it if they had had Christian presuppositions. But the real 'self' is first posited by the qualitative leap." [330] Thus, in Kierkegaard, self-knowledge separates itself from thought; *it denies thought existence,* except *qua* thought, *but it itself has been brought to this extremity by nothing other than the essential history of thought itself.* It perceives in thought, thought's pure formality, as rendered absolute by Hegel. But, because it lacks a knowledge of the essential history of thought itself, Kierkegaard's criticism is not to modern thought's intellectual foundations, but his criticism is that of an existent individuality with ' Christian presuppositions'—we may understand, it is the criticism of a man of clarified essence *directly* related to what is transcendent, without a *medium* of faith, since faith, for Kierkegaard, *is* this immediate relationship, the *how* of the Christian. The substance of the individual has in faith its own specific difference whereby it relates itself to its own thought as that immediate action which alone constitutes the truth of that thought. Kierkegaard writes in the *Concept of Dread:* "The content of freedom, intellectually regarded, is the truth which makes man free. But precisely for this reason is truth in such a sense the work of freedom that it is constantly

[330] S. Kierkegaard, *The Concept of Dread* (trans. W. Lowrie, Princeton, 1957), 70–71.

engaged in producing truth. It is a matter of course that I am not thinking here of the clever conceit of modern philosophy that the necessity of thought is also its liberty, which therefore when it talks of liberty of thought is only talking about the eternal immanent movement of thought. Such a conceit serves only to confuse, and to make more difficult communication between men. What I am talking about, on the other hand, is something quite simple and plain, that truth exists for the particular individual only as he himself produces it in action. If truth exists for him in any other way, and is prevented from existing for him in that way, we have there a phenomenon of the demoniacal. Truth has always had many loud preachers, but the question is whether a man is willing in the deepest sense to recognize truth, to let it permeate his whole being, to assume all the consequences of it, and not to keep in case of need a hiding place for himself, and a Judas-kiss as the consequence. In modern times there has been talk enough about truth; now it is high time for certitude, inwardness, to be asserted . . . in a perfectly concrete sense. Certitude, inwardness, which can only be attained by and exist in action, determines whether the individual is demoniacal or not." [331] Further in the *Concept of Dread* Kierkegaard writes: "The most concrete content consciousness can have is consciousness of itself, not the pure self-consciousness, but the self-consciousness which is so concrete that no author . . . has ever been able to describe such a thing, although such a thing is what every man is. This self-consciousness is not contemplation; he who thinks that it is has not understood himself, for he sees that he himself is meanwhile in the process of becoming and so cannot be a finished product as the object of contemplation. This self-consciousness therefore is a deed, and this deed in turn is inwardness, and every time inwardness does not correspond to this consciousness, there is a form of the demoniacal as soon as the absence of inwardness expresses itself as dread of its acquisition." [332] In modern thought the demoniacal reveals itself in the *pathetic*

[331] Ibid., 123–124.
[332] Ibid., 127–128.

elimination of the essential immortality of Aristotle's existent individual, in which *pathos*, absolute in Hegel, the individual falls totally within time. His formal immortality consists in his self-identification with the Spirit of the world-process, with the Divine Essence, no different than his own truth. But, in this identification, the individual, *qua* individual, is nothing. The demoniacal, then, is the individual, *qua* individual, contemplating his existence with that pure disinterest in which he withholds himself from existence itself, withholds himself in identification with his eternal essence. The conditions for a life lived wholly within this universe, set out absolutely by Hegel, may be seen then as exactly those conditions which, when appropriated by the individual, constitute him in the demonic passion which *prefers* its own nonexistence; prefers its own nonexistence to that liberating deed of self-conscious inwardness by which one relates oneself absolutely to God. In this way the individual would exist outside of himself in an objective order. In this connection, Kierkegaard writes in his *For Self-Examination,* where he is discussing the proper way to read the Gospel, "When thou readest about this, about the man upon whom Christ made an impression, but only such an impression that he neither could quite surrender himself, not quite tear himself loose, and hence chose the night time, chose to steal to Him by night—then thou shalt say to thyself, 'It is I.' . . . Thus it is . . . that thou shouldest read God's Word; and just as, according to the report of superstition, one can conjure up spirits by reading formulae of incantation, so shalt thou, if only thou wilt continue for some time to read God's Word thus (and this is the first requisite), thou shalt read fear and trembling into thy soul, so that by God's help thou shalt succeed in becoming a man, a personality, saved from being this dreadful absurdity into which we men—created in God's image!—have become changed by evil enchantment, into an impersonal, an objective something. Thou shalt, if thou wilt read God's Word in this way, thou shalt (even though it prove terrible to thee—but remember that this is the condition of salvation)—thou shalt succeed in the thing required, in beholding thyself in the mirror. And only thus is success possible. For if to thee God's Word is merely a doctrine, an impersonal, objective something, then there is no mirror . . . it is just as impossible to be mirrored in an objective doctrine as to be mirrored in a wall. And if thou dost assume an impersonal (objective) relationship to God's Word, there can be no question of beholding thyself in a mirror; for to look in a mirror surely implies a personality, an ego; a wall can be seen in a mirror but cannot see itself or behold itself in the mirror. No, in reading God's Word thou must continually say to thyself, 'It is to me this is addressed, it is about me it speaks.' "[333] If the individual is to exist, it will be possible only on condition that he assumes, *qua* individual, responsibility for himself. But this responsibility for himself, dreadful to the demonic passion that withholds itself in its essence, is possible only if the individual is related to the eternal. However, in a Hegelian universe, this is precisely the difficulty: The individual, *qua* individual, is *nothing* to the eternal Spirit of the world. The individual falls within an infinite succession of time; in this infinity he perishes. Hegel writes in his *Philosophy of Nature* I.1: "Only that which is natural, in that it is finite, is subject to time; that which is true however, the Idea, spirit, is *eternal.* The Notion of eternity should not however be grasped negatively as the abstraction of time, and as if it existed outside time; nor should it be grasped in the sense of its coming *after* time, for by placing eternity in the future, one turns it into a moment of time. . . . Time does not resemble a container in which everything is as it were borne away and swallowed up in the flow of a stream. Time is merely this abstraction of destroying. Things are in time because they are finite; they do not pass away because they are in time, but are themselves that which is temporal. Temporality is their objective determination. It is therefore the process of actual things which constitutes time. . . . The present makes a tremendous demand, yet as the individual present it is nothing, for even as I pronounce it, its all-excluding pretentiousness dwindles, dissolves, and falls into dust. It is the universality of these present moments which *lasts*. . . . The *present,*

[333] S. Kierkegaard, *For Self-Examination and Judge For Yourselves* (trans. W. Lowrie, Princeton, 1944), 67–68.

future, and *past,* the dimensions of time, constitute the *becoming* of externality as such, and its dissolution into the differences of being as passing over into nothing, and of nothing as passing over into being. The immediate disappearance of these differences into *individuality* is the present as *now,* which, as it excludes individuality and is at the same time simply continuous in the other moments, is itself merely this disappearance of its being into nothing, and of nothing into its being. . . . Incidentally, these dimensions do not occur in nature . . . for they are only necessary in subjective representation, in *memory,* and in *fear* or *hope.*" [334] If the inevitable progress of modern thought culminates, essentially, in Hegel's synthesis, wherein weight is of the very essence of matter, as the absolute pathos of self-determining reason in things themselves, then time is itself of the very essence of things in process. There is absolutely no possibility, therefore, of an individuality itself being related to eternity, when, indeed, it perishes in the higher continuity of its very present, which present, in turn, is simply an exterior moment of absolute reason's interminable occurrence, or becoming as time. This is the universe in which Leibniz' dictum that 'the things themselves should determine time and space' is absolutely complied with in Hegel's synthesis, yes, absolutely, in essence.[335] If this is a statement of absolute reason's purely intelligible universe, then, in the absence of a knowledge of the essential history of thought, Kierkegaard opposes to it that concrete self-consciousness of Greek thought with Christian presuppositions, wherein, in place of an Aristotelian essence clarified by the occurrence to it of the transcendental essence of existence itself, a finite existence relates itself in its inwardness to an absolute paradox: the Eternal made historical: the historical made eternal.[336] If we have been careful to formally define history (in its essential sense) as the identity of the storyteller

[334] *Philosophy of Nature* I, op. cit., 231–233.

[335] Cf. above, p. 219.

[336] *Philosophical Fragments,* op. cit., 76. Also, 89ff., where Kierkegaard subsumes 'coming into existence' under κίνησις; whereas Aristotle explicitly excludes it (*Physics* V.1, op. cit., 17, 19). But Kierkegaard, consistent with his own modernity, has no idea of the Aristotelian *essence.* He *begins* with *possibility.*

with the story of what has occurred to him in the course of his being in the world, and, if, further, we understand that history essentially exists in the appearance of the transcendental essence of existence itself, then, it should be noted that *Kierkegaard's modernity lies in his retrospective thought of history as process prior to the Moment at which the Eternal enters into time, or, he confuses history with time.*

Let us examine Kierkegaard's conception of time. He writes in his *Concept of Dread:* "When time is correctly defined as infinite succession, it seems plausible to define it also as the present, the past, and the future. However this distinction is incorrect, if one means by it that this is implied in time itself; for it first emerges with the relation of time to eternity and the reflection of eternity in it. If in the infinite succession of time one could in fact find a foothold, i.e., a present, which would serve as a dividing point, then this division would be quite correct. But precisely because every moment, like the sum of the moments, is a process (a going by) no moment is a present, and in the same sense there is neither past, present, nor future. If one thinks it possible to maintain this division, it is because we *spatialize* a moment . . . But even so it is not correctly thought, for even in this visual representation the infinite succession of time is a present infinitely void of content. (This is the parody of the eternal.) . . . On the contrary, the eternal is the present. For thought, the eternal is the present as an annulled [*aufgehoben*] succession (time was succession, going by) . . . Likewise in the eternal there is not to be found any division of the past and the future, because the present is posited as the annulled succession. So time is infinite succession. The life which is in time and is merely that of time has no present. . . . The present is the eternal, or rather the eternal is the present, and the present is full. The instant characterizes the present as having no past and no future, for in this precisely consists the imperfection of the sensuous life. The eternal also characterizes the present as having no past and no future, and this is the perfection of the eternal. If one would now employ the instant to define time, and let the instant indicate the purely abstract exclusion of the past and the future, and by the same token of the present also, then the instant precisely is not the present, for that which in purely abstract thinking lies between the past and the future has no existence at all. But one sees from this that the instant is not a mere characterization of time, for what characterizes time is only that it goes by, and hence time, if it is to be defined by any of the characteristics revealed in time itself, is the passed time. On the other hand, if time and eternity are to touch one another, it must be in time—and with this we have reached the instant. . . . Thus understood, the instant is not properly an atom of time but an atom of eternity. It is the first reflection of eternity in time, its first effort as it were to bring time to a stop. . . . The synthesis of the eternal and the temporal is not a second synthesis but is the expression for the first synthesis in consequence of which man is a synthesis of soul and body sustained by spirit. No sooner is the spirit posited than the instant is there. . . . nature's security is due to the fact that time has no significance for it. Only in the instant does history begin. . . . The concept around which everything turns in Christianity, the concept which makes all things new, is the fullness of time, is the instant as eternity, and yet this eternity is at once the future and the past. If one does not give heed to this, one cannot save any concept from heretical and treasonable admixtures which destroy the concept. One does not get the past as a thing for itself but in simple continuity with the future—and with that the concepts of conversion, atonement, redemption, are resolved in the significance of world-history, and resolved in the individual historical development. One does not get the future as a thing for itself but in simple continuity with the present—and with that the concepts of resurrection and judgment come to naught." [337] For Aristotle time is the measure of motion, the Now is the measure of time; nor is it possible to have actual time without the Now; nor the Now without time.[338] In Thomas Aquinas, we find that the Now is indifferently the Now of time, in which case it is imperfect, or, the Now of eternity, in which case it is perfect by

[337] *Concept of Dread*, op. cit., 76–81.
[338] Cf. above, p. 159.

virtue of the fact that as eternal Now it measures a thing that possesses itself always in the same way, that is, something essentially self-identical.[339] In Kierkegaard time is an infinite succession of moments, which is divided into past, present, and future only insofar as the eternal is reflected in time. The eternal is the present. The abstract instant is not the present; but, it is not time, for time is mere succession. The instant is where time and eternity touch one another; the instant is an atom of eternity, not of time. Nor are we totally surprised to discover the Now of eternity reflected in Kierkegaard's subjective consciousness (as Hegel would understand it) in such a way as to function as a hold on time's otherwise infinite succession. Indeed, it has been noted that Leibniz' continuity of being is, as a matter of fact, analogous to the time and space of sacred doctrine, as well as to the Now of eternity in Thomas Aquinas, that by analogy the time and eternity of transcendental reason coalesce in the unity (undifferentiated) of substance and function in Leibniz' system of monads.[340] This fact, together with Hegel's introduction of the synthetic function into things themselves,[341] thoroughly prepares us to see in Kierkegaard's understanding an entirely functional/substantial perception of Eternity as that without the reflection of which the present could not be, as that the reflection of which in the instant is the beginning of history. Kierkegaard, for the sake of being able to discriminate Christianity's concepts of conversion and resurrection out from their incorporation into world-history's continuity (essentially not history at all [342]), makes of time's possibility to be with a present, past, and future, a necessity of an eternal Now in time, or the instant as eternity; this paradoxical instant is then the beginning of history. But if this is the case, wherever I have a discrimination of the moments of time I will have history of one sort or another; or, wherever I have the thought of eternity I will have a discrimination of time in one way or another. But, perhaps, the most telling criticism of Kierkegaard's conception here is

[339] Cf. above, p. 185.
[340] Cf. above, pp. 206ff.
[341] Cf. above, p. 219.
[342] Cf. above, chapter on Leibniz.

that he shares a negative view of time with Thomas Aquinas and modern thought. It is a view in which time is a mere potentiality or determination of finitude, a view unmitigated, as was Aristotle's, by the thought of the Now's actually being time's continuity linking past with future. This is a view which we understand to be intelligible as a consequence of the appearance of the transcendental essence of existence itself, but which Kierkegaard presupposes as a condition of the appearance in time of the eternal. If, however, a distinction is made, on the ground of the appearance of the transcendental essence of existence, between essential history, as the identity of the storyteller with the story of what occurred to him during the course of his being in time, and the consequent isolation of the Nows of a potential time, then on this ground it is possible to perceive the history of being itself, together with the essential history of thought, a prospect denied to Kierkegaard, but one which clarifies the limits of his passionate existentialism. In his attitude to time, Kierkegaard shares, among other presuppositions of modernity, an antipathy to *limitations* as such.[343] But, if, in the light of the essence of history, it is understood that the appearance of the transcendental essence is the clarification of the essence of man, so as to make of him a manifestation of the glory of God, in light of this participated existence, man is redeemed from the negativity of time which exists as a matter of fact, not as a matter of necessity. On the ground of the transcendental essence of existence Thomas Aquinas held in distinction the Now of time and the Now of eternity. Existence itself is related indifferently to time and eternity, although, it is clear in Aquinas that *man's* state in this life is specifically limited by the sensible conditions of his nature (in this respect the Augustinian extension of man to God in this life as a matter of understanding is precluded). But in Kierkegaard, as a result of the assimilation of sacred doctrine's content to that of purely natural reason, begun by Descartes, completed, essentially, in Hegel, it being no longer possible to perceive, under the interdict of sacred doctrine, the transcendental essence of existence

---

[343] Cf. above, p. 204.

itself, there occurs in the individual's transcendent passion to exist a coalescence of eternity and time in the instant. The Moment filled with the Eternal is the *Fullness of Time.* Kierkegaard writes in his *Philosophical Fragments:* "This relationship of owing all to the Teacher [Christ] cannot be expressed in terms of romancing and trumpeting, but only in that happy passion we call Faith, whose object is the Paradox. But the Paradox unites the contradictories, and is the historical made eternal, and the Eternal made historical. . . . It is easy to see, though it scarcely needs to be pointed out, since it is involved in the fact that the Reason is set aside, that Faith is not a form of knowledge; for all knowledge is either a knowledge of the Eternal, excluding the temporal and historical as indifferent, or it is pure historical knowledge. No knowledge can have for its object the absurdity that the Eternal is the historical. . . . It is easy to see . . . that Faith is not an act of will; for all human volition has its capacity within the scope of an underlying condition. . . . How does the learner then become a believer or disciple? When the Reason is set aside and he receives the condition. When does he receive the condition? In the Moment. What does the condition condition? The understanding of the Eternal. But such a condition must be an eternal condition.—He receives accordingly the eternal condition in the Moment, and is aware that he has so received it; for otherwise he merely comes to himself in the consciousness that he had it from eternity." [344] For Kierkegaard, the entrance of the Eternal into time or history in the Moment is the beginning of the individual's consciousness that in himself he lacks an eternal determination. That time should be measured by the atom of eternity is an understanding of time not scientific, but grounded in Kierkegaard's happy passion of faith. Its significance is that, insofar as the Eternal touches time in the instant, the instant corresponds to the consciousness that the temporal is absolutely not, in its nothingness, a ground of an eternal reason. Its significance is also that the temporal or finite existence is absolutely in and for itself nothing whatsoever, but, paradoxically, at the

[344] *Philosophical Fragments*, op. cit., 76–79.

same time, is, in the instant, the ground of a decision by which one exists for the first time eternally. Kierkegaard writes in the *Concluding Unscientific Postscript:* "*Religiousness A* accentuates existence as an actuality, and eternity (which nevertheless sustains everything by the immanence which lies at the base of it) disappears in such a way that the positive becomes unrecognizable by the negative. To the eyes of speculative philosophy, existence has vanished and only pure being is, and yet the eternal is constantly concealed in it and as concealed is present. *The paradoxical religiousness* places the contradiction absolutely between existence and the eternal; for precisely the thought that the eternal *is* at a definite moment of time, is an expression of the fact that existence is abandoned by the concealed immanence of the eternal. . . . this is the breach with immanence. . . . If the individual is inwardly defined by self-annihilation before God, then we have *religiousness A*. If the individual is paradoxically dialectic, every vestige of original immanence being annihilated and all connection being cut off, the individual being brought to the utmost verge of existence, then we have the *paradoxical religiousness*. . . . even the most paradoxical determinant, if after all it is within immanence, leaves as it were a possibility of escape, of a leaping away, of a retreat into the eternal behind it; it is as though everything had not been staked after all. But the breach makes the inwardness the greatest possible. . . . The confusion of speculative philosophy . . . is due to the fact that it loses itself in pure being. Irreligious and immoral views reduce existence to a naught, a mere prank. Religiousness A makes the thing of existing as strenuous as possible (outside the paradox-religious sphere), but it does not base the relation to an eternal happiness upon one's existence but lets the relation to an eternal happiness serve as the basis for the transformation of existence. . . . it is true of all historical knowledge and learning that it is only an approximation, even at its maximum. The contradiction is: to base one's eternal happiness upon an approximation, a thing which can be done only when one has in oneself no eternal determinant (and that again is no more possible to think than how such a notion should occur to any one, since the Deity must provide the condition for it), and hence this again is connected with the paradoxical accentuation of existence." [345] Insofar as Kierkegaard, speaking as *the individual* having in himself no eternal determinant, at the utmost extremity of an existence in itself sheer temporality, at the same point refusing to acknowledge the reality of the absolute thought of existence, 'holds fast by the passion of inwardness to the objective uncertainty which repulses by virtue of the absurd,'—insofar as Kierkegaard expresses his faith in terms of a passion that is not able to reach that prospect from which just the validity of modern thought *as thought* is able to be criticized, and radically so, then, it is clear to what an extreme remove that thought has placed *the fact of its own origins*, namely, that it is, modern thought, a translation of the essential fact of history into terms of its own understanding, carried out absolutely, so that, in retrospect, it understands everything only with reference to the truth of appearances, or to its own self-realization.

---

[345] *Concluding Unscientific Postscript*, op. cit., 506–509.

Chapter 11

CLARIFICATION OF THE ABSOLUTE.
II: HUSSERL

Absolute truth is the truth of appearances. This truth of appearances is that of existence grounded absolutely in thought. The moments of this absolute thought of existence, namely, the individual, the species, and the universe, in and of themselves, stand without being; to these moments, in themselves, belongs finitude or nonbeing. In Kierkegaard, the individual, *qua* individual, is related, through its transcendent passion to exist, to its own eternal existence, an eventuality made possible, for the first time, by the entrance of the Eternal into history in the Moment. As a result, the individual to whom God has given the necessary condition is able by means of this faith to substantiate his existence before God as his proper thought of himself. To exist before God becomes the essential function of the individual. Outside of the individual, conceived of at the extremity to which he has been brought by the absolute pathos of modern thought, set out in its essential completeness by Hegel, there is, for Kierkegaard, for all practical purposes no relationship to God. He writes in his journals: "The law (which again is grace) which Christ has established by his life for man's life is that you have to do as a single person with God. Whether you are clever or simple, highly gifted or only slightly gifted, it does not matter in the least: relate yourself as a single person with God (oh divine grace, willing to have to do with an individual man, with each single man!); dare as a single person to have to do with him, he will adapt everything to your capacities and possibilities. So first have to do with God; not first with 'the rest'. But the fact is that to exist, existence, has something immensely frightening for a poor man. So he is afraid—he does not dare to set himself first in relation to God, the animal in him thinks, 'It is cleverer to be like the rest.' Every existence which thinks first of being like the rest is a wasted existence, and since from a Christian standpoint it has happened through one's own fault it is a forfeited existence." [346] Kierkegaard's delineation of Christian individuality is a spiritual monadology in which psycho-pathetic man is able to come into existence by virtue of a commitment of his entire temporality or finitude to the law of grace established by the life of Christ, according to which law God is related to this world, and not otherwise. This spiritual identity, if achieved, is that by which one's worldly being is placed at God's disposal absolutely. Thus, by becoming spirit a man is raised above his finite existence as a rational animal, which human existence as self-determination is, spiritually understood, nothing but sin. But this new spiritual identity, by which the historical is made eternal and the Eternal is made historical, is, as Christ's life makes clear, nothing but suffering in this world, although eternally it is exaltation. Kierkegaard writes in his *Training In Christianity:* "But is not this like a deceit on His part, that from on high He draws me in this wise unto Himself? Has he not suppressed something? To be the Truth, is it enough for Him to draw, is He not quite as much required to warn him who lets himself be drawn, reminding him constantly of the difference between them, that He the perfected One is in the environment of perfection, whereas the other is in the environment of actuality, of worldliness, of the temporal, where this loftiness must exhibit itself inversely as lowliness and humiliation, so that He draws from on high, and the man who feels himself drawn and follows finds himself, just in proportion as he follows heartily, in exactly the opposite case, of being in and sinking deeper into humiliation and lowliness? . . . No one

[346] S. Kierkegaard, *The Last Years: Journals 1853–1855* (trans. R. G. Smith, New York, 1965), 141.

can properly say that He has suppressed anything, for His life as he led it in humiliation and lowliness is surely well known. So it is not He who suppresses something, but perhaps it may be the individual who forgets something, who by looking only and with false passion upon the loftiness actually takes Him in vain, and therewith falls into forgetfulness of lowliness and humiliation, until it ends with his wandering too far off and finding himself where he least expected to be, where he lays the blame upon Him who from on high draws all unto Himself." [347] For Kierkegaard the exaltation of an eternal existence come upon in time is, as long as time continues, nothing other than *to be sacrificed,* which, insofar as the man lacks an eternal determinant in himself, either before God (or else the Eternal would not have entered into time), or before man (for then he would not have received the condition from Christ), is not to be confused with self-sacrifice. Self-sacrifice is a universal potentiality of human religiosity, or, perhaps, an ideal of world-historical speculative philosophy. But *to be sacrificed* is to suffer in time without the knowledge of an essential existence belonging to the self, or, with a passionate faith in an absurd fact, that is, one that is unintelligible, and, in modern thought, absolutely so, namely, that the Eternal entered time in such a manner as to make the individual, *qua* individual, eternal, in such a way that, in this life, he must suffer humiliation exactly proportionate to his future exaltation, must become *less* than nothing, since he is no longer in accord with reason itself; not in accord with his own humanity. The deed of faith may be understood to be to hold fast to this absurd Truth which repulses, and to hold fast in the examination that one's actual becoming in time is, in the midst of the only reality that a man knows: his own existence as temporality, as finitude. This holding fast, which must constitute what preeminently it means to be present to oneself, since it properly belongs not to time, but to eternity, means to a man throughout his life a continuing task. Since the deed of faith corresponds to the instant in which the eternal touches time, and since, as Hegel makes quite clear for modern conscious-

[347] S. Kierkegaard, *Training In Christianity* (trans. W. Lowrie, Princeton, 1944), 183–184.

ness, time is nothing other than the finitude of things, then, for Kierkegaard, *faith,* as this passionate being present to oneself, *is a measure of finitude*, that is, of the nothingness, absolutely considered, of worldly being. In this perspective we are able to appreciate the significance of a journal entry made by Kierkegaard only a week before he collapsed in the street (he was taken to a hospital where he died five weeks later): "The men in whom there is more spirit, and whom grace does not neglect, are brought to the point that life reaches the supreme degree of disgust with life. But they cannot be reconciled to this, they rebel against God, and so on. Only the men who are brought to this point of disgust with life and are able to hold fast by the help of grace to the faith that God does this from love, so that not even in the inmost recesses of their soul is there any doubt concealed that God is love—only these men are ripe for eternity. And it is these men whom God receives in eternity. . . . And what pleases him even more than the praise of angels is a man, who in the last lap of his this life, when God is transformed as though into sheer cruelty, and with the cruellest imaginable cruelty does everything to deprive him of all joy in life, a man who continues to believe that God is love and that it is from love that God does this. Such a man becomes an angel. And in heaven he can surely praise God. . . . But the apprentice time, the school time is also always the strictest time. . . . And every time . . . [God] hears praise from a man whom he brings to the uttermost point of disgust with life, God says to himself, This is the right note. He says, Here it is, as though he were making a discovery. But he was prepared, for he himself was present with that man, and helped him so far as God can help what only freedom can do. Only freedom can do it. But what a surprise for man to be able to express himself by thanking God, as though it were God who did it. And in his joy at being able to do this he is so happy that he will hear nothing about himself doing it, but he thankfully attributes everything to God. And he prays God that it may remain so, that it is God who does it. For he does not believe in himself, but he believes in God." [348] Faith is a measure of the absolute nothingness of a

[348] *The Last Years,* op. cit., 368–369.

man's existence in this world, which turns to dust before the eyes of the spiritual self he has become by God's grace. But this spirituality is *at once* a man's choice of God, in which choice a man's spirit is constituted. Thus, he sees nothing of himself, nor does he rest in any thought of his own freedom, but immediately, exclusively thanks God. This is the transparency of the self before God;[349] it is a self constituted in time to exist beyond time in eternity. It is the peculiar construction of modern consciousness (conceiving, as it does, for example, in Hegel, an essential identity of human and divine) that, in the event of a man's conversion, when the *fact of existence* [350] comes to be for him in absolute contradiction to the truth of appearances, he should experience his response to grace as *an exercise of absolute freedom of choice*. This is the negatively conditioned exhilaration at being liberated from that absolute weight of sufficient reason. Nor can a man rest in this *thought of absolute freedom,* lest the balance shift to its being the *absolute thought of freedom,* but he cancels thought absolutely to have his freedom only in praising God's love. In this most intimate communion of passionate awareness the simplicity of the transcendental essence of existence itself penetrates to modern man. Indeed, so long as modernity prevails there is perhaps no better understanding of what faith is in actuality. But, if modernity is man's work, and not God's work, then, *its passing is to be anticipated; this anticipation must take the form of the essential history of thought,* grounded in a knowledge of the transcendental essence of existence itself in its appearance, in a knowledge of the history of being itself.

If in Kierkegaard we encounter the individual, *qua* individual, as the first moment of the absolute thought of existence standing in opposition to that absolute thought, standing in its own finitude or nonbeing, and denying reality to the absolute outside of its thought, radically distinguishing thought from existence, so, similarly, in the phenomenology of Husserl, we encounter the species, *qua* species, as the second moment of the absolute thought of existence, standing in its own finitude or

[349] Cf. above, pp. 151ff.
[350] Cf. above, pp. 171ff.

nonbeing, rising to the form of an absolute transcendental subjectivity, and denying that the existence of the truth of appearances is capable of satisfying the sole criterion of a pure or absolute transcendental subjectivity, *self-evidence*. Husserl writes in his *The Crisis of European Sciences and Transcendental Phenomenology* IIIB: "An obscure dissatisfaction with the previous way of grounding in all science leads to the setting of new problems and to theories which exhibit a certain self-evidence. . . . This first self-evidence can still conceal within itself more than enough obscurities which lie deeper, especially in the form of unquestioned, supposedly quite obvious presuppositions. . . . We can understand, accordingly, that the history of transcendental philosophy first had to be a history of renewed attempts just to bring transcendental philosophy to its starting point and, above all, to a clear and proper self-understanding of what it actually could and must undertake. . . . As we know, transcendental philosophy appears in its primal form, as a seed, in the first Cartesian *Meditations* as an attempt at an absolutely subjectivistic grounding of philosophy through the apodictic ego; but here it is unclear and ambiguous, and it immediately subverts its genuine sense. . . . A true beginning, achieved by means of a radical liberation from all scientific and prescientific traditions, was not attained by Kant. He does not penetrate to the absolute subjectivity which constitutes everything that is. . . . All the transcendental concepts of Kant—those of the 'I' of transcendental apperception, of the different transcendental faculties, that of the 'thing in itself' (which underlies souls as well as bodies)—are constructive concepts which resist in principle an ultimate clarification. This is even more true in the later idealistic systems. . . . their ultimate incomprehensibility gave rise to profound dissatisfaction among all those who had educated themselves in the great new sciences. Even though these sciences, according to our clarification and manner of speaking, furnish a merely 'technical' self-evidence, and even though transcendental philosophy can never become such a τέχνη, this τέχνη is still an intellectual accomplishment which must be clear and understandable at every step, must possess the self-evidence of the step made and of the ground upon which it rests. . . . The great transcendental philosophies did not satisfy the scientific need for such self-evidence, and for this reason their ways of thinking were abandoned." [351] Husserl, then, on behalf of the species, *qua* species, that is, rational or scientific humanity, refuses to recognize as reality the purely hypothetical constructions that underlie the truth of appearances. But, unlike Kierkegaard, who, *qua* individual, dismissed modern thought as 'airy idealism,' while turning to inwardness in the relationship to God (though constricted in his understanding of his faith by his modernity), Husserl demands on behalf of the *species* a clarification of the philosophic grounds for modern science. Indeed, he demands an ultimate clarification of modern philosophy itself as itself scientific. Husserl's modernity is quite deliberate: He will begin again with Descartes' *Meditations,* in which the latter attempted to found a new science on the basis of the presuppositionless thought of his own existence.[352] Husserl, like Kierkegaard, takes up his opposition to the thought of existence within that Moment of Descartes' thought, but this time on behalf of the species in opposition to the generic identity of the individual with the universe in existence. Not by relating, *qua* individual, to God, but, rather, *qua* species, to pure transcendental subjectivity, is the man, the rational animal, to be liberated from the absolute weight of reason. He is to be elevated to a position of pure disinterested vision wherein he sees himself as an individual within the world as a *possibility* to be radically distinguished from his transcendental ego for whom both world and individuality have a merely exemplary or essential scientific significance. Let us listen to Husserl as he thinks over again Descartes' thoughts. He says in his *Paris Lectures:* "We can no longer accept the reality of the world as a fact to be taken for granted. *It is a hypothesis that needs verification*. Does there remain a ground of being? . . . Is not 'world' the name for the totality of all that is? Might it not turn out that the world is not the truly

[351] E. Husserl, *The Crisis of European Sciences and Transcendental Phenomenology* (trans. D. Carr, Evanston, 1970), 198–201.
[352] Cf. above, chapter on Descartes.

ultimate basis for judgment, but instead that its existence presupposes a prior ground of being? Here, specifically following Descartes, we make the great shift which, when properly carried out, leads to *transcendental subjectivity*. This is the shift to the *ego cogito,* as the apodictically certain and *last basis for judgment* upon which all radical philosophy must be grounded. . . . We can no longer say that the world is real—a belief that is natural enough in our ordinary experience—; instead, it merely makes a claim to reality. . . . the entire concrete world ceases to have reality for me and becomes instead mere appearance. However, whatever may be the veracity of the claim to being made by phenomena, whether they represent reality or appearance, phenomena in themselves cannot be disregarded as mere 'nothing.' On the contrary, it is precisely the phenomena themselves which, without exception, render possible for me the very existence of both reality and appearance. . . . I no longer judge regarding the distinction between reality and appearance. I must similarly abstain from any other of my opinions, judgments, and valuations about the world, since these likewise assume the reality of the world. But for these, as for other phenomena, epistemological abstention does not mean their disappearance, at least not as pure phenomena. This ubiquitous detachment from any point of view regarding the objective world we term the *phenomenological epoché*. . . . Everything in the world, all spatio-temporal being, exists for me because I experience it, because I perceive it, remember it, think of it in any way, judge it, value it, desire it, etc. It is well known that Descartes designates all this by the term *cogito*. For me the world is nothing other than what I am aware of and what appears valid in such *cogitationes*. *The whole meaning and reality of the world rests exclusively on such cogitationes*. My entire worldly life takes its course within these. I cannot live, experience, think, value, and act in any world which is not in some sense in me, and derives its meaning and truth from me. . . . I certainly do not discover myself as one item among others in the world, since I have altogether suspended judgment about the world. I am not the ego of an individual man. I am the ego in whose stream of consciousness the world itself—including myself as an object in it, a man who exists in the world—first acquires meaning and reality." [353] In its inevitable progress modern thought, which, beginning with Descartes, is the thought of existence, comes to the realization, absolute in Hegel, of its original intention to be for universal contingency its reason for existing. Indeed, modern thought's intention is to be the essence of everything that exists: its substantial reality—the truth informing appearances, apart from which determination appearances in and of themselves are nothing. Hegel writes in his *Science of Logic:* "Actuality is the *unity of essence and Existence;* in it, *formless* essence and *unstable* Appearance, or mere subsistence devoid of all determination and unstable manifoldness, have their truth. Existence is, indeed, the immediacy which has proceeded from ground, but form is not as yet posited in it. In determining and forming itself it is Appearance; and when this subsistence which is determined only as reflection-into-an-other is developed further into reflection-into-self, it becomes *two worlds, two totalities* of the content, one of which is determined as *reflected into itself,* the other as *reflected into an other*. But the essential relation exhibits their *form relation,* the consummation of which is the *relation of inner and outer* in which the content of both is only one *identical substrate* and equally only *one identity of form*. By virtue of the fact that this identity is now also identity of form, the form determination of their difference is sublated, and it is *posited* that they are *one* absolute totality. This unity of inner and outer is *absolute actuality*. . . . this actuality is, in the first instance, the *absolute* as such. . . . Secondly, we have *actuality* proper. *Actuality, possibility* and *necessity* constitute the *formal moments* of the absolute, or its reflection. Thirdly, the unity of the absolute and its reflection is the *absolute relation,* or rather the absolute as relation to itself—*substance*." [354] It is precisely this *substantiality* of the world that appears that Husserl means to bracket, or, to put in suspense, by means of his phenomenological epoché. But substance, beginning with the transcendental form of natural reason in sacred doctrine,[355] is

[353] E. Husserl, *The Paris Lectures* (trans. P. Koestenbaum, The Hague, 1970), 6–8.
[354] *Science of Logic*, op. cit., 529.
[355] Cf. above, p. 63.

formally determined. Since the transcendental form of reason was brought into being by the impression made upon it by the transcendental essence of existence itself, it was in its very being capable of beholding the substantiality of its object with self-evidence.[356] But, with Descartes, natural reason takes as its own, as if it were its original form, transcendental subjectivity, so that it attempts to ground the world self-evidently in its own substantiality. With the shift from the equipoise of self-evidence and substantiality in the epistemology of the transcendental form of natural reason (an equipoise grounded *not* on an original capability of human reason, but, explicitly, on the fact of the appearance of the transcendental essence of existence) to the preponderance of substance over self-evidence, or, of the thought of existence over the appearance of things, begins the progress of modern thought. The latter, as Husserl says, is continually beset with questions of its own self-evidence. Indeed, it could hardly help but be, since, as we understand in examining Aristotle's understanding of the relation between knowledge and being, substance and self-evidence never coexisted in a true metaphysics, although, indeed, knowledge and substance most certainly identified themselves in actuality, *beyond the formality of reason*.[357] It should be immediately clear that the phenomenology of Husserl is nothing other than the shift, within the radical *imbalance* of modernity, in direct opposition to Descartes, to the preponderance of self-evidence over substance. But, the transcendental subjectivity of Husserl is clearly distinguished from the transcendental form of natural reason since it must, in order to have its self-evidence, refuse to assent to the substantiality of what appears. It lacks the transparency of the transcendental form of natural reason. Rather, *it abstracts from the formality of absolute reason* to a purely *inner* determination, by a pure ego, of appearances as simply for itself. In other words, in place of the obscurity of modern thought which *posits* the intelligibility of the totality of all things, transcendental subjectivity reserves to itself *the sole function* of determining for itself what is *possibly* real purely on the basis of what

[356] Cf. above, pp. 65ff., 69ff.

[357] Cf. above, chapter on Aristotle. Also, above, pp. 76ff., 83ff.

self-evidently appears to it. Husserl says further, in *The Paris Lectures:* "Descartes was thoroughly sincere in his desire to be radical and presuppositionless. However, we know through recent researches . . . that a great deal of Scholasticism is hidden in Descartes' meditations as unarticulated prejudice. But this is not all. We must above all avoid the prejudices, hardly noticed by us, which derive from our emphasis on the mathematically oriented natural sciences. These prejudices make it appear as if the phrase *ego cogito* refers to an apodictic and primitive axiom, one which, in conjunction with others to be derived from it, provides the foundation for a deductive and universal science, a science *ordine geometrico*. . . . It is not true that all that now remains to be done is to infer the rest of the world through correct deductive procedures according to principles that are innate to the ego. Unfortunately, Descartes commits this error, in the apparently insignificant yet fateful transformation of the ego to a *substantia cogitans,* to an independent human *animus,* which then becomes the point of departure for conclusions by means of the principle of causality. . . . We must regard nothing as veridical except the pure immediacy and givenness in the field of the *ego cogito* which the *epoché* has opened up to us. In other words, we must not make assertions about that which we do not ourselves *see*. In these matters Descartes was deficient." [358] For Husserl, then, Descartes took the wrong turn at the point where he failed to distinguish his presuppositionless phenomenon from his being a thinking substance in the world. This confusion becomes the basis for the inconsistent inclusion of historical, mathematical, and metaphysical elements that do not properly belong to the pure transcendental subjectivity conceived by Husserl in its absolute priority to any world. But, since it lacks a knowledge of the essential history of thought, phenomenology is oblivious to the fact that the world or totality, to the reality of which it refuses its assent, is not only grounded in a constructive conceptualization in principle unable to be clarified, but that *it endures, this essential denial of the history of being, in the abstraction, that phenomenology itself is* (on

[358] *The Paris Lectures*, op. cit., 8–9.

behalf of the species, *qua* species), *from the absolute doubt of Hegel,* in which appearance emerges, *qua* appearance, as *a totality for another.*[359] Phenomenology is the end of reflection in essence, or, it is a transcendental reflection on the finitude of the species whereby a pure description is able to be understood to be the essence of science; Husserl continues in his *Paris Lectures:* "The independent *epoché* with regard to the nature of the world as it appears and is real to me—that is, 'real' to the previous and natural point of view—discloses the greatest and most magnificent of all facts: I and my life remain—in my sense of reality—untouched by whichever way we decide the issue of whether the world is or is not. To say, in my natural existence, 'I am, I think, I live,' means that I am one human being among others in the world, that I am related to nature through my physical body, and that in this body my *cogitationes*, perceptions, memories, judgments, etc. are incorporated as psycho-physical facts. Conceived in this way, I, we, humans, and animals are subject-matter for the objective sciences, that is, for biology, anthropology, and zoology, and also for psychology. . . . the phenomenological *epoché*. . . . eliminates as worldly facts from my field of judgment both the reality of the objective world in general and the sciences of the world. *Consequently, for me there exists no 'I' and there are no psychic actions, that is, psychic phenomena in the psychological sense.* To myself I do not exist as a human being, (nor) do my *cogitationes* exist as components of a psycho-physical world. But through all this I have discovered my true self. I have discovered that I alone am the pure ego, with pure existence and pure capacities (for example, the obvious capacity to abstain from judging). Through this ego alone does *the being of the world,* and, for that matter, any being whatsoever, make sense *to me* and has possible validity. The world—whose conceivable non-being does not extinguish my pure being but rather presupposes it—is termed *transcendent*, whereas my pure being or my pure ego is termed *transcendental.* Through the phenomenological *epoché* the natural human ego, specifically my own, is reduced to the transcendental ego. This

[359] Cf. above, p. 256. Also, above, p. 135.

is the meaning of the phenomenological reduction."[360] By means of the phenomenological reduction, a man, *qua* man, that is, *qua* species, is elevated above the psycho-pathetic understanding of human nature characteristic of the sciences of the world, and of the thought of existence in which they are grounded, raised to *a purely rational existence* as himself the ultimate judge of the world's validity. This 'true self' of Husserl's meditation would remind us of the Aristotelian essence or true self of the man, which is his absolute priority to the order of the universe, except that that 'essential immortality' of Aristotle is not self-evident to reason. It would remind us of Hegel's absolute reason which is *ex hypothesi* self-evident, except that that absolute genus is ground to a universal synthetic judgment of existence constituting the actuality of the world in the divine essence,[361] indeed, just that from which phenomenology abstracts itself *qua* species. Nor is the true self or pure ego of Husserl to be confused with Christianity's immortal person, whose essence is first constituted interpersonally within worldly being by the appearance of the transcendental essence of existence, and whose worldly being is not even the same worldly being that Husserl envisions, since *his* is abstracted from that of modern thought for which the fact of history is *originally* unknowable. For an immortal person the nonbeing of his world, in the sense of unreality, would imply his unreality, except if we imagine that finitude is nonbeing (as Hegel does). In this event, however, immortal personality is best represented by Kierkegaard's infinitely *'interested* self-consciousness, in sharp contradistinction to the pure *disinterest* of Husserl's transcendental ego. The latter is a disinterest that extends to whatever exists within the world, including, therefore, commitment, belief and knowledge, including that of the man himself within the world. Now, in Aristotle, essential immortality constitutes a man as a person.[362] With the appearance of the transcendental essence of existence, that person's essence is so clarified as to convert the person, *qua* person, to immortal existence.[363] Similarly, in

---

[360] *The Paris Lectures*, op. cit., 9–10.
[361] Cf. above, pp. 127ff.
[362] Cf. above, p. 49.
[363] Cf. above, pp. 192ff.

Husserl, we find that *the discovery of the pure transcendental ego provides a foundation for a definitive rational understanding of human personality,* in which the only 'interest' is self-realization, *qua* man. Husserl writes, in his *Philosophy as Mankind's Self-Reflection; the Self-Realization of Reason* (*Appendix IV* of *The Crisis of European Sciences*): "One has expected the same objectivity from psychology as from physics, and because of this a psychology in the full and actual sense has been quite impossible; for an objectivity after the fashion of natural science is downright absurd when applied to the soul, to subjectivity, whether as individual subjectivity, individual person, and individual life or as communally historical subjectivity, as social subjectivity in the broadest sense. This is the ultimate sense of the objection that one must make to the philosophies of all times—with the exception of the philosophy of idealism, which of course failed in its method: that it was not able to overcome the naturalistic objectivism which was from the beginning and always remained a very natural temptation. . . . only idealism, in all its forms, attempts to lay hold of subjectivity as subjectivity and to do justice to the fact that the world is never given to the subject and the communities of subjects in any other way than as the subjectively relative valid world with particular experiential content and as a world which, in and through subjectivity, takes on ever new transformations of meaning; and that even the apodictically persisting conviction of one and the same world, exhibiting itself subjectively in changing ways, is a conviction motivated purely within subjectivity, a conviction whose sense—the world itself, the actually existing world—never surpasses the subjectivity that brings it about. . . . *Reason* is the specific characteristic of man, as a being living in personal activities and habitualities. This life, as personal life, is a constant becoming through a constant intentionality of development. What becomes, in this life, is the person himself. . . . Human personal life proceeds in stages of self-reflection and self-responsibility from isolated occasional acts of this form to the stage of universal self-reflection and self-responsibility, up to the point of seizing in consciousness the idea of autonomy, the idea of a resolve of the will to shape one's whole personal life into the synthetic unity of a life of universal self-responsibility and, correlatively, to shape oneself into the true 'I,' the free, autonomous 'I' which seeks to realize his innate reason, the striving to be true to himself, to be able to remain identical with himself as a reasonable 'I'; but there is an inseparable correlation here between individual persons and communities by virtue of their inner immediate and mediate interrelatedness in all their interests—interrelated in both harmony and conflict—and also in the necessity of allowing individual personal reason to come to ever more perfect realization only as communal-personal reason and vice versa. The universally, apodictically grounded and grounding science arises now as the necessarily highest function of mankind . . . namely, as making possible mankind's development into a personal autonomy and into an all-encompassing autonomy for mankind—the idea which represents the driving force of life for the highest stage of mankind."[364] Here, in Husserl, by means of the phenomenological reduction, is discovered the true bearer of reason's absolute autonomy: the human person (inseparably related to the community of persons). What is prior to this human personality is the transcendental sphere of the pure ego. This itself constitutes the transcendental intersubjectivity or ground upon which the person and persons stand, *qua* species, in their freedom over against the claim of the thought of existence to, itself, be the absolute reality of the world. This claim is held in radical suspense by a personal mankind that comes in the course of history to realize that the world is structured from time to time immediately in accordance with structures of its own subjective intentions. But the radical possibility of autonomous personal life is the pure ego of transcendental life for whom the world as it exists in itself is a matter of disinterest compared to what it manifests *essentially*. Husserl writes, in his *Experience and Judgment:* "This world *as the correlate of lived experiences* always belongs to the lived experiences which the psychologist meets with entirely as a matter of course and which he studies, but from these lived experiences he has no way of going back to the *origin of this world itself*—a world which is what it is because of the

[364] *The Crisis of European Sciences*, op. cit., 337–338.

subjective operations, cognitive activities, and pursuit of scientific methods through which it stands before us as determined in such and such a way and as in principle infinitely determinable with regard to its true being. . . . The revealing of these intentional implications and with them the history of the world itself, in which the subject of psychology finds himself as in one ready-made, *also* means, therefore, a retrogression to what is subjective, since it is through the intentional activity of the subject that the world has obtained this form; but it is *a retrogression to a hidden subjectivity*—hidden because it is not capable of being exhibited as present [*aktuell*] in reflection in its intentional activity but can only be indicated by the sedimentations left by this activity in the pregiven world. . . . Moreover, such regressive inquiry does not involve seeking the factual, historical origin of these sedimentations of sense in a determinate historical subjectivity. . . . Rather, this world which is ours is only an example through which we must study the *structure and the origin of a possible world in general* from subjective sources. We would not be able to understand this definite historical origin of productions of sense in historical subjects if we did not reaccomplish them ourselves, if we did not re-experience this origination of the operations of idealization from original life-experience. . . . We then understand ourselves . . . *as a subjectivity bearing within itself, and achieving, all of the possible operations to which this world owes its becoming. . . . as transcendental subjectivity*, where, by 'transcendental,' nothing more is to be understood than the theme, originally inaugurated by Descartes, of a regressive inquiry concerning the ultimate source of all cognitive formations, of a reflection by the knowing subject on himself and on his cognitive life, the life in which all scientific formations valid for him have been purposefully produced and are preserved as available results."[365] Transcendental subjectivity is, then, a *reappropriation* of this world by a subjectivity whose distinctive feature is exactly that it is a reappropriation of the original appropriation of this world undertaken by modern

[365] E. Husserl, *Experience and Judgment* (trans. J. S. Churchill and K. Ameriks, Evanston, 1973), 48–50.

thought, beginning with the flawed subjectivity of Descartes. To *reappropriate* is to bring to clarity, step by step, the operations of subjectivity by which this world originally came into being. Now the reason for the lack of clarity or self-evidence in the foundations of modern philosophy is, for Husserl, due to the fact that this *original* appropriation lacked that disinterest or detachment achieved by bracketing the reality or substantiality of this world in the phenomenological *epoché*. Indeed, it lacked precisely that clarity about itself as in itself pure ego *indifferent* to the question of the nonbeing of the world. As pure transcendental ego, it is this world's absolute presupposition whether this world exists or does not exist. For Husserl, modern thought's predicament is the result of a lack of clarity, *to begin with,* about itself and, therefore, not, primarily, but secondarily, about this world's reality. We may understand, further, in terms of our analysis of the essential history of thought, that when Descartes introduced his *animus* or *substantia cogitans* into his thought of existence immediately thereupon he reflected upon his finitude as a creature existing in a merely potential succession of time. Leibniz eliminated this *real* time as inessential to the thought of existence. That is, it is an obstacle to reason's appropriation of the contingency of the world. But Leibniz retained *substantiality* in the continuity of being. It appears in Kant's *noumena* and finally, in Hegel's synthesis of subjectivity with actuality. In the concept of substance, human subjectivity located itself outside of itself, since it failed to perceive its transcendental character radically. It continued to identify itself with nature or natural reason, as it learned to do at the knees of sacred doctrine, even after it had taken for its own original form a transcendental perspective. And although to Husserl it was a mistake on Descartes' part to introduce his substance into modern thought's inception, it is, nevertheless, quite clear that this substantial interest of the thinker in his science is *of the essence of the practical* that characterizes modern science from Descartes on. The phenomenological *epoché* is, then, *the antithesis of the appropriation of this world.* Or, *the reappropriation of this world* is actually *the appropriation of itself by the human species,* which self-appropriation is the personal act whose absolute autonomy is such that it seeks the reappropriation of the origination of this world in *no factual, historical subjectivity*, as if it belonged to this world as a reality. Rather, in Husserl, this reappropriation studies the history of this world's origination only insofar as it may exemplify, for radical personal autonomy, the 'origin of a possible world in general.' Thus phenomenology studies the structure and origin of this world simply to learn how a subjectivity, such as it is, builds up worlds in general; it asks what the essence of a world is: what it means to be a world in general. *It does not ask what the essence of this world is,* as if it simply granted this world's claim to be real; this knowledge of essence that it seeks is a knowledge of its subjective intentions. The ultimate object of the pure transcendental ego is its own constitution; it is in view of this that its interest is restricted to looking for the essence of what appears. Husserl writes in *The Paris Lectures:* "Every one of us who has been guided back, through phenomenological reduction, to *his* absolute ego discovered himself, with apodictic certainty, as an actual existent. Looking around, the ego discovered diverse classes—classes which can be fixed descriptively and developed intentionally—and could soon proceed to the intentional disclosure of its own ego. But it was no accident that the expressions 'essence' and 'essential' [*wesensmäßig*] escaped me repeatedly. These expressions are equivalent to a definite concept of the a priori, a concept clarified only by phenomenology. It is clear that if we explain and describe a class of *cogitata,* such as perceptions, *qua* class—such as the perceived, retention and the retained, recollection and the recollected, assertion and the asserted, seeking and the sought etc.—then we are led to results which persist regardless of how we abstract from actual facts. The individuality of the instantiating actuality—as for instance the present flux of perceptions of the table—is completely irrelevant to the class. Equally irrelevant are the general impressions which I—the actual ego—acquire in my experiences of this class. The description does not depend on discovering individual facts or establishing their existence. The same holds for all egological structures. An analysis of the class of sensible and spatially objective experiences may serve as example. I proceed systematically to the constitutive problem of how such experience would have to unfold itself consistently so that one and the same object might disclose itself completely, *i.e.*, with all its intended attributes. It is then that I hit upon the great realization that before anything can be, for me, a truly existing object, it must fulfill certain necessary a priori conditions. It must appear in the form of a specific and relevant structural system dealing with experiential possibilities. . . . Evidently I am quite free in what I may imagine my ego to be. I may view the classes as pure ideal [*ideale*] possibilities of the now merely possible ego, or of any possible ego whatsoever (as a free interpretation of my actual being). In this manner I reach *classes of essences, a priori possibilities, and corresponding essential laws of being* [*Wesensgesetze*]. The same applies to general structures of the essence [*Wesensstrukturen*] of my ego insofar as it can be thought of at all. Without these structures I can neither conceive of my self in general nor a priori. . . . We have reached a methodological insight which, next to the genuine method of phenomenological reduction, is the most important in phenomenology: *the ego,* to use traditional language, *possesses an enormous inborn a priori.* . . . The actual facts of experience are irrational, but their form—the enormous formal system of constituted objects and the correlative formal system of their intentional a priori constitution—consists of an inexhaustibly infinite a priori. Phenomenology explores this a priori, which is nothing other than the essence [*Wesensform*] of the ego *qua* ego, and which is disclosed, and can only be disclosed, by means of my own self-examination." [366] Perhaps it will help us to understand the significance of this apparently unbridled freedom of the pure transcendental ego to imagine itself now as existent, now as merely possible, now as any ego whatsoever, that is, as an ego in general, if we recall Aristotle's understanding of the act of knowledge.[367] For Aristotle, the possibility of knowledge is first the priority of essence to reason in such a way that reason's ultimate constitution is an identity of its own but not self-

[366] *The Paris Lectures*, op. cit., 27–29.
[367] Cf. above, chapter on Aristotle.

evident to reason. It is an essential identity without which it understands that its knowledge would remain forever purely potential. Aristotle understands, further, that in the *act of knowledge* the mind is identical with its object. Again, in a manner not self-evident to reason, *qua* reason, pure intellectual knowledge is identical in existence with its object, since a knowable essence is always that of an existent which, *qua* existent, is inseparable from its essence which is its existent individuality. Knowledge, then, in Aristotle is always, if it is actual knowledge, or the knowledge of an individual man in the act of knowing, *substantial* identity with what is known. Indeed, we may see retrospectively, that *in the absence of self-evidence,* which does not emerge until we encounter the transcendental form of natural reason in sacred doctrine,[368] *it is quite impossible,* as the understanding of Aristotle's epistemology makes vividly clear, *to distinguish out, in the act of knowing, substance, knowledge or particularity*. In sacred doctrine, to the contrary, as we noted earlier, reason, in the light of revelation, beheld its substantial object with self-evidence. Moreover, we emphasize here, it beheld this object as a *particular* existent. But, beginning with Descartes, the particularity of the object is subject to an increasingly successful attempt on the part of the thought of existence (a substance with a transcendental perspective!) to appropriate that particularity. Or, in an unparadoxical or un-Aristotelian manner, it simplifies the situation by denying a separate reality to the object apart from its being known (this in abstract imitation of the transcendental form of natural reason). This appropriation is, of course, essentially complete in Hegel. [369] As a result of this appropriation by absolute reason, *particularity came to be identified with the universal substance of thought itself.* When Husserl's phenomenology, on behalf of the species, *qua* species, abstracts itself from the absolute thought of existence, when it refuses, *qua* rational ego, to acknowledge the claim of the formality of reason to be substance, when the balance of modernity shifts to the weightless essence of the pure transcendental subjectivity,

[368] Cf. above, chapter on Aquinas.
[369] Cf. above, chapters 3,4,5,6, passim.

then, this pure ego discovers itself to exist prior to all particularity whatsoever including the universal specificity of this world. There is for the pure ego neither substance, nor *preexistent* formality or specificity. The liberty of the pure ego consists in its knowing originally in general; Husserl writes in his *Experience and Judgment:* ". . . for the acquisition of pure concepts or concepts of essences, an empirical comparison cannot suffice but . . . by special arrangements, the universal which first comes to prominence in the empirically given must from the outset be freed from its character of contingency. Let us attempt to get a first concept of this operation. It is based on the modification of an experienced or imagined objectivity, turning it into an arbitrary example which, at the same time, receives the character of a guiding 'model,' a point of departure for the production of an infinitely open multiplicity of variants. It is based, therefore, on a *variation*. . . . Thus, by an act of volition we produce free variants, each of which, just like the total process of variation itself, occurs in the subjective mode of the 'arbitrary.' It then becomes evident that a unity runs through this multiplicity of successive figures, that in such free variations of an original image, e.g., of a thing, an *invariant* is necessarily retained as the *necessary general form*, without which an object such as this thing, as an example of its kind, would not be thinkable at all. While what differentiates the variants remains indifferent to us, this form stands out in the practice of voluntary variation, and as an absolutely identical content, an invariable *what,* according to which all the variants coincide: *a general essence*. . . . The essence turns out to be that without which an object of a particular kind cannot be thought, i.e., without which the object cannot be intuitively imagined as such. This general essence is the *eidos,* the idea in the Platonic sense, but apprehended in its purity and free from all metaphysical interpretations, therefore taken exactly as it is given to us immediately and intuitively in the vision of the idea which arises in this way." [370] This arbitrary variation of the free imagination of the pure ego of transcendental subjectivity is that process in

[370] *Experience and Judgment,* op. cit., 340–341.

which emerges in *absolute self-evidence* the a priori knowledge in general governing every possible special knowledge. Husserl says in *The Paris Lectures:* "Phenomenological procedure possesses no antecedent realities or conceptions of reality, but instead, from the very beginning, creates its concepts through original acts—which in turn are fixed in original concepts. Furthermore, phenomenology, because of its necessity to disclose all horizons, governs all differences of range and all abstract relativities. Consequently, phenomenology must arrive from within itself at those conceptual systems which determine the fundamental meaning of scientific constructions. These are the concepts that trace out all formal demarcations of the idea of a possible world, and, consequently, must be the genuine foundation-concepts of all knowledge. For such concepts there can be no paradoxes." [371]

What remains radically unclear in phenomenology, that is, what is not self-evident in the constitution of transcendental subjectivity, is the fact of its *essential historicity.* In the first instance, it abstracts from the absolute reason of modern thought as the pure ego of a transcendental and phenomenological monadology which grounds the autonomy of the human species, *qua* species. In the second instance, it begins again with Descartes, not in a merely external proximity to 'scholasticism,' but in a blind but ultimate dependence on the fact of the transcendental form of natural reason brought into being by the appearance of the transcendental essence of existence itself. But, closer to hand, if this worldly existence is for Kierkegaard, *qua* individual, *nothing* at all, it is, to the contrary, for the species, *qua* species, in the phenomenological reduction to transcendental subjectivity, *a possibility* for the pure ego in whose inner determination is reflected man's freedom to be for himself the author of his world.

[371] *The Paris Lectures,* op. cit., 37.

Chapter 12

CLARIFICATION OF THE ABSOLUTE. III: HEIDEGGER

The essential modernity of Kierkegaard resides in his taking up the position of the individual, *qua* individual, within the Moment of the thought of existence, in opposition to absolute reason's claim that it itself constitutes the individual's reality through the universality of the particular. The individual relies, instead, in its own nonbeing or finitude, upon its coming to exist in the course of time by its coming to be related, *qua* individual, immediately to God by means of a passionate faith in face of faith's inherent absurdity. In Husserl, the species, *qua* species, similarly takes up a position within the absolute thought of existence in opposition to that thought's generic identification of the individual with the universe in existence. It takes to itself, *qua* species, in its finitude or nonbeing, a ground in its own transcendental subjectivity before the pure ego of which, constituting an infinite a priori, the world, together with the individuality of which it is the principle, must *first* present its claim to substantial existence as self-evident, so that, in the nature of phenomenology's modernity, this world's reality is radically reduced to *a possibility*. (Whereas, for Kierkegaard, this same reality became in the eyes of faith *nothing* at all apart from its being an examination of the individual by God.) If we hold these two moments of radical opposition to the Hegelian synthesis before the mind, we notice how essentially restricted they are by virtue of *the necessity of beginning with their own finitude or nonbeing*. That is, they are radical only in opposition to the absoluteness of the synthesis of modern thought essentially complete in Hegel. If we so hold these moments, the thought of Kierkegaard and the thought of Husserl, we begin to notice that what these moments of absolute thought's self-opposition actually portend is the *fragmentation of reality itself.* Indeed, as abysmal a thought as it may be, they portend the *multiplication of realities:* The fruit of the absolute thought of existence is absolute relativity in the event of refusing to thought its absolute determination of its own moments.

In the thinking of Heidegger we come upon the third moment of the absolute thought of existence standing in its own finitude and nonbeing in opposition to the claim of that thought to be for the universe its substantial being. In Heidegger, the universe, *qua* universe, opposes itself to the specification of the individual as the immediate determination of what-is-in-totality. The universe, *qua* universe, or, what-is-in-totality, while at the same time refusing to acknowledge its dependence upon absolute reason, nevertheless, demands to be beheld in its immediacy. In Kierkegaard, the individual, *qua* individual, came under the requirement that it be beheld in that immediate consciousness of its own becoming which is the deed of inwardness beyond all scientific reckoning, a clarity of self-perception actually achieved by the individual only in relating himself immediately to God as the power in which he exists transparently. In Husserl, however, the species, *qua* species, demands of this world a clarification of its origination and structure in self-evident science. The content of this is nothing other than the self-examination of transcendental subjectivity's infinite a priori, made possible by the phenomenological reduction of the self to the pure ego, in turn the point of departure of a transcendental monadology as ground of both this world and the sciences of the world.[372] In Heidegger we encounter the same demand, this time on behalf of the totality-of-what-is, to be made *visible* in its *as such,* or *as totality* to be *seen,* but, not, as in Kierkegaard, in its nothingness for the individual, not, as in

[372] Cf. above, first two chapters on the "Clarification of the Absolute."

Husserl, in its being a possibility for the transcendental subjectivity that lies beyond the world as its horizon in potentiality. In Heidegger, the universe, *qua* universe, demands that its totality be visible *in the midst of the world,* as this world's *actuality* beheld as such by being related immediately to *its own Being,* therefore, *as a totality in its independence of the thinking beholding it.* But the background against which the *totality*-of-what-is might be *seen as such* must be Nothing, or what is left over from the totality-of-what-is. Is it possible in the midst of the world's actuality to experience this Nothing with respect to which alone the universe, *qua* universe, may become visible in its own Being? Heidegger deals with this question in his *What Is Metaphysics?:* "As certainly as we shall never comprehend absolutely the totality of what-is, it is equally certain that we find ourselves placed in the midst of what-is and that this is somehow revealed in totality. Ultimately there is an essential difference between comprehending the totality of what-is and finding ourselves in the midst of what-is-in-totality. The former is absolutely impossible. The latter is going on in existence all the time. . . . However fragmentary the daily round may appear it still maintains what-is, in however shadowy a fashion, within the unity of a 'whole.' Even when, or rather, precisely when we are not absorbed in things or in our own selves, this 'wholeness' comes over us—for example, in real boredom. . . . Real boredom comes when 'one is bored.' This profound boredom, drifting hither and thither in the abysses of existence like a mute fog, draws all things, all men and oneself along with them, together in a queer kind of indifference. This boredom reveals what-is in totality. There is another possibility of such revelation, and this is the joy we feel in the presence of the being—not merely the person—of someone we love. . . . The affective state in which we find ourselves not only discloses . . . what-is in totality, but this disclosure is at the same time far from being a mere chance occurrence and is the ground-phenomenon of our *Da-sein.* . . . Yet, at the very moment when our moods thus bring us face to face with what-is-in-totality they hide the Nothing we are seeking. We are now less than ever of the opinion that mere negation of what-is-in-totality as revealed in these moods of ours can in fact lead us to Nothing. . . . Does there ever occur in human existence a mood of this kind, through which we are brought face to face with Nothing itself? This may and actually does occur, albeit rather seldom and for moments only, in the key-mood of dread (*Angst*). . . . And although dread is always 'dread of,' it is not dread of this or that. 'Dread of' is always a dreadful feeling 'about'—but not about this or that. The indefiniteness of *what* we dread is not just lack of definition: it represents the essential impossibility of defining the 'what.' . . . All things, and we with them, sink into a sort of indifference. But not in the sense that everything simply disappears; rather, in the very act of drawing away from us everything turns towards us. This withdrawal of what-is-in-totality, which then crowds round us in dread, this is what oppresses us. There is nothing to hold on to. The only thing that remains and overwhelms us whilst what-is slips away, is this 'nothing.' Dread reveals Nothing. . . . In the trepidation of this suspense where there is nothing to hold on to, pure *Da-sein* is all that remains. . . . Only in the clear night of dread's Nothingness is what-is as such revealed in all its original overtness (*Offenheit*): that it 'is' and is not Nothing. . . . The essence of Nothing as original nihilation lies in this: that it alone brings *Da-sein* face to face with what-is as such. . . . insofar as *Da-sein* naturally relates to what-is, as that which it is not and which itself is, Da-sein *qua Da-sein* always proceeds from Nothing as manifest. *Da-sein* means *being projected into* nothing. . . . Projecting into Nothing, *Da-sein* is already beyond what-is-in-totality. This 'being-beyond' . . . what-is we call Transcendence. Were *Da-sein* not, in its essential basis, transcendent, that is to say, were it not projected from the start into Nothing, it could never relate to what-is, hence could have no self-relationship. Without the original manifest character of Nothing there is no self-hood and no freedom. . . . Nothing not merely provides the conceptual opposite of what-is but is also an original part of essence (*Wesen*). It is in the Being (*Sein*) of what-is that the nihilation of Nothing (*das Nichten des Nichts*) occurs. . . . 'Pure Being and pure Nothing are thus one and the same.' This proposition of Hegel's ('The Science of Logic,' I, WW III, p. 74) is correct.

Being and Nothing hang together, but not because the two things—from the point of view of the Hegelian concept of thought—are one in their indefiniteness and immediateness, but because Being itself is finite in essence and is only revealed in the Transcendence of *Da-sein* as projected into Nothing." [373] The visibility of the third moment of the absolute thought of existence, namely, the universe or totality-of-what-is, *qua* totality, this clarification, like those of the two previous moments, takes its stand in opposition to absolute reason. It is therefore, under the circumstance of its *point of departure,* subject indirectly to the interdict of sacred doctrine whereby it is oblivious of the appearance of the transcendental essence of existence itself. In this oblivion the universe, *qua* universe, holding itself in its own essence, is seen to be radically related, through *Dasein's* transcendent existence (its essence), to Being, itself *finite.* This grounding of the universe in the finitude of Being or Nothing is, at the same time, the primordial event out of which emerges Dasein's own self or identity as Being-in-the-world-beyond-the-world. This is human *existence,* in its actuality, as being projected into Nothing, out of which emerges simultaneously the distinction of world and self. The finitude of Being itself, as it appears in Heidegger's thinking, manifests itself, in its *essence,* as *Being-related.* The Being of the universe, *qua* universe, in its finitude, is revealed only in and through man's transcendent being in the world. This existent essence of man is prior to scientific thinking, that is, essentially prior so that it may never be comprehended in thought. It is, therefore, essentially prior to the faith of the Kierkegaardian individual, whose transcendent passion to exist presupposes, in its holding fast to the absurd, the validity of the scientific understanding, or of human self-knowledge. The absolute and relative distances, achieved by Kierkegaard and Husserl, respectively, on behalf of the individual and the species, disappear altogether in Heidegger where the *essence* of *Dasein* is that through which the universe, *qua* universe, comes into being in the immediacy of

[373] M. Heidegger, *Existence and Being* (Introduction by W. Brock, Chicago, 1949), 333–346.

Being before thought in a relationship of absolute nearness, to which man, in his essence, is brought near. This intimate indwelling of man in the world, in the way of the world, is the actuality of the essence of *Dasein*. Heidegger writes in *What Is Metaphysics?*: "The old proposition *ex nihilo nihil fit* will . . . acquire a different meaning, and one appropriate to the problem of Being itself, so as to run: *ex nihilo omne ens qua ens fit:* every being, so far as it is a being, is made out of nothing. Only in the Nothingness of *Da-sein* can what-is-in-totality—and this in accordance with its peculiar possibilities, i.e., in a finite manner—come to itself. . . . Man's *Da-sein* can only relate to what-is by projecting into Nothing. Going beyond what-is is of the essence of *Da-sein*. But this 'going beyond' is metaphysics itself. That is why metaphysics belongs to the nature of man. . . . Metaphysics is the ground-phenomenon of *Da-sein*. It is *Da-sein* itself."[374] It was Hegel who, on behalf of the thought of existence, understood that the 'nonbeing of the finite is the being of the absolute.'[375] Also, it was Hegel who understood the coming to be and passing away of beings in time as nothing other than the objective determination of the finitude of things in the world.[376] This negativity of the finite world derives ultimately from reason's determination of Christianity's doctrine of 'creation *ex nihilo*' according to terms of its own comprehension, so that it was unable, in the absence of the knowledge of the transcendental essence of existence itself, to distinguish between the *relation* of the creature to its creator, on the one hand, and the *essence* itself of that creature, on the other.[377] We are now in a position to perceive that Heidegger's equation of the *ens qua ens,* being *qua* being, with *nihil,* with the self-contradictorily finite being grounded in its own nonbeing (*ex nihilo fit*), is, in essence, and in explicit opposition to the being of the absolute thought of existence, the greatest extenuation of modernity possible. The world is no longer grounded in the *infinite* essence of man, as in Hegel, but, now, in Heideg-

---

[374] Ibid., 346–348.
[375] Cf. above, p. 223f.
[376] Cf. above, pp. 239ff.
[377] Cf. above, pp. 195ff., p. 204.

ger, in the *finite* essence of man. With this there is a shift from the infinite substantiality of universal thought to Being itself as finite; within the limitations of modernity, man, in Heidegger's thought, continues in his essence to be indispensable to the coming to be of the universe, *qua* totality. This *indispensability* of man is the *extenuated understanding* of what in modern thought proper is originally taken to be man's essential identity with the absolute reason of the universe in its infinite determination. But it is now understood, in the event of the refusal of the universe, *qua* universe, to acknowledge the absolute reality of thought, to be man's belonging together with Being itself in the origination of the totality-of-what-is in the manner of its finitude. If the essence of *Dasein* is that transcendence by which man is projected beyond what-is into Nothing, and, if this being-beyond is metaphysics, that is, if the essence of man is, in Heidegger, what metaphysics is, then, in this incarnation of the Logos of Hegel, metaphysics itself is reduced to finitude. It is reduced to possessing its truth in its being together with Being itself. Here we may begin to glimpse how it is that in Heidegger truth itself is radically relativized, as a result of that modernity of his thinking in which it is impossible to distinguish man's essence from metaphysics or the thought of true being itself, both, here, brought to essential temporality, to a radical scepticism. The peculiarity of this scepticism is that, in the absence of a knowledge of the transcendental essence of existence itself, that is, in the absence of the knowledge of truth itself, it relates itself, in its finitude, to the Being of the totality-of-what-is in the midst of which it discovers itself to be essentially. It looks to Being itself for what truth it might come to think, that is, the essential priority of its thought is the finite Being of the totality-of-what-is, the truth of which essentially is yet to be thought. Heidegger writes in his *The Question of Being:* ". . . a thoughtful glance ahead into this realm of 'Being' can only write it as ~~Being~~. The drawing of these crossed lines at first only repels, especially the almost ineradicable habit of conceiving 'Being' as something standing by itself and only coming at times face to face with man. According to this conception it looks as if man were excluded from 'Being.' However, he is not only not excluded, that is, he is not only encompassed into 'Being' but 'Being,' using the essence of man, is obliged to abandon the appearance of the for-itself, for which reason, it is also of a different nature than the conception of totality would like to have it, which encompasses the subject-object relationship. The symbol of crossed lines can, to be sure, according to what has been said, not be a merely negative symbol of crossing out. Rather it points into the four areas of the quadrangle and of their gathering at the point of intersection. . . . The being present as such turns towards the essence of man in which the turning-towards is first completed, insofar as the human being remembers it. Man in his essence is the memory of Being, but of ~~Being~~. This means that the essence of man is a part of that which in the crossed intersected lines of Being puts thinking under the claim of an earlier demand. Being present is grounded in the turning-towards which as such turns the essence of man into it so that the latter may dissipate itself for it."[378] Being itself, in being present, reveals itself as being for another. This is made possible only by the essence of man which is transcendence, in which projection-into-Nothing there appears to man the Being of beings, the Being of the totality-of-what-is, as that which appropriates man for this revelation of its own essence. Thus, by virtue of his unique participation in the totality-of-what-is, or, by virtue of his *meta-physical* essence, man is face to face with Being in its presence, for the sake of which man *exists.* That is, he is *in* the world as the transcendent Being *of* the world. In the dissipation of man's essence he comes to an identity with Being itself, with the Being itself of the totality-of-what-is, *qua* totality, not as it is for thought, an identity of subject with object (a universal identity in substance), but, rather, in the actual finitude of the world's Being, an identity of Being the Same in essence, that is, man remembers that he is in essence the memory of Being, that, therefore he *belongs* to Being. This *belonging* is the peculiar form of identity within the limitations of finitude or temporality, from which absolute

[378] M. Heidegger, *The Question of Being* (trans. J. T. Wilde and W. Kluback, New Haven, 1958), 81, 83.

being-related nothing is excluded in the thinking of Heidegger. Indeed, as we have seen, the exclusion of Nothing from the totality-of-what-is is the ground-phenomenon of Dasein's transcendence in which finite Being itself is manifest as being present to man, or, by the exclusion of Nothing man is reminded of Being itself, of its *inapparent essence* (~~Being~~), of its finitude as the ground of the universe (instead of the absolute thought of existence). Therefore, we may understand that Heidegger's thought is an attempt to restore the essence of man, or meta-physics, to its original calling as the opening for the manifestation of Being as presence. This attempt is made, we must notice, within the same 'this world' that Husserl examines the structure and origin of, namely, the world created by modern thought as the totality-of-what-is. For Husserl, it must be reappropriated by pure transcendental subjectivity as a possibility. But for Heidegger, it is seen from *within,* is seen as a *metaphysical* concern of man's *actual existence.* It follows, then, that whereas Husserl calls into question the truth of appearances, presented in modern thought, and does so with ubiquitous detachment from the question of this world's existence, Heidegger's critique of modern thought, situated in the midst of this world, and without the faith in God which only the individual, *qua* individual, can have in this *totality* of being (as Kierkegaard makes clear), is directed not at the truth of appearances, but, in the first instance, it seeks out the *inapparent essence* of the apparent totality-of-what-is. By keeping this in mind, we will be constantly aware of Heidegger's extenuated modernity, even while he claims to return to the Greeks for a fresh start (in the manner of Husserl's beginning again with Descartes). Heidegger's thinking everywhere presupposes modern thought's actual development (as does Husserl's). Indeed, what is really crucial is that Heidegger's understanding of the history of being presupposes the history of being in precisely that form in which neither modern thought nor the three moments of its extenuation are capable of knowing it, the latter simply because they are operating out of a world brought into being by man or demon, but not by God. Heidegger's interest, then, is in metaphysics, because his interest is in man's being in the midst of the totality of the universe. It is clear also that Heidegger's understanding of metaphysics must take its point of departure, in actuality, not from Aristotle, but from Leibniz. In his *The Question of Being* he writes (speaking of the lecture, *What Is Metaphysics?*): "The lecture closes with the question: 'Why is there being at all and not rather Nothingness?' Here 'Nothingness' is intentionally and, contrary to previous procedure, written with a capital. According to the wording the question is, to be sure, broached which Leibniz posed and Schelling took up. Both thinkers understand it as the question about the highest reason and the first existing cause of all being. Present-day attempts to restore metaphysics have a special liking for taking up the designated question. But the lecture *What Is Metaphysics?* in accordance with its differently constituted way through another area, also thinks this question in a transformed sense. It is now asked: what is the reason why everywhere being is given precedence, why the negative of being, 'this nothingness,' that is, Being in regard to its essence, is not rather considered?" [379] Leibniz intended that the question should lead to the sufficient reason of the *existence* of the world. Heidegger reinterprets the question within the limits of the universe, *qua* universe, so that it leads to the *essence* of the world itself, to its nothingness, at the same time, to that of human being in the world, which essence *is* the existence of the world in its finitude. It is not a question of discovering a being outside the world to account for its existence. Rather, for Heidegger, it is a matter of discovering or uncovering Being itself, which traditional metaphysics is understood to be oblivious of, preoccupied as it is with itself. This preoccupation closes off, in advance, Being itself in oblivion. This is the origin of metaphysics, namely, the oblivion of Being (~~Being~~). The restoration of the latter is the restoration of the former. All the while we must keep in mind that thinking with Heidegger we move within the limits of Leibniz' question, remembering that this question arises in modern thought's attempt to appropriate the contingent universe for itself. Leibniz takes for granted the contin-

[379] Ibid., 99.

gent existence of this world. So does Heidegger. But Heidegger resolves upon a metaphysical leap off the ground of existence into the Nothingness of his essential identity with the finite Being of the world. Let us listen to Heidegger in his *Identity and Difference:* "Man and Being are allocated to each other. They belong to each other. It was owing in the first place to this belonging-together (to which thinkers have not paid much attention) that Man and Being have acquired those determinations in essence by which they are comprehended metaphysically in philosophy. This predominant *belonging*-together of Man and Being we fail to acknowledge stubbornly so long as we are looking upon everything in orderly arrangement and mediation, be it dialectically or without dialectics. In such situations we then find regularly nothing but connections which have been brought about either on the part of Being or on the part of Man and represent the belonging-together of Man and Being as an intertwining of the two. Thus far we have not entered into *belonging*-together. The question may be asked how such homing may be accomplished. The answer is by keeping aloof from the attitude of representational thinking. This keeping aloof is a positing in the sense of a leap. It is a bounding away from and a leaving behind of the familiar concept of Man as the *animal rationale,* the rational animal, who nowadays has become the subject for his objects. The bounding off is at the same time a getting away from Being. However, beginning early in occidental thinking, this Being has been interpreted as the ground in which every Existent, as Existent, has its authentication. In what direction does the leap from the ground point? Is it heading for an abyss? It does so, indeed, so long as we picture ourselves the leap and do so more specifically within the scope of metaphysical thought. It does not head for an abyss in so far as we leap and let ourselves go. In what direction? We are going to where we are already imbedded, into the fold of Being. Yet, Being itself is our very own, for only in Man can Being be domiciled, that is to say, can it *be* present. Hence we must leap in order that we may experience in our own person the *belonging*-together of Man and Being. This leap is the precipitous homing without benefit of bridges into that belongingness which alone Man and Being as mutually related and, hence, their pattern can provide. The leap is the sudden return into the realm where Man and Being are already found together in their essence, because both were assigned to each other in a sufficiency. It is not until the entry into the sphere of this mutual assignation is effected that thought experience becomes attuned to it and determined." [380] Man leaps into his very own Being, letting go the support of traditional metaphysics, that is, his conceptualization of being itself as a reason for existence, as well as his thought of himself as a specific individuality.[381] Abandoning all thought of legitimacy, he steps suddenly into the freedom that arises from his own Being to meet him in mutual concern, stripped of all previous conceptions. He enters into the security of the deep well of his own Being. If modern thought looked for the sufficient reason of the existence of the universe, then, in Heidegger, this immediate return of man into the realm of his identity with the Being of the totality-of-what-is, that is, with his own Being, reveals itself as that *original mutuality* of the essence of Man and Being which *relationship* itself, prior to all thought, constitutes a *sufficiency*. Indeed, we may understand, that *what Heidegger uncovers in the remembrance of Being is nothing other than the self-supporting contingency of the universe,* qua *universe, that deep within the abyss of the oblivion of Being lies waiting for Man the resource of his survival in the world.* But that self-assurance that characterizes modern thought as the thought of existence, that, in the clarification effected by Heidegger, is transformed into a matter of the *essence* of Being, considered in its difference, *qua* difference, from existence. The relationship between Man and Being is one of their own essence. The guarantee whereby history is thought through to its completion in man, that guarantee, absolute in Hegel, is, in Heidegger, *suspended in essence.* That is, it is suspended in the finite Being of the actual world, from the midst of which Heidegger looks out, as it were, to see neither God, nor pure ego, but Nothing. It is a Nothing recalling Man

[380] M. Heidegger, *Essays In Metaphysics: Identity and Difference* (trans. K. F. Leidecker, New York, 1960), 22–24.

[381] Cf. above, p. 271.

to a remembrance of the difference that Being is from what-is, reminding him of his *essential* freedom. Heidegger writes in his *Identity and Difference:* "What we learn by way of the modern world of technology in the frame-work as a pattern of Being and Man, is merely *prelude* to what we call con-cern. Con-cern, however, does not necessarily persevere in its prelude. For in con-cern we are persuaded of the possibility of developing the mere sway of the frame-work into a more primitive solicitude. Such a development of the frame-work from concern to solicitude would bring about the eventful reduction (never initiated by Man himself alone) of the world of technology from lordship to servitude within the realm in which man more properly involves himself in con-cern. Where has our road taken us? It has led our thinking back home into that simplicity which we have called con-cern. . . . the nearest of the near is entreating us to share immediately what we are already domiciled in. In fact, that was what we wanted to express by the word con-cern. For, what could we consider more intimate than what brings us closer to where we belong, wherein we are corporate members, that is, con-cern? Con-cern is the internally oscillating realm through which Man and Being touch each other in their essence and attain their essential nature by divesting themselves of these determinations which metaphysics imputed to them."[382] If we understand that, in Hegel, absolute reason thinks the *identity of the two as such,*[383] or, *identity in difference,* thereby thinking the Platonic *duad* or matter itself as the essence, which essence, in turn, reveals itself in the *thing* as an internalized presupposition of the coincidence of matter with form, Heidegger's thinking, in contrast, is an attempt to be receptive to *not* the two *in identity,* but rather, *the difference between* the two. It is a receptiveness to that forgotten essence of Being, which as difference itself transcends the essence of existence thought by Hegel. This difference yields the distinction of Being and Existence to thought, from out of the primordial simplicity in which Man and Being touch each other in essence.

---

[382] *Essays In Metaphysics,* op. cit., 27–28.
[383] Cf. above, pp. 128ff. Also, above, p. 137.

But this difference when thought absolutely determines the two as *grounds* one for the other, or, absolute thought is not the thought of *difference as such.* Heidegger says in *Identity and Difference:* "Still with an eye to difference, yet releasing it in 'backtracking' into that which is *to*-be-thought we can assert the following. Being of Existence means the type of Being which Existence is. The 'is' in this case is to be taken transitively, as implying passage. Being asserts its nature here in the manner of a transition to Existence. However, Being does not go over toward or into Existence by leaving its place or position, as if Existence, previously devoid of Being, could be contacted by Being for the first time. Being transcends and covers, while revealing itself, what is encountered in open presence by such enthrallment. Encounter means seeking refuge in open presence, thus, being in sheltered presence, being an Existent. Being exhibits itself as revealing enthrallment. Existence as such manifests itself as an encounter fleeing into unmasked presence. Being in the sense of revealing enthrallment and Existence as such as in the sense of refuge-seeking encounter have their being as elements that have been differentiated from the Same, that which underlies difference. What underlies distinction is what originally is responsible for yielding and keeping apart the between, wherein enthrallment and encounter are conjoined and mutually supported in their fluctuating relationships. The difference of Being and Existence as the ground of the distinction between enthrallment and encounter lies in the *unmasking-enshrouding issue* of both. Light is shed throughout the issue on the self-enshrouding occlusion. It is this pervading luminosity which is responsible for the reciprocity of enthrallment and encounter."[384] We may understand that the self-enshrouding occlusion is the oblivion of difference in which the elements, in this case, Being and Existence, support each other in the absolute thought of existence, indeed, together are grounded in that thought, the thought of Existence *as such,* which is ultimately the thought of an eternal Being. But the ground, that absolute thought is, is an infinite ground or Being

[384] *Essays In Metaphysics,* op. cit., 57–58.

as ground. The restoration of metaphysics would be the restoration, *to begin with,* of the oblivion of difference; *to begin with* difference *as* difference is to return to that original *belonging-together* of Man and Being in their shared essence. Indeed it would be *to begin with Nothing.* Absolute thought's solution to the problem of the beginning is considered by Heidegger, in his *Identity and Difference,* as follows: "In order to gain a comprehensive view of the Hegelian metaphysics in our seminar, we chose expediently a discussion of the section with which the first book of the *Science of Logic,* 'The Doctrine of Being,' begins. Even the title of the section furnishes enough food for thought in every word. It says: 'Wherewith must we start science?' Hegel's answer to this question consists in the demonstration that the beginning is of a 'speculative nature.' This means that the beginning is neither something immediate nor something mediate. The nature of this beginning we then endeavored to express in one speculative statement: 'The beginning is the result.' According to the plural interpretation of the dialectic, the 'is' in this sentence indicates several things. For once, it conveys that the beginning is—taking the *resultare* literally—the rebounding of thought reflecting upon itself from the perfection of dialectical movement. The perfection of this movement, the absolute Idea, is the unfolded closed Whole, the fullness of Being. The rebound from this fullness results in the emptiness of Being. Science (the absolute self-knowing knowledge) must begin here. Beginning and end of the movement, and prior to them the movement itself, remains everywhere Being. Being asserts itself as internally gyrating movement from fullness to extreme alienation and from alienation to self-perfecting fullness. Object of thought is, hence, for Hegel thought thinking itself as internally gyrating Being. Were we to reverse the speculative statement concerning the beginning, which is not only justifiable but necessary, it would read: 'The result is the beginning.' Actually, we ought to start with the result in so far as the beginning evolves from it. This says the same as does the remark which Hegel inserts casually and parenthetically in the section dealing with the Beginning, toward the end: '. . . (and it would seem *God* has the indisputable right that we should begin with him).' "[385] The problem of the beginning, then, is solved by Hegel, on behalf of modern thought as a whole, by beginning with the *result.* The *result* in its profoundest meaning is that of the internally gyrating Being of thought thinking itself, which for Hegel is God. Heidegger, in contrast, will not begin with the *result,* because he will not begin with *thought.* Ultimately, he will not begin with *something* (as Leibniz does), but, rather, with *Nothing.* Now as we have pointed out, in beginning with *Nothing,* Heidegger remains within the limits of modernity, or, specifically within the alternatives posed originally in Leibniz' question, by means of which Leibniz desired to be led to the sufficient reason of the universe (which becomes in Hegel the absolute thought of existence, or God). And, although Heidegger quite consciously transforms the question so as to point it in the direction away from thought towards the abyss, although he points to the *Nothing* away from *something* whose contingency demands a sufficient *reason* for its existence, he, nevertheless, in the leap from the *ground* into the *abyss* is borne up, as it were, in the *pure sufficiency* of his identity with Being *in that mutual contingency which is their essence.* In short, as I peer over Heidegger's shoulder as he looks at modern thought, I see that what escapes his notice in his beginning with Nothing as opposed to beginning with something (even God) is the fact that he shares with Hegel (therefore, with modern thought essentially) the *pure beginning with.* Only now, this *to begin with* lacks the content of a universal substantial particularity; it is *to begin with* nonbeing. That Hegel begins with the result signifies something more than Heidegger realizes. It signifies something *not originally alternated with nothing.* Indeed, we first discover this *alternation* or *contrariety* of *something with nothing* in the *Meditations* of Descartes.[386] There it is clear that neither God nor truth are presupposed, both existing merely in humanity's eventuality, in which eventuality existence itself is grounded in doubt. That is, eventually in absolute thought. This latter is a result of

[385] Ibid., 45–46.
[386] Cf. above, pp. 93ff.

the progress that begins with Descartes. Now, we understand that in fact Descartes' *cogito ergo sum* represents a *presuppositionless phenomenon* [387] with which to begin modern science. (This is true despite a merely apparent conflict with Husserl's understanding that Descartes begins with certain 'scholastic' presuppositions, etc.,[388] for which Husserl has the authority of historians of ideas. On the level of this naiveté, neither Husserl, quite consistently with his modernity, nor his authorities are able to perceive the spiritual dimensionality of modern science which *begins with what is not able to be presupposed,* but *begins with it nevertheless,* namely, the *novitas mundi* or created universe of sacred doctrine.) If we understand this *presuppositionless* beginning of modern thought to be the necessity of thought that it is in the event of its own determination to appropriate for itself the universe of contingent existence, then, we realize that thought's identification of God and truth with its own eventuality is the necessity intelligible only if God and truth were of such a nature that it was impossible for them to stand in front of the scientific enterprise because the latter is mankind's self-determination. Then, God and truth would have to be what was made manifest in the appearance of the transcendental essence of existence itself as the historical fact *par excellence,* the shining forth of the *fact* of existence.[389] Otherwise, no cause would necessitate locating God and truth within humanity's eventuality, as is clear, for example, in Aristotle. But, it is clear then that the nothing which alternates with existence in Descartes' *Meditations* is there because it is, in the transcendental form of natural reason (sacred doctrine), that which *immediately goes before* existence. In this event, the event of creation, it is not the *contrary* of something existing, but, rather is its *contradiction,* itself absolutely *before* the fact of existence, unthinkable as it might be, but not to be confused with what it is *thought* to be *after* the fact of existence. Heidegger, then, true to the extenuated modernity of this thinking, fails to perceive the dis-

---

[387] Cf. above, chapter on Descartes.
[388] Cf. above, p. 258.
[389] Cf. above, pp. 193ff.

continuity of the transcendental form of reason with a true metaphysics, and, of modern thought with them both. This is quite clear in his treatment of the nothing of sacred doctrine in his *What Is Metaphysics?:* "Christian dogma . . . denies the truth of the proposition *ex nihilo nihil fit* and gives a twist to the meaning of Nothing, so that it now comes to mean the absolute absence of all 'being' outside of God: *ex nihilo fit—ens creatum:* the created being is made out of nothing. 'Nothing' is now the conceptual opposite of what truly and authentically (*eigentlich*) 'is'; it becomes the *summum ens,* God as *ens increatum.* Here, too, the interpretation of Nothing points to the fundamental concept of what-is. Metaphysical discussion of what-is, however, moves on the same plane as the enquiry into Nothing. In both cases the questions concerning Being (*Sein*) and Nothing as such remain unasked. Hence we need not be worried by the difficulty that if God creates 'out of nothing' he above all must be able to relate himself to Nothing. But if God is God he cannot know Nothing, assuming that the 'Absolute' excludes from itself all nullity (*Nichtigkeit*). This crude historical reminder shows Nothing as the conceptual opposite of what truly and authentically 'is,' i.e. as the negation of it. But once Nothing is somehow made a problem this contrast not only undergoes clearer definition but also arouses the true and authentic metaphysical question regarding the Being of what-is. Nothing ceases to be the vague opposite of what-is: it now reveals itself as integral to the Being of what-is." [390] So Heidegger attributes to sacred doctrine a vagueness in its conception of the Nothing that, as a matter of fact, is more in keeping with modern thought's ingestion of the Nothing into the essence of what-is, than it is consistent with the precision of Thomas Aquinas' understanding of the Nothing, of which he says in the *Summa Theologica* I,45,1 ad 3: "When anything is said to be made from nothing, this preposition *from* (*ex*) does not signify the material cause, but only order; as when we say, *from morning comes midday* [*ex mane fit meridies*]—*i.e.,* after morning is midday. But we must understand that this preposition *from* (*ex*) can comprise the ne-

[390] *Existence and Being,* op. cit., 345–346.

gation implied when I say the word *nothing,* or can be included in it. If taken in the first sense, then we affirm the order by stating the relation between what is now and its previous non-existence. But if the negation includes the preposition, then the order is denied, and the sense is, *It is made from nothing—i.e., it is not made from anything*—as if we were to say, *He speaks of nothing,* because he does not speak of anything. And this is verified in both ways, when it is said, that anything is made from nothing. But in the first way this preposition *from* (*ex*) implies order, as has been said in this reply. In the second sense, it imports the material cause, which is denied."[391] It is evident, then, that the transcendental form of reason introduces a *beginning* of the world, a *novitas mundi,* and, in this, is radically different than Aristotle's conception of the 'eternity' of the universe. Nevertheless, it is only with modern thought that we come upon a *beginning with,* which, as is clear in Hegel, is a *beginning with the* resultant *thought of existence, resultant* in an essential way that escapes the notice of Heidegger. Namely, it is a *thought* of existence *resulting from the introduction of the fact of existence* in the new form of transcendental reason. Modern thought *results* from the essential fact of history; *it leaps back* from the rift in time. It *re-echoes* the fact of existence, that is, this world's contingency, in its *rebounding* upon *itself.* Hegel *begins with the result* simply because modern thought is nothing other than a *result* to begin with. The transcendental form of reason begins, but *neither with something, nor with nothing.* Perhaps we may say, *it begins from above;* its beginning comes after nothing. It does not begin with the world, but the world simply begins. In any event, contrary to Heidegger's perspective on it, what relation the world can be said to have to Nothing (as Aquinas says, it is purely a relation of *order,* not of matter), Nothing is related to *creation,* not to God, in the transcendental form of reason that is sacred doctrine. Indeed, as Thomas Aquinas also points out (*S.T.* I,45,3),[392] creation itself is a *relation,* real in the creature, but in God a relation of reason. Therefore, there is no prece-

[391] *Summa Theologica,* op. cit., 233.
[392] Ibid., 234–235.

dent for *beginning with Nothing* in sacred doctrine. It is first *possible* in the resultant thought of existence which first thinks something in potential alternation with nothing. It is first *actual* in the moment when the universe, *qua* universe, stands in its own nonbeing or finitude in opposition to the reality of the absolute thought of existence, as it does in the thinking of Heidegger, which *begins with* Nothing. This extenuation of modern thought, then, clearly depends upon the *result* reached in Hegel, if only in *self-denial to deny the result,* to refuse to take a thing from the absolute thought apart from its own finitude. In Heidegger's thinking we encounter the *final* result in which man denies himself within the world in essential communion with the Being of this world, by which he enters into that primordial *concern* wherein Man and Being touch each other in their essence.

This *concern* in which Man is embedded with the finite Being of the world is also identified by Heidegger as *Appropriation.* He speaks of this primordial Appropriation in *On Time and Being* as follows: "When we say 'It gives Being,' 'It gives time,' we are not making statements about beings. . . . we . . . must consider the possibility that . . . in saying 'It gives Being,' 'It gives time,' we are not dealing with statements that are always fixed in the sentence structure of the subject-predicate relation. And yet, how else are we to bring the 'It' into view which we say when we say 'It gives Being,' 'It gives time'? Simply by thinking the 'It' in the light of the kind of giving that belongs to it: giving as destiny, giving as an opening up which reaches out. Both belong together, inasmuch as the former, destiny, lies in the latter, extending opening up. In the sending of the destiny of Being, in the extending of time, there becomes manifest a dedication, a delivering over into what is their own, namely of Being as presence and of time as the realm of the open. What determines both, time and Being, in their own, that is, in their belonging together, we shall call: *Ereignis,* the event of Appropriation. . . . One should bear in mind, however, that 'event' is not simply an occurrence, but that which makes any occurrence possible. . . . We now see: What lets the two matters belong together, what brings the two into their own and, even more, maintains and holds them in their belonging together—the way the two matters stand, the matter at stake—is Appropriation. The matter at stake is not a relation retroactively superimposed on Being and time. The matter at stake first appropriates Being and time into their own in virtue of their relation, and does so by the appropriating that is concealed in destiny and in the gift of opening out. . . . (Insofar as the destiny of Being lies in the extending of time, and time, together with Being, lies in Appropriation, Appropriating makes manifest its peculiar property, that Appropriation withdraws what is most fully its own from boundless unconcealment. . . . We catch sight of the other peculiar property in Appropriation as soon as we think clearly enough what has already been said. In Being as presence, there is manifest the concern which concerns us humans in such a way that in perceiving and receiving it we have attained the distinction of human being. Accepting the concern of presence, however, lies in standing within the realm of giving. In this way, four-dimensional true time has reached us. . . . Appropriating has the peculiar property of bringing man into his own as the being who perceives Being by standing within true time. Thus Appropriated, man belongs to Appropriation. This belonging lies in the assimilation that distinguishes Appropriation. By virtue of this assimilation, man is admitted to the Appropriation. . . .). . . . The task of our thinking has been to trace Being to its own from Appropriation—by way of looking through true time without regard to the relation of Being to beings. To think Being without beings means: to think Being without regard to metaphysics."[393] For Heidegger, in the Appropriation the history of Being, that is, the destiny of Being as metaphysics, comes to an end. Man and Being as Appropriated are transformed from the hard edges of their metaphysical shapes as *rational animal* and *finite universe* into *mortals in the fourfold of the world.* In his *The Thing,* Heidegger writes: "Thinging, the thing stays the united four, earth and sky, divinities and mortals, in the simple

[393] M. Heidegger, *On Time and Being* (trans. J. Stambaugh, New York, 1972), 18–24.

onefold of their self-unified fourfold. . . . The mortals are human beings. They are called mortals because they can die. To die means to be capable of death as death. . . . Death is the shrine of Nothing, that is, of that which in every respect is never something that merely exists, but which nevertheless presences, even as the mystery of Being itself. . . . Earth and sky, divinities and mortals—being at one with one another of their own accord—belong together by way of the simpleness of the united fourfold. Each of the four mirrors in its own way the presence of the others. . . . The appropriative mirroring sets each of the four free into its own, but it binds these free ones into the simplicity of their essential being toward one another. . . . None of the four insists on its own separate particularity. . . . This appropriating mirror-play of the simple onefold of earth and sky, divinities and mortals, we call the world. The world presences by worlding. That means: the world's worlding cannot be explained by anything else nor can it be fathomed through anything else. . . . causes and grounds remain unsuitable for the world's worlding. As soon as human cognition here calls for an explanation, it fails to transcend the world's nature, and falls short of it. The human will to explain just does not reach to the simpleness of the simple onefold of worlding. The united four are already strangled in their essential nature when we think of them only as separate realities, which are to be grounded in and explained by one another."[394] With Heidegger, the essential fragmentation of modern reality into the discrete moments of the finite continuum is complete. The third moment: the universe, *qua* universe, standing in its own finitude, beyond the parameters of science, in the primordial Event of Appropriation: the world, *qua* world, gathered together in its unique resource—no longer withheld.

---

[394] M. Heidegger, *Poetry, Language, Thought* (trans. A. Hofstadter, New York, 1971), 178–180.

Section C

EPILOGUE: THE ESSENTIAL ANTICIPATION OF THE FINALITY OF THE FACT

In essence the fact is neither to be affirmed nor to be denied. Nor is the fact in doubt. It is not possible to demonstrate that the fact itself in essence exists; it fact-evidently, in its very essence, triumphs over simple nonexistence. In essence the fact is simply the fact of existence. To ancient, true metaphysics the fact in essence is simply unknown; the known fact is the fact of thought, apart from which there is no fact of existence. The fact of thought's existence is known in the potentiality of matter. (This is *essentially* so with Aristotle; but with the Stoics merely *ideally* so; with the Skeptics only *accidentally* so.) It is presupposed in essence by Neoplatonism. In this presupposition of the fact of existence in thought's essence, Neoplatonism expresses, in the form of pure negativity, its opposition to the appearance of the transcendental essence of existence itself in time. But it is only with that appearance that existence itself is an issue for thought. Therefore Neoplatonism, in its presupposition of the fact in essence, places the fact of existence itself absolutely beyond thought. But it is clear that the Neoplatonic Absolute, the One, is just that: the appearance of the transcendental essence of existence itself denied. This, in essence, is what that formal reconstruction of ancient metaphysics is. It is a denial of the appearance of the fact in essence (not a denial of the *fact,* essentially impossible; but a denial of the *appearance* of the fact), or, a formal denial of the essence of history. But the essence of history is able to be denied only so long as it itself is not a fact for thought. That is, it is able to be denied so long as appearance, *qua* appearance, is something impenetrable to thought and the appearance of the transcendental essence of existence itself is not able to be known to thought in a way comparable to thought's knowledge of the fact of its own existence. Under this same condition, which amounts to the essential incompatibility of knowledge with spatiotemporal conditions, the essence of history is able to be affirmed as not itself a fact for thought, yet issuing in the fact of existence in essence for thought. The thinking that begins with the appearance of the transcendental essence of existence itself affirms the existence of the essence of history, taking to itself as the primary fact of its thinking the fact of existence itself. Thereby it transforms itself for the first time into the transcendental form of reason. The fact of existence is its *formal* presupposition, since, as Neoplatonism clearly demonstrates, this same fact *essentially* presupposed is absolutely beyond thought (involving thought in absolute self-transcendence). Modern thought hypothesizes that the fact of existence belongs to thought apart from its affirmation of the existence of the essence of history. The fact of existence becomes the *material* presupposition of thought or the thought of existence. Modern thought, then, does not deny the essence of history as Neoplatonism does. Rather modernity appropriates the essence of history as indistinguishable from its own thought of existence. It banks upon its identity in thought with this world's apparent essence to avail itself of a formally infinite power. Essentially inhibited by that absolute weight that is its own abstraction or world-determination, it breaks down internally into moments of self-opposition which, in the absence of the knowledge of the essential history of thought, but faced with the evident vacuity of the modern enterprise of power, withdraw from the radically hypothetical thought of existence to the inapparent essences of their own discrete modes of Being, beyond the specificity of thought or history, but not beyond the hypothetically predictable limits of opposition. Indeed, the essential hypothesis that modern thought is proves itself to be correct; that is, within the limits of its own ambition it is true enough. It is so much so that, although knowledge of the essential history of thought shows modern thought to be mistaken in its failure essentially to recognize another upon whom its identity with the apparent essence depended (an essentially impossible recognition for modern thought), it continues to be the case that every form of *opposition* to modernity is to have imputed to it a share in its essence as fashioned by Being for its own purposes. There is here no turning back. It is not of the essence of the fact. In this new thinking now occurring for the first time in history (*novitas mentis*), the essence of which is the absolutely evident appearance of the transcendental essence of existence itself in its effect in thought, a *new* fact appears, history itself, in terms of which it is for the first time demonstrable that this world exists in fact. So, far from denying to modern thought its *material* presupposition of the fact of existence, it is demonstrably evident that it belongs to it as the issue of history converted to power's essence: the thought of existence. At the same time it is of the essence of this demonstration that this thinking is essentially discontinuous with past thinking, that in it history itself now appears as the new fact: this fact's existence is no longer at issue. It is therefore *no longer necessary to presuppose the fact of existence.* Neither essentially, is it necessary to deny history, nor formally to affirm history, nor materially to appropriate history. The presupposed fact of existence of previous thought is, in the inversion that is transcendental historical thinking, shown to be simply the fact of this world's existence reflected by different points of view in time, each *essentially* unhistorical, though each with its own peculiar attitude toward history, due to the fact that the essence of history was evident to none prior to this thinking now occurring. This thinking now existing clearly perceives that modernity in its appropriation of history eliminated absolutely every material-formal distinction separating its thought of existence from the essence of history. It identified history with its own point of view or the point of view in essence. It demonstrates thereby that, in terms of its own will to power, in terms of this world's apparent essence, nothing but history's essence accounts for the fact that it itself does not occur before now to thought, that is, nothing but the essential incompatibility of *to einai katholikon* with the point of view in essence, since, as a matter of fact, modernity's clarification of its absolute truth (effected within the limits of self-opposition) has taken away every substantial barrier to the intelligibility of the appearance of the transcendental essence of existence itself in thought at this time. In light of the *novitas mentis,* the new fact, this is the *irreversible historical function* of modern thought: the *preparation absolutely* of the necessary state of affairs for the *advent of history itself in thinking:* what now in fact occurs in thought. *The world is introduced in essence to the fact of its historical existence.* There is now in fact no staying behind in the forms of past thought. There is no possibility of updating in fact past modes of apprehending this world's reality in thought, that thought essentially unhistorical, an opacity, if not to the fact of existence, to the historical essence of existence itself. Nowhere in past thought, most certainly not in modern thought (whose very essence is thinking-in-the-past), is to be found this transparency now occurrring to this world's essence, in which *novitas mentis* history's being in thought as the appearance of the transcendental essence of existence itself manifests the historical structure of being as *occurring truth,* a structure erected upon being essentially indifferent to thinking (not Being beyond the thought of this world, but): ἀλήθεια, *loving of truth,* the essential fact of being disclosed in the essence of history now manifest in thought itself. We too easily fall back into past thinking's habit, if we understand *even this* being essentially indifferent to thinking (*alētheia*) as in itself absolute. *The essential fact of being is indifferent to the frame of reference absolute-relative.* The historical structure of being is erected without reference to this framework, the latter belonging essentially to the point of view in time. Indeed, so perfectly indifferent to the frame of reference is the essential fact of being that it is indifferently the historical structure in the exaltation of its being ἀγάπη, *truly loving.* This exaltation of the essential fact of being, its historical structure, is *love's revelation in truth,* ἀποκάλυψις, in which structure the essential fact of being abides indifferently to its manifestation in the appearance of the transcendental essence of existence itself as truth occurring before now in time, now as the occurring truth or essence of history absolutely evident in its effect in thought. The revelatory, historical structure of being (*apokalypsis*) manifests that the essential fact is indifferently love (*agapē*)—truth (*alētheia*). The abiding indifference of the historical structure of being (*apokalypsis*) to truth occurring before now in time—now occurring in thought is not that in fact existence itself has not appeared in the transcendental essence. (Indeed the transcendental essence is the essence of the fact of being in its exaltation.) It is to say that the historical structure of being is not a determination of the appearance of its essence (no more than the fact of existence is a determination of this world's essence), but rather that the appearance terminates in existence itself as the essence of this world's existence in fact, that is, as the fact of creation. In the beginning the fact of creation appeared to thought reflecting on the fact of nature as matter for a thinking itself quite natural (where this fact appears not, there matter is potentially the fact of thought's existence beyond nature). In time the fact of creation appeared to thought trusting in the man's historical identity as the formal presupposition of a thinking itself thoroughly conscious of its dependence on this trust (where, then, the fact of creation appears not, the form of thought is actually an identification with history's appearance in the form of this world). In this thinking now occurring the fact of creation appears in the absolute evidence of the essence of history, that is, of the appearance of the transcendental essence of existence itself, as essentially demonstrable. Here, now, the fact itself in essence appears. There is now no alternative to thinking the fact of creation: the *novitas mentis* is the end of the mystery in thought. This thinking now occurring is without resource, without Being beyond itself. It knows transcendentally that beyond itself every-thing exists in its historical essence (*to einai katholikon*). It knows that beyond this fact of being is the exaltation of the essential fact, its manifestation terminating in this fact of existence. It knows that beyond itself the fact itself in essence appears in its utter finality anticipated essentially in this thinking now occurring for the first time in history, but in knowing this it is not beyond itself.

Nothing of the essential fact of being is hidden from the *novitas mentis* in its inspection of its historical essence. Beyond its knowledge there is only the historical structure of being in its essential indifference to the manifest appearance of the essence, in which revelatory structure of being *truly loving of truth* (*apokalypsis*) abides the fact in its utter finality, essentially indifferent to thinking. This thinking now occurring is utterly impoverished of every perspective beyond the fact in essence of the historical, that is, demonstrable existence of this world. But in essence the fact is no perspective. This is the wealth of transcendental historical science: Taken up with the transcendental essence of existence itself, it accounts for every-thing's existence (*ho logos katholikos*) without in the least discounting what every-thing is in itself. It is of the essence of history that the *novitas mentis* reaps where it itself has not scattered. In this thinking now newly existing thought appears to itself a moment to have risked every-thing. In fact in essence it has ventured nothing. Not a moment apart from the existent fact of its essence, it comprehends the promise of existence itself in the essential fact of being's indifference to thinking, manifest in the essentially new fact that it exists in radical *intellectual* discontinuity with modernity's irreversibly absolute truth. Venturing nothing in essence it gathers, in the fact of history, the promise of every-thing's existence, indeed, in essence the promise of existence itself, a transcendent occupation immanently beginning at this time. What is evident to the *novitas mentis* in its inspection of the appearance of the transcendental essence of existence itself now in thought is that the *essential fact of being*—essentially indifferent to time—to thinking—to history itself—is *the promise kept.* Upon this essential fact of being—*the promise is kept*—is erected indifferently, that is, no less-no more the fact, being's historical structure (*truly loving of truth: apokalypsis*), the exaltation of the kept promise. The disclosure of the essential fact is the kept promise open through the essence of the historical fact, which appeared before now in time, but now appears manifestly in an essentially new thinking. The essence of history (*to einai katholikon*) is manifestly the *promise kept open,* now in thought, to every-thing, to this world. The transcendentally perceived utter finality of the fact is approached in the recognition that the essential fact has never been anything but open. But now in this new thinking, every reflective distance overcome in essence, thought comprehends this fact in its existence. Every barrier of time or space is taken away in modernity's essentially *resulting* thought or Being beyond thought. Taken up with the essence of history, we stand face to face with existence itself in thought: an objectivity essentially indifferent to every unhistorical perimeter of essential Being or Being beyond essence which rationalizes the object's schematic inconsistency as the difference attributable in essence to the point of view, which therefore betrays its essential incomprehension of the historical essence of the object itself. If transcendental historical thinking perceives every object's intelligibility to be its historical essence, then it comprehends in essence that absolutely objective science (science of the fact of existence) is life-science. It knows nothing of death or inertia, except that they belong to the radical inconsistency of the schema of past thinking, in which, adhering to difference itself, life perishes in the contradiction of its own subjectivity. Indeed, absolutely so, in the event of Nothing. Here, now, in thought, the promise of existence is fulfilled; only the existence of the Nothing (the factuality of which is demonstrated in the essential history of thought) separates this thinking now occurring from the utter finality of the judgment of the fact itself. But this Nothing is not to be eliminated by the contemplative judgment of the *novitas mentis.* It silently testifies to the essential priority of the transcendental thought of existence itself to previous forms of thinking. Now there is a new understanding of an essentially new fact in its essence, not only a new form of thinking (as in Aquinas), nor just a new matter for thinking (as in Descartes), but a new essence: a new thinking radically conscious of its own historicity together with everything comprehended in the historical essence of the fact. In comprehension of the fact in essence, choice disappears. Actually it disappears in the moment before history itself appears in this thinking now occurring. If it does not disappear, choice is confronted with Being itself, or it has its freedom as Nothing. But the fact in essence is no sooner seen than decided upon. This is the essential identity of historical existence. In recognition of the essential indifference of the revelatory, historical structure of being to time, this thinking now occurring in essence *leaves the future behind,* as that mode of reckoning history belonging to the past. The authority for this indifference comes to the *novitas mentis* in the inspection of its own essence. Therein it discovers that before now it was not to be: Neither, in terms of its own historical essence, to be sooner than it now is in fact, nor, in terms of the essence of thinking-in-the-past, to be at all. Here is the *novitas mundi,* the new state of the world now occurring for the first time in thought, transcendently comprehended in its historical essence (*to einai katholikon*) by a thinking transcendentally anticipating the essential fact of history itself; to which manifestation of the *kept promise* thinking is essentially nearer than, before now, it ever was.

Section D
APPENDICES

Appendix α

THE REALITY OF TRANSCENDENTAL HISTORICAL THINKING

> Finally we shall not seek the reason for natural things from the end which God or nature has set before him in their creation; for we should not take so much upon ourselves as to believe that God could take us into His counsels.
>
> Descartes, *Principles of Philosophy* I.28

> It was the stone rejected by the builders that proved to be the keystone; this is Yahweh's doing and it is wonderful to see. This is the day made memorable by Yahweh, what immense joy for us!
>
> *Psalms* 118:22–24

To that thinking essentially in the past which took it upon itself to be, in effect, at the foundation of the universe, to it, the knowledge of God, and consequently, of the end of creation, was what in the nature of things remained to be seen. Modern thought saw not what remained to be seen but the thing itself, in its essence. That is, that thinking saw the *thing itself* existing essentially prior to the *conception* of its nature, just as it essentially saw itself existing prior to its conception of God. That thinking, which constructed the universe now about to be left behind, simply undertook what was sufficient for its purposes (in this consists its modernity), setting aside as essentially inappropriate, and correctly, too, in the form of its conception of

God or nature, *what remained to be seen.* But to conceive of the universe now about to be left behind is to have been ushered into the perception that *what remains to be seen* exists, now, for the first time, in place of the thing itself. It is to see that the end of creation exists now not in the form of past conceptions of God or nature, but now in the form of the appearance itself, in and for thought, of what is required by thought, namely, a complete change in being, which, in the very essence of appearance itself, is now seen to have occurred prior to thought, not, therefore, as a result of its being what thought now requires. It is seen to be, in its effect, the appearance of the essence of transcendental existence itself for the first time in and for thought perfectly. Therefore, neither is it seen to be a result of its own being prior to thought. Its priority to thought is seen in the sheer immediacy of its appearance in place of the thing itself as what has occurred in fact. The appearance itself is what essentially remains to be seen to have occurred not before now.

The appearance itself of the essence of transcendental existence, that is, existence itself essentially now, is the history of its essence now in thought for the first time. It is the fact-evidence that its essence exists prior to its coming to be in the course of time. The essentially new form of *time itself,* in its now-evident priority to the indifferent matter of previous change, is the consciousness of the radical transformation of the matter of past thought into an essentially new matter. This transformation is effected by the appearance itself of that *essentially tautologous* thinking now occurring. The essence of this thinking is the perfectly unnecessary repetition of existence itself, therefore, of the appearance itself, as thought's object. The appearance itself is the perfectly new matter of thought now existing for the first time. This is *an essential change in the very foundation* of things, brought about subsequently to thought's erection of the universe in time. This is *an essential modification of the structure* of things, brought about subsequently to thought's radically self-critical discovery of the essentially incomplete nature of the product of its own enterprise. This is *an essential transformation of the totality* of things, the universe itself, therefore, into being historical, a transformation brought about by thought's perception of the fact that not before now, in the very essence of things, was its completion in sight, but that now, in full view of thought, appears the end of creation in the form of existence itself immediately intelligible. This new immediacy is the new matter or the essence effectively of history itself, that is, of the essentially tautologous change of the conditions of existence themselves. The matter of the past was effectively the essence of the thing itself, or the merely *formal* tautology that constituted that thinking comprehensively as *appropriation* of the conditions of existence (so that while appearance was understood to be the ground of thoughtful existence, in the event of disappearance Nothing remained unchanged—nothing remained to be seen). Now, however, not the thing but its history is the stay of the world, except that there is no staying of the world. Behind the essentially tautologous change of the very conditions of existence is *the revelatory structure of being itself.* Not the thing, but its history is now seen to be the resolution of the contradiction of existence, except that there is no resolution (there is Nothing), but there is history's contradiction of the essence of the thing: the perfect resistance of history itself to that attempt of the thing itself in the infinite form of past thought (or in the finitude of its own Being) to impose itself upon being. History's resistance is accomplished, in essence, in that thinking now occurring, effortlessly beholding the fact that the contradiction belongs to the appearance of the *simple* essence of the thing, that is, belongs to the fact that the thing *itself* in the *simplicity of its construction* lacks *essentially tautologous* being, that is, not to the fact that the thing itself is constructed, but to the fact that it is *not essentially constructed,* that is, *in existence itself.* Insofar as past thinking recognized its own essential limitations, it attempted to construct its own essence, but not essentially, or in existence itself, but rather, *formally,* by reconstruction. Thereby it remained within the essential circularity of its own determinations, seeing not the appearance itself but *mere phenomena.* In this fact of its constitution consists the essential blindness of the phenomenological enterprise to its historicity.

The object of transcendental historical thinking is the thing itself transformed into being essentially constructed. In this thinking now occurring the essential circularity of the thinking of the past is ruptured. The thing itself is seen to terminate in existence itself in and for thought, to be essentially historical *qua* object. Not that the thing itself *has* a history (as if it should itself be an individual, a species, or a world, or, indeed, thought itself), but that now the thing itself *is* its history. Thus, the object is no longer simply *what is new* (the formal construction of that thinking whose essence it is to be in the past), but the object of this essentially new thinking is the essential construction, *what is now new,* wherein is seen for the first time in history the *new form* of the thing, *qua* new. Now-transformed *what is new* is seen to be *simply the appearance* of the substantial change of the object effected by history itself without mediation in essence, or without either the mediation of *time itself* (seen now for the first time in its tautologously essential distinction from appearance itself and therefore in contradiction to its necessary being in the self-objectification of the essence of past thought), or the mediation of *distance* (seen now for the first time in its tautologously essential identity with time itself in its distinction from the appearance). The truth is no longer *the truth of appearances* within the horizon of which modern man found it not only possible, but, at the same time, necessary to be the subject of a merely transcendental knowledge of existence in one or another form of reflection (either in essence or in existence), or, in the event of Nothing (reflection's end in non-existence), found the necessity of existence to be self-contradictory in one or another form of finitude (either the suspension of thought in essence or of the will in existence). The truth is now *the truth of history,* or *time itself,* essentially without horizon, wherein, for the first time, appears a transcendent object for thought itself in sheer immediacy. Time itself now being its essential form (there being for an essentially tautologous existence no distance apart from the time of its appearance), the self-limiting forms of the contradiction embodied in the thing of the past no longer constitute thought's horizon. The change of the object is no longer bound to the entirely formal indifference of the space-time continuum. The essentially tautologous identity of time itself with distance is time's no longer being able to be measured by any frame of reference or distance-analogy. Indeed, being essentially distance, it is, as such, dis-analogous to whatever was previously thought about time or in time. Essentially, therefore, the measure of time itself is the object of transcendental historical thinking. That is, the measure of time itself is the change itself of the object essentially prior to time itself. Or, the measure is changing substance, which, in its appearance, is essentially distinguished from time itself as *pure motion.* In this thinking now occurring for the first time in history, *the appearance itself* is essentially constructed; that is, *its very conception in thought is its existence in fact.* Therefore *the appearance itself* is now for the first time conceived with perfect clarity as the essentially tautologous *identity in existence of elemental concepts* now for the first time perfectly distinct: motion *itself,* time *itself,* change *itself.* This perfect conception of the appearance itself is *perception.* The *perfection of thought itself* is its perception of the change in essence prior to the motion now, its perception that the object now beheld in its essence exists prior to time. *The thing is not thought's contemporary.* The reality of the thing is what transcendental historical science contemplates as being its fully appearing essence, not at all its unreality but its being essentially constructed in existence. The reality of the thing is its very appearance. It is to reality that imperfection is essentially conceived to belong, but not to thought itself, which, in contemplative judgment, perceives that what now appears is what before now it was not. That is, thought perceives perfectly its object to be *what is now new:* a complete change in essence. The reality of thought itself is its perfect perception of the fact that before now it itself did not exist in essence. Rather, it was *essentially reflection* upon the thing itself, beginning with the apparent essence of what was to be thought. The object of this essentially new science is radically unpredictable or pure motion in its essential appearance. Conceived to be now, it is what remains to be seen in consequence of a complete change in being. It radically remains to be seen what it is, so essentially new is it. In the perfectly clear conception of appearance itself, *motion* is perfectly distinguished as the *measure of change, potentially,* while *time* is *actually* the *measure of change itself, consequently, of pure motion.* Transcendental historical science distinguishes itself from its object by understanding that *motion* properly belongs to *the object,* while *time* belongs to essentially tautologous *thought.* Time itself, so conceived, is seen, in its perfectly unnecessary repetition, to belong essentially to history itself, but formally to the object of history as motion; as that transformation through which the merely apparent essence is realized in thought. Nothing is historical by virtue of its being in time alone, but by virtue of a change in its being now comprehended essentially. Nor is something historical by virtue of the fact that it simply moves, as, for example, modern thought takes itself to be. The latter takes itself to be the historical object *par excellence.* It involves itself, therefore, in that essential contradiction, its characteristic mark, that as its own object it knows itself to belong to the past. It takes itself as essentially in-the-past. (This is a contradiction which may well be life before now, or life formally considered, but which is clearly not history. It is, furthermore, a contradiction not essentially resolved by the suspension of the question of self-relation in the pure form of time.) But something is historical if its motion is seen to be a measure of a change in its being now comprehended essentially. (This, indeed, has now happened to modern thought in the comprehension of the history of being that actualizes the essential history of thought as knowledge, radically incorporating modern thought in that history as the irreversible self-conscious appropriation of the merely apparent essence of change. It is, then, only now that modern thought is historical, not in its essence [as an *essentially* tautologous thinking is], but *formally* historical. It is essentially incapable of rising to the perception that the object, *qua* historical, is not in essence identically thought, but *is* what appears to have occurred to thought now perfectly conceived.) The essence of history is the appearance itself, the new matter of thought now existing. What appears to have occurred to this thinking now occurring *is* what has occurred (the appearance itself is the essence of transcendental *existence*) perfectly repeating itself in thought—its new matter: *The essential identity of transcendental historical science belongs to its object.* In and of itself this thinking now occurring for the first time lacks identity.

Mere transcendental subjectivity taken up with the appearance itself is raised to being actually transcendental objectivity. In and of itself this thinking now occurring does not exist; it belongs to the past as essentially inconceivable. This is its transcendence in fact to the past essence of thought. Its essence is an objectivity *beyond imagination. This is the transparency of thought itself to the distinction in existence of its object, in which distinction its own existence consists formally.*

The perfect objectivity of thought's essentially constructing the appearance itself is not to be confused with that thinking which, after a lapse of time, identifies the truth with what has occurred to thought, evidencing thereby its essential distance from truth, its priority to truth. The truth of transcendental historical science is not truth apart from its now occurring without reference to what has occurred. This sheer immediacy of perception in an essentially dis-analogous moment is *knowledge, the identity of appearance itself with reality* (what has occurred in essence—the fact). Thought is now no longer, as it was in the past, an essentially redundant (merely *formally* tautologous) judgment upon the truth of appearances, an absolute intuition. The thinking now occurring is direct contemplation of the reality of the appearance itself, the form itself of this judgment (*essentially* tautologous) being the truth of history. It is not a judgment upon truth, but true judgment itself. In the very essence of the reality now occurring in and for thought, only historical judgment is true judgment. Essentially false judgment is the identification of appearance with reality in which the judgment's Being is distanced from its so-called occurrence or event by extrinsic time (transcendental idealism), or by intrinsic time in the pure form of the finitude of thought itself (transcendental realism). But truth is neither an *ideal* form of the appearance in judgment (in which event there is a series of *relatively* true statements about reality), nor is truth the *reality* appearing in the form of the judgment (in which event there is a series of relatively *real* truths transcendent to the judgment's Being). Truth itself is the truth of history; essentially it is judgment itself, historical if true. Transcendental historical thinking faces the fact that what appears now to have occurred is what has occurred. It faces squarely up to the fact that *the universe, constructed by thought before now, now exists* in and for thought. *The truth is that thinking now conceives itself to exist in transcendence to the essentially complete construction of its own Being.* It now appears that thinking itself is something more than it ever appeared to itself to be before now: *the constructor, in essence, of things that exist beyond thought left behind in the form of time itself.* It appears now to thought that it wields an infinite responsibility. Its every judgment is a *world* in existence. The question of truth appears to be the question of the being of every-thing. Thought is now transformed in its essence into being the judgment of history, the maker of what before now remained to be seen. Thought now makes what remains to be seen to have occurred. Through its perfect perception of the reality of the appearance itself it shares in essence the finality of the fact it faces. The word it utters is its final word. The thinking now occurring for the first time in history is the thoroughgoing consciousness of the fact that now there is no room for error, that what is at stake is not its own Being or the Being of the world but truth itself. Truth itself is the transcendental essence of existence itself appearing now essentially in thought's judgment upon the reality of past thought, completing it in the form of time itself or the truth of history. Transcendental historical thinking shirks not the heavy weight of this responsibility to which, in essence, it is more than equal. Truth itself is in the balance. Nothing accrues to truth. But truth itself is the dispensation of existence itself to everything otherwise bound in the self-contradictory schema of a merely formal tautology. It perfectly liberates every-thing from the conception of its own weight. (This conception of its own weight is true enough within the horizon of Being fashioned for its own purposes, within the circle of past thought's understanding of the transcendence of difference itself to its own identity, to its being in essence the-same-as-past-thinking.) Everything remaining essentially unchanged in the conception of its matter, the judgment of past thinking is distant, in essence, from its event. It survives it, as it were, living upon its interest in a world of its own making, or, in the event of radical self-criticism, its interest in its own constitution of essence. Invariably what is encountered in these false forms of judgment is but a *form* of appearance (either *substantial* or *phenomenal.*) The changeability of this form is merely a matter of *experience,* or a matter of *pure reflection* prior to experience. *But in neither event is it a matter of thought itself changing in essence in its judgment upon the reality of the appearance itself,* that is, of thought itself essentially changed by its experience of something beyond itself (encountering the essence itself of existence). False judgment encounters, then, merely a form of appearance, not appearance itself. That is, it encounters not what, in essence, remains to be seen now, but what is there to be seen in the first place—an idea, a sign, a symbol, etc.—or what is there to be seen *to begin with.* Nor is the essential falsity of the form of appearance changed by its being known to be constructed by thought itself, since the construction *to-begin-with* essentially precludes the *realization of the object itself in existence* and because the thinking essentially belonging to the past does not comprehend the fact that, in the perfect dynamism of truth itself, true judgment is essentially transformative of thought itself. Instead, false judgment, essentially *apprehensive,* or self-relational (no less so in the form of self-denial than it is in the form of self-assertion), apprehends its object; it takes it to itself too soon, that is, essentially before now. In so doing it masks the manifest evidence of the appearance itself with a semblance of identity or, it may be, a dissemblance of perception. The appearance itself of what has occurred is obscured either through self-mockery (the agreement in appearance) or through a different perception of nothing (the disagreement in appearance). In the perfect conception of appearance itself thought takes nothing to itself, therefore, nothing too soon (before now). Radically unapprehensive, transcendental historical perception is not at all a matter of agreement or disagreement. But, in light of the clear perception of appearance itself, the truth of false judgment is silence.

The radical inconsistency of the schematic identity of the object of past thought is manifest at that point where it recognizes itself as changing its object in the very act of apprehending it. In the instrumentality that is of the very essence of the modern enterprise of power (the essential distance of thought from its event), precision varies inversely with accuracy. To the degree to which the appearing distance of apprehension is overcome, to that degree the framework of apprehension is obscured, and vice versa. Past thought's conception of the merely apparent essence is its extension of its object's appearance beyond itself in essence. *The change in the object apprehended by past thought is not, therefore, a change in essence, but it is a function of the object's exteriority to thought itself, or, of past thought's essential inability to conceive of the appearance itself.* The subject-object reciprocity of theoretical/(so-called) empirical science is a mere surface of modern thought's self-contradictory essence, its *Being in itself before now,* its Being *a priori* in its conception of a universe of its own making, therefore, inhibited, in essence, in its approach to its object. The tautology of past thought is merely formal. What the reality of its object is has not occurred to it. Therefore, it is synthetic *a priori* conceptualization (no less so in its form of radical naiveté: positivism) of an object the reality of which is a *determination* of thought, but essentially not what remains to be seen. For that thinking now occurring for the first time in history the reality of the object is the appearance itself. That is, now, for the first time, there exists a conception of perception itself as the reality beyond the merely apparent essence of past thought, as the essential construction of the object by thought itself perfectly *distinct in existence.* Truth itself is no longer a matter of perception. It is the essential synthesis of the fact. It is the new essence of thought now occurring, transforming thought into the pure activity of contemplative judgment. That is, it is historical judgment or judgment now in essence after the fact. It is not judgment after the fact in time, for *time itself* is now seen to be the truth of history, or, the form itself of judgment. As this essentially tautologous thinking sees it, modern thought or thinking-in-the-past is *afterthought,* not after-the-fact in essence but *in time.* Radically unable to comprehend the fact in essence, thinking-in-the-past reflects upon what it takes to be truth's merely apparent essence in time. Thinking before now is not essentially synthetic, but it is *formally synthetic, essentially recollective.* As such, modern thought is essentially oblivious of the history of being. It knows that history merely formally, either *infinitely, in the idea,* as the absolutely constructive activity of thought itself, or *finitely, in the nothingness* of the idea as the form of the history of Being beyond itself. For modern thought the form of history is its idea or it is nothing. The object (which in essentially tautologous thinking is, *qua* object, *pure motion*) is, in essentially recollective thinking, a motion accruing time in the form of its past position. That is, it is a moving mass in the form of a series of events, or a mass moving with a certain direction. This direction is essentially *internal.* That is, time is the apparent essence of the object remaining *the same* within a precise specification of its mass. *Time changes* together with its mass *within a framework* of other moving objects, that is, by being included within a larger field of energy, wherein it itself is an essentially *inaccurate* measure of the times of other things. There is a difference between times, measured only by accruing time, by a mass moving in a certain direction, so that this difference is overcome only in the form of the accumulation of past time. But since the apparent essence of the object remains, in itself, the same, it accrues this past time only inessentially, or, outside of itself. This exteriority of temporal change to the essence of the object of past thought is this recollective thinking's self-restriction whereby it constructs its object as mere appearance or pure form. Within the essential circularity of this synthesis it judges the truth of appearances to be essentially relative. The unlimited accumulation of past time is impossible. There is a definite limit (the velocity of light) with respect to which every motion is equidistant. The object is precluded from a change in being essentially by past thinking's premature substitution of a *framework* of motion for *motion itself* (that is, essentially tautologous motion). *Through the framework motion acquires time and direction;* through the framework motion acquires identity in the form of *being-the-same as-in-the-past. Change* remains a difference beyond identity in the direction of other moving objects, if it is not, in the essence of apprehension itself, a function of the appearance of essence to thought. In neither event is there change in essence, since in thinking-in-the-past there is no conception of *change itself.* But change before now is an inessential transformation of matter wherein identity recedes essentially into the past. This matter is in a constant state of transformation acquiring new identity indefinitely, but definitely not an essentially new identity, such as occurs to matter now for the first time in the form of an essentially new thought. The acquisition of new identity by the matter of recollective thought is at the same time retrieval of identity in the past. Otherwise, there is death, by which identity comes so thoroughly under the domination of change as to vanish under a highly accurate perception of its essential mutability, that is, its being *something that moves.* Total perception, in the thinking that belongs to the past, essentially is the *disappearance of identity in essence* or the convertibility of matter to energy within the framework of a merely *formal* tautology, so that energy remains necessarily an ideal essence of things, the immediate perception of which requires the formal repetition of time. *Total perception is the ideal of a thinking-in-the-past, founded upon the mistaken notion of a merely apparent essence in the form of something that moves.* This is the common point of departure of those two fundamentally opposed perceptions of the reality of the universe (relativity/quantum mechanics, each, on its own scale, conscious of an unachievable simplicity in its conception of its object) by which modern thought is brought face to face with itself, with the self-contradictory essence of a merely apparent or formally tautologous existence. In the self-objectification of recollective thinking motion involves time necessarily. Time is *included* in the essence of motion as a *thing that moves.* Time is a *clock* or a thing changing through formal repetition at a constant rate in relation to other things. To move absolutely, then, is to cease to exist, to be completely in the past, to be Nothing, to have brought the totality of time to stillness, to be included within the absolutely essential limitation of the world of the thing, absolutely involved in a world without measure. But such an absolute motion is absolutely impossible for past thought to conceive. It is the very essence of nonexistence, from which *thinking* in the past preserves itself by its being in *two* places at one time, in *two* things at once. That is, it is essentially, in thought (itself) *and* in reality (the thing). This is a 'simultaneity' in essence purely formal, an existence as *unrealizable* in past thought as its continuing to be thought in the event of its nonexistence. (The merely apparent conception of the essentially inconceivable state of absolute relativity is poetry.) The object as the absolutely formal appearance is the locus of that contradiction by which thinking-in-the-past denies to reality its own essence in an absolute form in order to preserve itself from nonexistence. The essential nonexistence of the object in itself is the condition of its intelligibility, in which it is clearly perceived to exist in distinction from other things but not in distinction from thought itself.

The essential conception of perception itself exists in that thinking now occurring for the first time in history. Before now, perception itself is absolutely inconceivable. Now, in fact, the absolutely inconceivable is conceived, but, therefore, not as mere appearance, nor as merely apparent conception, but as the appearance itself in and for thought of the transcendental essence of existence. In this light the absolute itself is no longer the merely apparent essence of what has occurred, an essentially self-constricted, purely formal construction of reality. Modern thought is unable to conceive a change that is not a motion in essence. It knows nothing of the essential priority of change itself to motion now occurring in thought. Therefore, change appears to be essentially a motion before now involving identity with something in time. *Change* is known to thinking-in-the-past, but known *formally;* not *change itself,* but some-*thing* is seen to have changed. What *it* is remains to be seen, or, *clearly perceived it disappears, but it disappears* (as it does in the event of the clarification of modern thought effected in the forms of its extenuation corresponding to the essential moments of thought) *in a conception whose essence is itself mere appearance.* Such is the essential circularity of a *formally* tautologous perception that its clarification of the thing itself is seen to be a form of its being in time. That it is a fact that the conceptions apparently clarifying what has occurred in modern thought are in essence temporal points to the fact that a radical criticism of the absolute pretension of Being, prior to the appearance itself of transcendental historical thinking in time, cannot be *intellectual* criticism in essence. Criticism, before now, is, in time, a transcendent passion to exist, or, a preoccupation with transcendental essence, or, in the last resort, poetic retrieval of the simplicity of the universe. For the individual, in its resort, time is an absolute matter of consciousness suffering the contradiction of not-being in itself. This absolutely material consciousness, essentially incapable of terminating in existence, is transcendentally preoccupied in repeating itself in the only form it knows. Shaping its existence in accord with its idea of itself, it *transforms* absolutely the motion of thought into nothingness, bringing about the end of reflection in thought itself. This is the relativity of thought itself in substance. For the species, time is an absolute form of that subjectivity which dispenses, in essence, with thought's substantiality, resorting to a pure phenomenology, to the formal relativity of the thinking that constitutes its own essence. In this resort the absolute form of time *eclipses* the motion of thought; it becomes the locus of the contradiction in the shape of the invariable essence: its intentional object. For the universe, in the final resort, time is the absolute essence of its *Being beyond* motion in unpretentious stillness, the end of contradiction. In this merely apparent conception, time is thought to be the resort of Being from contradiction (not the end of reflection in essence [phenomenology], but its end in nonexistence). In the final resort, time ceases to be in itself together with existence. *There remains the interlude between times; in this interlude a formally tautologous thought is not being itself, but,* due to the essentially coincident disappearance of time in essence in this merely apparent attempt to see what it is that has occurred, *is not being beyond the difference.* This *transcendent difference* is the evidence that, in the final resort of the universe to Being beyond motion, thought, essentially belonging to the past, *merely appears to disappear in essence.* Under cover of the perfectly transparent, or, essentially inconceivable, form of absolute motion (where no one before now would think to look for it) it remains, but not to be seen, essentially true to form, to the inconceivability of change without motion. It remains inconspicuously in the form of the essence into which time has disappeared (the form, identically, of absolute motion): the difference (being itself) beyond which is no/thing or the thing luminous in its absolutely unreflective nonexistence. The interlude is the difference between epochs being itself. *It is the absolute retention of direction in the epochal pause,* the stoppage itself in the passage of time, the blockage of thought in the form of its inapparent essence, coincidentally that of motion and time. In the final resort, the universe is suspended in essence; it contracts absolutely to not being beyond the transcendent difference of this worldly existence in order to prevent its self-destruction, *to arrive ahead of time* by a different route. In the *stoppage* of time, bound up, as it is, in past thought's inapparent essence with motion, looms *the passage itself* as the absolute essence of finitude, the passage *indifferently,* being *the same* to time or thought, *the same* apart from time's involvement with motion in thought's inapparent essence, *the same* in which, ahead of time, the *self,* the self-contradictory essence of past thought, is dismantled in the grip of an absolute apprehension of its own end as the direction in which it is essentially moving. But then *Nothing is something that has changed;* its appearance is essentially motivated by the dreadful experience of a world of its own making on the part of something no longer visible, but in essence absolutely *the same being.* Such is the elaborate subterfuge of self-denial to which thinking-in-the-past resorts when brought face to face with itself in time. The index of the absolute absence of the concept of *change itself* is that, in the merely apparent conception of Being beyond itself, in essence the difference remains *transcendently* the same. That is, there is no essentially tautologous difference (essentially indifferent to absolute motion), but difference changes from some-one-thing to being no-other-thing (the same). The difference does not remain *the same transcendentally,* or, *an essential construction,* seen now for the first time, but, the fool of time in thinking before now (essentially), difference *remains* the same *by becoming* the same. It remains, in the essentially redundant formality of past thought, by changing. *Change* (formally constructed as difference) *changes.* This is the essential impasse to which past thought comes in its being merely apparently the thought of what has occurred—*change itself,* now seen for the first time in history. The essentially redundant formality of modern thought is pervasively evident as the *meaning* of things, or, in the event of self-denial, the *belonging* of everything to Being. In that thinking now occurring for the first time in history everything is conceived *without meaning,* that is, essentially conceived, or, conceived *in existence.* Every-thing in transcendental historical thinking exists essentially *without belonging* (either to Being or to itself, therefore), essentially because existence is no longer conceived as a thing of the past that itself has come into being by changing. But, now, for the first time, thought sees the appearance itself of the transcendental essence of existence. In this perception itself of the *novitas mentis* it sees *change itself* for the first time, not *in time* (a redundancy by which it would be returned to the essence of past thought), but *in history,* in an occurrence that *brings time into existence for the first time,* that makes time *essentially* new, that makes it *time itself,* or *now,* the form of an *essentially new appearance. This is the essence of an essentially new universe,* seen now for the first time *in thought,* the essence of which is the essence of history (a thought which, therefore, is identically what it says has occurred to it in the course of time). Change itself is the fully apparent essence of the construction of existence itself, what now essentially appears to have occurred (what, therefore, has occurred in fact). Change is the fully apparent essence of a new universe that transcends the epochal essence of past thought. It therefore, transcends as well that thought's merely formal perception of change as a difference transcendent to identity. The essentially tautologous transcendence of transcendence issues in the perfectly clear perception of the fact that the *difference itself is transcendental,* that is, *the same as an essentially new identity.* (Before now it was neither something nor nothing. It was what is seen to have occurred not before now. Its priority is not temporal but prior to *time itself.*) In transcendental historical thinking (now essentially transcendental thought) it is not that transcendence *is* immanence (as if it were a case of either/or, or two modes of the same essence), but that transcendence *begins* now for the first time, or, immanently, so that there *essentially exist no past epochs* of thought or time. (Their existence *is* a matter of *pure* potentiality. The merely formal construction of history in modern thought, confusing in essence time with history, understood its history to be a series of epochs at the end of which emerges Nothing in essence, or, to be the essence of time as direction, essentially something changeable, actually real in the Idea.) For transcendental historical thinking there exist in the past essentially only misconstructions of the essence of existence, whose common essence was a misconstruction of history itself. Only in this thinking now occurring for the first time in essence is there history. Prior to the appearance itself of the transcendental essence of existence there is no conception of *essential change,* that is, prior to the *change itself* now occurring in thought.. That *this* change is not itself simply a new form or epoch of thought is manifest in the fact of its *transcendent identity,* or, *its being the same as what now appears for the first time,* in the fact that it constructs existence without identity. Transcendental historical thinking now exists in essential distinction from itself as the transcendent identity of appearance itself. (Thus, as it were, two essences share one existence, except that the analogy is essentially imperfect. *Existence* is neither some-thing nor no-thing; neither is it to be domiciled in nor to be taken apart. Existence in essence is the perfectly unnecessary repetition of itself in thought, thereby giving to thought an identity that it lacks to begin with.) Beyond the transcendent difference of past thought, this transcendental objectivity exists. The transcendent difference of past thought with its absolutely transcendental identity, with its merely formal existence (or merely apparent nonexistence, in the event of Nothing), is the construction of past thought itself. Now, in this thinking occurring for the first time in essence, as before now, it remains the same as it was. The different is not transcendent to transcendental historical (objective) thinking; past thinking remains in the past by virtue of its own essence. Essentially it is a result of its self-determination, apparently the result of the passage of time (indifferently construed as the passage of thought). Now thought itself is taken up with the transcendental essence of existence itself. It receives an identity essentially not there to begin with. Thought itself, and consequently the entire universe of its previous construction (having been brought, in the meantime, to the impasse implicit in its building with identity—*being incomplete in essence*), is essentially transformed. It is *brought to completion in essence* through a perfect perception of what has occurred in essence, the appearance itself, and through the distinct conception of the fact itself of existence.

The universe contemplated in an essentially transcendental science is not the universe that holds meaning for past thought. It is not the universe that belongs essentially to the thinking of the past. But it is that universe now existing in an essentially new form. Objectively this new universe exists in a state of pure motion, *essentially distinct in existence, essentially constructed by that thinking now occurring.* Pure motion, then, or *motion itself* as the essence of the object, *qua* object, is not to be confused with the absolute motion of past thinking. This latter, insofar as it existed essentially, was absolutely not distinguished from thought itself, while, insofar as it was constructed by past thought, it was constructed in *nonexistence.* Or, it existed, through the inapparent essence of past thought, surreptitiously, in the form of an essentially inconceivable Being beyond thought. *Motion itself,* clearly, distinctly *perceived* now for the first time in history, is not to be confused with that motion which, in the final analysis, is seen to have been the essential presupposition of the *permanence* of past thought. The latter was a motion which, insofar as it existed, existed essentially not without reference to time, not without stopping, therefore, in essence. It was a motion constituting the *essential presupposition of transcendental identity itself,* now seen for the first time in history to be in essence a mere appearance, to be itself not in fact what has occurred not before now—the appearance itself of a transcendent identity to thought, in essence, *transcendental existence itself.* The *motion itself* of the object of transcendental historical thinking exists in and for that thinking now occurring essentially beyond transcendent difference, beyond that universe belonging perfectly to past thought, as well as beyond *every conceivable* universe belonging to whatever *epoch* of past thought. Through the mediation of the mere appearance of modern thought the *motion itself* of the object existing beyond its universe exists by extension beyond *every previous universe known to thought,* in whatever way, including that universe in which motion exists remote in essence from thought's transcendent identity. In *pure motion* an essentially new thinking conceives its object existing without the limitation in essence of its past position, existing without the horizon in essence of direction, *without temporality in essence.* It exists now as the appearance itself of a change complete in essence prior to its pure motion in the form of *time itself.* Motion in an essentially transcendental thinking is not some-thing's motion, *presupposed* to be absolute in essence, but conceived to be, in the merely formally tautologous construction of past thought, actually relative in essence to a Being or beings beyond it. But *motion itself* is now seen to be for the first time in history, in the perfect conception of the appearance itself, *essentially relative.* It is under no necessity whatsoever to involve itself in the framework of previous motion (to involve itself with the continuum of space-time, with mass convertible to energy within the limitations of self-restriction) in order to acquire an identity in the final analysis merely transcendental. Motion itself is the appearance itself, *qua* object, of a *transcendent* identity, for the first time in history unmediated by anything whatsoever, except the conception of transcendental existence itself. But this latter is essentially no-thing, but the fact itself of existence now evident in thought, not before now. The inapparent but essential presupposition of absolute motion to the universe of the past is now left behind in the essence of past thought (perpetually drawing near, it nowhere arrived, precluded in essence by its being everywhere at once). In the thinking now occurring, the transcendental identity of the object itself no longer exists in place of what remains to be seen, but is itself seen to be a merely formal construction of thought itself, a symbolic or merely apparent essence, obscuring the reality of what has occurred, which is now seen to be the appearance itself of *change itself,* transforming every-thing so that it moves in the form of pure motion objectively before thought now beholding it. This is the fact-evidence of the transcendent identity of the object in and for thought: the fact that transcendental subjectivity exists simply in essence and lacks existence in fact, but that transcendental thinking exists in fact knowing in its own essence the transcendent identity of its object conceived essentially. The object of an essentially new thinking is itself so essentially new that it now remains to be seen. No longer is it, as it was a moment before now, a change effected through the essential instrumentality of transcendental subjectivity, within the limits of its imagination, so that it is apprehended as a change complete in appearance. It is no longer a change formally complete in its appearance to thought itself which takes it to be merely this appearance of an essence, to the distinct constitution of which appearance in existence thought itself remains transcendentally indifferent (the object's existence essentially presupposed, or, in the event of a radical self-clarification of the structure of scientific consciousness, a matter of indifference). But now the object remains to be seen in and for an essentially transcendental thought that presupposes nothing whatsoever (that understands the existence of Nothing to be essentially a mere appearance). This transcendental thought conceives the existence of an object which is the appearance itself of a complete change in essence. The appearance itself presupposes no matter whatsoever, neither does it presuppose the *abstract identity* of thought itself (by the pure potentiality of which past thought is essentially precluded from the perception of *change itself*), nor does it presuppose the *indifferent matter* of formally relative motion, the absolute essence of which remains to be seen in thought's inability to conceive a transcendent identity in place of its conception of itself. The object of trancendental historical thinking, essentially tautologous, repeats itself now in an essentially different matter. Now, not a moment before now (without, therefore, *momentum;* without the *sameness of time,* that is, the passage of time, essentially; without being something produced by past thought: a *mass* in which that thought stored its energy for future use; without being essentially a product of thought's consciousness of the limitation essential to being its own resource), now what is new essentially is thought at the end of thinking in the past, the end which thought does not survive without a change in essence, without separating itself from itself in existence, without recognizing its object to be an other in and for thought now for the first time in history, immediately its conception itself distinct in existence, its appearance itself in thought unconditionally real. Now thought comprehends in its object the transcendent difference of past thought, now, in fact, transformed, its own identity. In its object thought now comprehends *the historical essence in motion:* the essence of history transforming the dual matters of essentially recollective thinking into the *novitas mundi* now occurring in the *novitas mentis* for the first time as the intelligibility of the fact in essence, the synthesis of the fact itself. In this pure act thought itself surmounts the barriers. It rises above the rule. It sings of the glory of creation itself. Conscious now for the first time of the essential difference between permanence and identity, it sees that *change itself* is the essence of transcendence, not *continuous,* but *essential* change. The transcendent identity of the appearance itself of its object now occurs to thought in place of the stoppage of time, in place of the blockage of thought. This is a change in essence now comprehended as motion itself. This perfect unity of distinct elements constitutes, for the first time in history, an essentially transcendental conception of *energy itself.*

In the course of time, the appearance of the transcendental essence of existence itself was first comprehended as the *transcendental form of the conversion of the world from being nothing to being something.* Subsequently the objectivity of this world's existence appeared to thought in the *form of transcendental natural reason* (in which, for the first time, thought conceived of the *material* conditions of existence itself). Modern man appropriated the natural form as the form of thought itself in which the existence of the world apart from this thought is *essentially* doubted, in which, that is, transcendental reason conceives of the *formal-material* conditions of existence as being of the *essence, indifferently, of thought and of the thing itself. Something existing* was, in that thinking belonging essentially to the past, penetrated with thought's identity so thoroughly that the difference between things remained a transcendent temporality (as such, essentially inconceivable), and energy remained a radically conditioned essence. Now, for the first time in history, the appearance of the transcendental essence of existence itself is understood to be the *transcendent essence of the conversion of thought itself* from being something essentially conceived to exist as an absolute form (self-conception in essence) to being essentially existence itself in the form of thought identifying itself wholly with its object. This thinking essentially conceives its object's existence as wholly unconditioned, its appearance itself. In this conversion of thought itself to being formally transcendent, now, for the first time, the universe is transformed essentially from being *something new* to being *what is now new,* its *appearance itself.* The universe is, in essence, *its very appearance* (it being of the essence of that thinking now occurring that *everything is now itself,* that is, *its appearance,* nothing is hidden in essence, the *merely apparent essence of things* reduced to nothing whatsoever, or to the recollection of Being upon itself). Furthermore the *acquisition* of energy by past thought in the form of its universe of things (an acquisition involving, of necessity, thought's identification of itself with the condition for the appearance of energy, that is, thought's self-conception of its essential priority as the constructor of the forms of appearance, ultimately thought's self-identification with the essence of energy merely appearing: the self-constructing formal essence of the universe), *this acquisition is now seen, in accord with the irreversible historical function of modern thought as the preparation absolutely of what occurs not before now, to be transformed into thought's identity with energy itself.* That is, thought is identified with *its appearance itself,* transcending the transcendent difference of merely formally tautologous thinking wherein energy alternated, essentially without end, with thought in its identification with an essentially indifferent matter. *Energy itself,* concretely, *is the transcendent comprehension of the identity of what remains to be seen without the passage of time in essence,* without, therefore, an alternative route outside of thought. *Energy itself is the end of the alternative to thought* (the end of the merely apparent essence of things) *occurring now to thought* for the first time in history as *the appearance itself of the unconditioned reality of its object.* In this thinking now occurring *it is the same as an essentially new identity: to be conceived in essence, to appear in existence.* The form of thought's comprehension of its identity with the transcendent energy of the universe is *time itself,* through which thought comprehends the appearance itself of what has occurred, a complete change in essence now appearing. Through the form of *time itself* the object of the past is rendered transparent, transformed, *relativized in essence,* which may be expressed as:

$$eT = mc^2,$$

so that, $$T = \frac{mc^2}{e} \quad \text{or} \quad T = mc.$$

The essentially new matter is the transcendent form of essentially transcendental thought. Or, a merely formal tautology ($E = mc^2$), conceived essentially ($eT = mc^2$), appears in existence ($T = mc$), that is, without the essential limitation of being merely apparent. Energy itself appears as pure motion now occurring to thought, more precisely, as an essentially transcendental thinking's *transcendent form* of comprehending *change itself. Energy itself appears as time itself,* through which pure form thought understands what remains to be seen to have occurred without mediation in essence, to have been conceived in existence, therefore, to belong in essence not to the past (not to be something essentially different before now), but now to belong to thought as its transcendent identity in existence. This is in consequence of thought's perfect conception of appearance itself, transcending the merely formal relativity of past thought which *distances* the object from itself *in essence.* For this thinking now occurring there is *no distance in essence apart from time itself.* The *constant thing* of past thought is now seen to be essentially no different than *time itself; no longer is the thing measured by its own standard:*

$$m = \frac{E}{c^2},$$

as if it belonged to itself in essence apart from its being apprehended by a thinking itself essentially transcendent (essentially unconscious of its historical essence). But *insofar* as the thing of the past is able to be contemplated in that thinking now occurring *as* a thing (something essentially impossible: the thing continuing now as a pure convenience):

$$e^2T = mc^2,$$

then,

$$T = m,$$

that is, *time itself is the transcendent unity of indivisibles, or, indivisible motion's transcendent actuality in thought; time itself is the measure of the absolute tenuity of the thing.* (Nor is there any danger of truth itself becoming a matter of convenience, since, in that thinking now occurring, it is evident that truth itself is the contradiction of the truth of appearances, that truth itself is *not something arrived at by thought.* Rather it is the *form of the historical judgment itself now taking place in thought* with the appearance itself of the transcendental essence of existence now for the first time comprehended as the essence of history in its effect in thought: a complete change in essence.) In the event of this tenuity of the thing, time itself, as the essentially tautologous measure of energy itself, is the form by which the *appearance itself* transcends the impasse in essence to which thinking in the past comes face to face with itself, and from which it resorts, *qua* thinking, to the positing of the transcendental identity of two different things (to a 'higher continuity' of one sort or another). The *appearance itself,* then, is the unblinking perception of thought's essential *discontinuity* in the course of time (both with respect to itself and to its object). But, *at the same time,* that is, at the time itself of an essentially new identity, it is the perception that thought itself remains *formally continuous* with the fact itself of history and that the very conception of its object's existence depends upon that fact, upon the perception of the object's historical essence. The *formal continuity* of an essentially transcendental thinking is *time itself* as the comprehension of the *essentially new state of being every-where, ἡ ἐνέργεια καθολική,* not at one, essentially indifferent, continuous time, but every-where at the same time, that is, through the form of time itself. This universal and perpetual energy is an essentially new motion in thought itself of what is now essentially different, namely, an identity in time itself transcending the passage of time, continuous *not with other times* (as if time itself consisted of a number of things-in-essence, or as if time itself were a common essence of things-in-existence), but, comprehending its continuity with the *change itself* manifest in the essence of history, this identity is conscious of its being *energy itself,* therefore *prior in essence to the very conception* of world-*ordering* or *reordering.* This transcendent identity now knows itself to be, in itself, an essentially new locus for the pure occurrence of new being. The route to this location, existing every-where as the conception (*hē energeia katholikē*) of an essentially new universe, existing every-where as the perception of light itself as thought's brand-new shining object, the route, is history itself, now occurring in and for thought as the revelation of the essential priority of its object in existence, by which transcendental thinking is, for the first time, taken up with the recognition that what it conceives in essence is what in fact appears, but that, before now, it has conceived nothing in essence. Now it conceives life itself.

Appendix β

THE NOW EXISTING THOUGHT OF FAITH

> And we, with our unveiled faces reflecting like mirrors the brightness of the Lord, all grow brighter and brighter as we are turned into the image that we reflect; this is the work of the Lord who is Spirit.
>
> *2 Corinthians* 3:18

In history before now faith is tempted to be thought to be other than itself, to be for another, to be, in essence, a mere appearance. Before now faith is tempted to think of itself as the preexistent ground of the world in history and in this way to deny the ultimate reality of its appearance, turning, as it were, stones into bread by conceiving of the world as imbued absolutely with life-giving power available to a self-realization of the transcendent origin of all things. This essentially theurgic conception of self appears in history as Neoplatonism's denial of history in essence, which it accomplished by placing the fact itself of existence absolutely beyond its transcendent thought of existence, in the form of its transcendental identity with all things, that is, by making the fact of existence itself the presupposition in essence of thought itself. As a result, there was to have been in history no change in substance, but the transformation of the appearance of the essence of transcendental existence itself into the image of a self-transcending identity. This first temptation of faith in history to prove itself to thought by identifying itself in essence with the appearance of thought itself is overcome in Augustine by faith's acknowledgment of the essential priority of transcendental existence to its conception of its own self-transcendence. Thus, the latter has the merely negative function of illuminating the region of the distinct knowledge of the existence itself of all things. In this way the ultimate reality of faith's appearance in history is seen immediately, although it is, immediately, borne up beyond the temporal distention of the soul.

Before now in history faith, taken up, as it were, to the pinnacle of the temple, is tempted to hurl itself down, to demonstrate the fact itself of its existence, to prove its transcendence to thought by the self-contradictory motion of relying upon the conditions of contingency to necessitate the appearance of existence itself, tempting God to be thought in essence other than itself. This was the second temptation of faith before now in history, specifically, to be beside itself in an attempt to transform the appearance of the transcendental essence of existence itself into an accident in essence. This temptation was to make that appearance dependent upon a state of transcendent self-realization, that is, to render it, *qua* contingent, intelligible. This materially theological, that is, metaphysical conception of existence as that existence, other than self-existence, the fact itself of which is dependent upon experience, appears in ibn Sina's doctrine of the unification of all things in the Necessary Existent. Therein, in fact, it is impossible to conceive of existence itself apart from a contingent order of being. In ibn Sina the essence of history is transformed into being the subject matter of a metaphysics incipiently theoretical, not actually so. It is not historical essence itself that is thought, but strictly its intelligibility *qua* contingency, that is, its being understood to be in terms of a being beside itself. In Thomas Aquinas this second temptation of faith in history to prove itself to thought, to a thought itself other than itself in essence, and to fashion itself in the likeness of a Necessary Existent by placing its appearance itself in the hands of the angels is overcome by the *affirmation* of the unconditional contingency of the appearance of existence itself. It is overcome by the affirmation that thinking in essence is *to begin with what is transcendental,* and that it is *not to begin with what is transcendent.* In Thomas' formulation of sacred doctrine in essence thinking begins immediately, with faith, that is, without self-transcendence. A materially new form of science comes to be in history, namely, the transcendental form of natural reason. It understands its object to be what is transcendent, known to be an actual existent without being thought to be other than itself, that is, without the *necessity* of referring its existence to essence. The object is known to exist immediately as a matter of fact. The fact of existence is the formal presupposition of that thinking that begins precisely with the indemonstrable truth of faith.

Before now in history faith is tempted a third time. This third temptation is no longer cast in the form of faith's *proving* itself to thought. Faith is not tempted to exist theurgically, nor to exist theologically. In the third temptation of faith in history, in modern thought (beginning with Descartes), faith is tempted to exist theoretically. That is, it is tempted to exist precisely without being tempted to prove its existence. Faith is tempted to conform itself absolutely to thought, to bow down, as it were, in worship of thought's theoretical conception of its own purposes, to itself prove nothing, but to be proven to exist for another. Modern thought *appropriates* the appearance of the transcendental essence of existence itself as the form of its self-conception. Appropriation is the essence of the proof of existence offered to faith by modernity's transcendental subjectivity. Faith is tempted to see itself reflected in thought's appropriation of the contingency of the world (standing, as it were, upon a towering peak) as heir apparent of the essence of power. Existence itself is the material presupposition of modern thought; of faith it requires nothing, but conformity. At what is historical irony's very summit, faith itself, whose ontological priority is the *conditio sine qua non* of thought's essential transformation from being metaphysical to being historical, and whose final result in history before now *is* modern thought, faith itself is tempted *to exist simply for that thought.* Or, simply, faith, without which the devil is inconceivable, is tempted to play the devil—by the devil! It is tempted to see itself reflected in the substantiality of thought as the absolute actuality of this world. In the absolute self-relation of modern thought the appearance of faith in essence is retrospectively anticipated. Modern thought, is bound in its essence to the past, but bound in such a way as to be essentially unaware of the factuality of its existence. This is due to the fact that apart from the form of its own substantiality, modernity has no idea of existence itself, from which, in its pure factuality, it lies cut off, by virtue of its appropriation of the transcendental form, as from the absolutely unintelligible. Everywhere in modern thought existence is a matter of precedent possibility. As a result, although this thinking, *qua* thought, in its essential originality is transcendental substance, when it is opposed arbitrarily, or inessentially, by a rational materialism, God himself is liable to appear to exist as the indispensable actuality of a universe of relative self-realization. Faith is tempted to believe, but for faith to be tempted to believe is to be tempted to believe in itself. Kierkegaard, by standing in *essential* opposition to the *substantiality* of modern thought, anticipated this possibility in the dark clarity of his transcendent passion to exist, and *resolved* in faith not to believe in himself, but to believe in God. In this resolve of faith Kierkegaard set up a clear limitation to the possibility of a precedent existence in essence by opposing this same in the form of the transcendental identity of thought itself with the absolutely unintelligible actuality of his private existence, *qua* private. He resolved in faith upon the absurd. But, insofar as Kierkegaard opposed in essence the *substantiality* of modern thought, he did not in essence oppose *the essentiality* of the third temptation of faith before now in history. That is, he did not in fact accomplish the repetition of existence that he willed; he saw, but with blind eyes. His God was other than himself, like Kierkegaard constrained to conform himself to the thought of existing for another. Well might it appear to Kierkegaard, as it did, that there was something in the thought that Christianity was the invention of the devil! Kierkegaard willed the essential repetition of existence, but precisely because he willed it, he had no conception of it. He willed it in the form of what had in fact become an absurdity: self-transcendent existence. (Theology after Kierkegaard, when it shares his antipathy to the substantiality of thought, but not his passion for the absurd, understands existence to be the *pure determination of an other;* that is, when its antipathy, like his, is essential, not arbitrary, opposition.) Kierkegaard appropriated the existence of thought itself, bringing reflection to an end in substance. For him the substantiality of thought is impossible apart from this individual appropriation. The absolute extremity of the existential paradox is that Kierkegaard became the embodiment of the *theoretical* conception of faith itself. He experienced in the flesh the third temptation of faith. Unable to conceive in essence the essential repetition of existence, Kierkegaard suffered the fact that modernity is, before now, the logical implication in essence of faith itself, without being able to conceive of the essentially unnecessary existence of the thought itself of existence. He conceived it to be the unnecessary *form* of existence to which in essence no alternative existed but his passion. In his passion Kierkegaard knew himself to have suffered the loss of his identity, but it could never have occurred to him that, thereby, he had suffered the loss in essence of the identity of another's existence. This remained a matter of the pure formality of thought, inconceivable in essence before that thought now occurring for the first time in history as *the logical explication of faith itself in essence.* Bereft of faith itself, Kierkegaard resolved in faith to believe in another, that is, by self-contradiction to overcome the contradiction of faith itself (that by existing for thought it would accomplish something for God). He did this by positing the contrary (that before God it accomplishes nothing). Thereby, Kierkegaard's act of faith implicitly accorded with thought's essential supposition that what is absolutely impossible is in fact nonexistent; his opposition was essentially immaterial to thought itself. What now occurs for the first time in history is absolute opposition to modern thought. (It is not but *formally* essential opposition to the substantiality of modern thought, not, that is, essentially juridical or intramural opposition, which leaves it *essentially* uncontradicted, but an opposition that transforms *in essence* the substance of thought from being absolute to being explicitly historical.) This is a material opposition to thought itself appearing in thought's essence, so that it is clearly perceived that the substance of thought is not there to begin with in essence for thought, but is, in essence, what appears in existence in thought. There now exists, then, for the first time in history, an essentially new thinking wherein is conceived the *essential* repetition of existence, *essentially* tautologous, without self-relation, and *essentially* discontinuous. That is, it exists without being other than itself, occurring itself in essence, therefore, without necessity: The appearance itself in essence of transcendental existence. In this thinking now occurring for the first time in history the third temptation of faith in the form of thought is overcome. In the essence itself of the absolute pretension of thought itself faith is perceived to exist immediately, there existing now no immediacy that is not thought, no necessity for the essence to appear to exist for a thought other than itself. Faith mirrors itself in the absolute formality of the temptation as existence in essence at the disposal of another. Faith thereby overcomes in essence the negativity of thought itself by coming to be itself in the form of thought, bringing reflection to an end in existence. Now faith actually thinks for itself. A breakthrough in essence: *The word is spoken in essence.* This is the same word through which, in the beginning, every-thing (without exception) came into existence, and upon which faith, before now in history, relied in response to its temptation to theurgic self-conception. In time, it became flesh, and with this word faith, before now in history, identified itself in response to its temptation to conceive of itself metaphysically. That same word, now for the first time in history, and in response to its temptation to look upon itself theoretically, is spoken through faith itself in essence. The thoroughgoing implication is accomplished, in essence, by thinking, before now conceiving itself to be absolute, now conceived in essence to be essentially at variance with itself through the fact itself of the transcendental appearance in essence of the essence of existence itself. No sooner than the word is spoken in essence is thought's perfect immediacy conceived in essence. Now *thought itself is the framework.* Thought itself is the universal, perpetual constant. Thought itself is, as it were, the speed of light through which now is conceived not simply the variability of *things* in essence, but the variability, in essence, of *being itself*. What is now conceived for the first time in history is the essential change in existence, comprising the negation as nothing at all in essence, except it be implicitly faith in a universe of thought now seen to belong to the past. Nothing at all escapes this essential change in existence, not the different itself. Before now the difference between things was conceived to be in essence transcendentally identical. But now it is conceived to be in essence the transcendental difference of an essentially new identity, transcendent in form. The different is different than itself in essence without being other than itself. This is the essentially perfect inclusiveness of the new form of thought brought into being by a complete change in essence from transcendent to transcendental existence.

Before now thought itself fashioned itself for its own purposes in the form of the condition for the appearance of faith itself in history. Now faith accepts the condition as the essence of its formal appearance. At a stroke, faith itself, appearing in essence, conceives the essential. What is the essential in essence that it is able to be conceived in fact? What is the essential condition of the appearance of faith itself? Or, to repeat essentially: What, in essence, is the essence of history? It is *being itself at the disposal of thought itself*. It is the dissolution in essence of the distinction underlying the essential negativity of thinking-in-the-past, namely, the distinction in essence, self/other (therefore, also, the dissolution of the distinction in essence, determinate/indeterminate being), through the appearance in essence of what is distinct in existence. Through the essential distinction in existence, through change itself in essence comes into being *the essence of being in its exaltation* in the form of thought itself now for the first time in history. It is the essential repetition of the essence of history in thought itself. Before now, in the life of Abraham, history is to be comprehended as the logical concomitant of faith. Implicit in the deeds of his life, in his obedience to God's word, is Abraham's faith in Yahweh's creation. Time separates Abraham's faith from history itself. The implication in essence of Abraham's existence in time is the essence of history, that is, in time, Jesus the Nazarene. He is the appearance of the transcendental essence of existence itself, who, in response to a question about his identity, says in *John* 8:56–58: " 'Your father Abraham rejoiced to think that he would see my Day; he saw it and was glad.' The Jews then said, 'You are not fifty yet, and you have seen Abraham!' Jesus replied: 'I tell you most solemnly, before Abraham ever was, I Am.' "[395] The essential implication of Abraham's existence necessitates the *expectation* of the appearance of the transcendental essence of existence itself, but the appearance itself is in essence transcendent to time. The time of the appearance itself is essentially unexpected. With the appearance of the essence of history in time, *essentially* the creation is repeated, but the essential implication of Jesus the Nazarene is a completely new universe. This implication, just as the implication of Abraham's existence, upon which in essence it depends, is elaborated, before now, inessentially, but necessarily, in the course of time. Time alone separates Jesus the Nazarene from a completely new universe; he together with it now exists in essence transcendentally. The elaboration of the implication of the essence of history, before now, in time is essentially a function of the inessential repetition of the appearance itself of the transcendental essence of existence in one or another form of transcendent existence. This is to say that before now in history there has been no thinking *essentially* transcendental. So Augustine, in response to the theurgic temptation of faith, immediately apprehends in the essence of history the essential repetition of creation. That is, he recognizes the fact itself of existence in the form of the transcendental conversion of the universe to existence *ex nihilo*. The essence of history, *qua* transcendental appearance itself, is *not* repeated in Augustine's essentially self-transcendent knowledge of existence itself. It remains a form of transcendent apprehension of existence itself. Thomas Aquinas, in response to the theological temptation of faith (that the essence of history be thought to be other than itself: a manifestation of the necessity of existence), repeats the essence of history *materially* as the transcendental form of natural rea-

[395] *The Jerusalem Bible: N.T.*, op. cit., 166.

son apprehending transcendently an object existing as a matter of fact. In this thinking of Thomas, existence itself is not apprehended, as in Augustine, but is the *formal* presupposition of the transcendent knowledge of a created universe. The presupposition of existence itself is in the form of the truth of faith. In modern thought, the theoretical temptation of faith to exist in essence for another, the essence of history is *formally* repeated as the necessary appearance of the essence of thought itself in time. In this essential indistinguishability of history from thought a complete change in essence is impossible in essence; that is, a complete change is *absolutely* impossible. Although for the first time in history modern thought conceives a complete change in appearance, it remains an ideal possibility of a thinking formally transcendental but essentially transcendent. Modern thought is a similitude in essence of the essential implication of Jesus the Nazarene. That is, it is the purely logical implication of the essence of history. As such, it is at once faith's final temptation, and *formal* condition of its appearance in essence in thought itself. But the essential appearance of faith itself in thought is the *essential* repetition of the essence of history. It is the occurrence *in fact* of what is absolutely impossible. It is a complete change in essence. In its sheer immediacy as fact synthesized (there being in this thought now occurring for the first time in history no immediacy that is not thought) it is being itself now for the first time in history at the disposal of thought itself. In essence no time at all elapses in the creation of this essentially transcendental thinking. Now for the first time the implication of the appearance of the transcendental essence of existence itself, namely, a completely new universe, is elaborated, not in time, but essentially in thought itself. The transcendent form of this thinking is time itself, and this thinking is itself essentially transcendental. There now occurs in fact, in contradiction to the similitude modern thought is in essence, the further approximation in essence of the implication of Jesus the Nazarene. As the appearance of the essence of history in time *essentially* repeats the creation, so now the repetition in essence of the appearance of the transcendental essence of existence itself constitutes creation's *formal* repetition. An essentially new time is conceived in essence as the form of an essentially new universe, anticipating in essence its utter finality in fact. The implication of this *formal* repetition of creation now occurring is the formal repetition of the essence of history, but not, as it appears in modern thought, belonging in essence to the past, a necessity imposed upon faith itself. Rather, it is an essentially unnecessary transformation of everything existing here and now transcendentally into a transcendently new identification with time itself so as to make possible for the first time in history the furthest approximation in essence of the implication of the essence of history, that is, the *material* repetition of the creation. This last, together with its possibility, although now for the first time *formally* comprehensible, remains *in essence yet to be conceived.* For the time being it recurs to the fact itself in its essential indifference to everything apart from its being itself the promise kept. This thinking now occurring is, however, the essential retrieval of faith itself from the form of its acceptance at the hands of modernity. That is, it is the retrieval of faith itself from Being diverted to its own end, from being waylaid by the mere appearance of existence, or by the thought of existing for another. This retrieval of faith itself is *essential* insofar as it involves no loss of time in faith's further realization of the implication of the essence of history (in this consists the essential historicity of this thinking now occurring for the first time). Indeed, it is now clear that modern thought's irreversible historical function (whatever other benefits it may have brought forth within the circuit of its own intention) is to have been the occasion of the appearance in thought of faith itself. The essential condition is Jesus the Nazarene, in essence being itself at the disposal of thought itself. The essential actuality of the appearance of faith itself is *being at the disposal of another* (not being *for* another). This, in terms of this thinking now occurring for the first time in history, is faith's thinking for itself in *essential* independence of that thinking which belongs in essence to the past, that is, modern thought. This explicit conception of faith itself essentially transcends as well theological modes of opposition, adaptation, or accommodation to modern thought that themselves, unhappily or happily, despair of possessing essential intellectual originality, modes of theology ultimately justifiable in terms of the *intellectual* desperation of Kierkegaard's faith. In this thinking now occurring, thinking is *essentially* conception, or a thinking in which what is conceived in essence appears in existence. It is opposed to modern thought, a thinking *formally* conception, in which what is formally conceived is *thought* to appear in essence. It is a thinking in terms of which, in opposition to Hegel's essentially modern formulation, to wit, "*Logic . . . coincides with Metaphysics, the science of things set and held in thoughts,*—thoughts accredited able to express the essential reality of things," *logic is understood for the first time to coincide with existence itself.* Logic is the knowledge of the essential identity of the reality of the thing with the appearance itself. In this thinking the dependence of the knowledge of reality upon the absolute self-relation of thought itself is *severed* in essence (not merely phenomenologically suspended in existence or in essence). The essential condition of this thinking is the essence of history's essential repetition, the appearance of *an other none other than itself in essence* in and for thought for the first time in history. This essentially transcendental thinking now exists essentially as a matter of history in transcendence to every previous form of thinking that remains in the past essentially a result of other/self-determination. It now, in fact, exists in what is essentially a previously impossible transparency to the essence of history itself. In this thinking *the essentially unconditioned actuality of the appearance of faith itself is conceived to be the acceptance in essence of being itself in the word,* the essential synthesis of the fact itself of the word's being itself spoken in essence: the word made flesh in essence. This is the essentially new substance not of thought itself, but of its essentially implicit object, a completely new universe. (Its explicit object is what now exists conceived to be in essence historical.) This essentially new substance is the eucharist of existence itself now for the first time in history perfectly conceived. This conception itself incorporate in the substantiality of faith itself as the moment of thought's being *reception* itself. It constitutes the essentially further approximation of the implication of the essence of history, the moment of absolute objectivity in thought itself. In this thinking now occurring for the first time in history thought undergoes the time itself of its conception. Under the impact of the substantial fact of faith itself, the very conception of which essentially comprises thought itself as a transcendentally differentiated identity with the appearance itself of a transcendent object, the very conception of which *is* (immediately) thought's existing in essence transcendentally, thought suffers essentially the death of its thinking in the past. It suffers the death of its essentially transcendent existence, to be taken up now for the first time with the appearance of the transcendental essence of existence itself in essence. This thinking now occurring suffers the death in essence of its transcendental identity. It is actually no longer the case, neither is it necessary nor possible, that thought *posits* its identity in essence with what appears (an identity essentially inapparent). Thought undergoes this death, suffers the time itself of its conception of the appearance of an object none other than itself in essence, and submits itself to the absolutely objective perception of itself as *formally* conception, but *essentially preconception,* that is, *essentially immediacy or transcendent existence.* In this purely active contemplation of itself to begin with (effected in essence through the grace of the fact itself of existence), thought, which, before now (through a series of maneuvers undertaken in essence by Descartes, Leibniz, and Kant, and culminating essentially in Hegel's synthesis) has attempted to divert the appearance in essence of faith itself to being fashioned by Being for its own purposes, thought, which has, before now, thereby taken up a position from which retreat is absolutely inconceivable (modern thought having made of thought *a position* in essence), and absolute self-extrication an impossibility, now, without itself moving from the spot, indeed, rooted to the spot, this thought *passes through in essence* its being essentially a position taken up in existence in transcendence to the fact itself of existence. It, thought itself, is converted on the spot: it passes through its essential immediacy to transcendental existence in essence: it now conceives immediately in essence its previous supposition of transcendent existence to be without foundation in essence. Now, for the first time in history, *thinking begins in essence.* Nothing including thought in essence is now inconceivable. The conception itself is in fact the appearance of the transcendental essence of existence itself in essence, in fact, being itself at the disposal of thought itself. Nothing excluding thought in essence is now conceivable except as belonging in essence to the past. But what belongs in essence to the past, that is, what begins not in essence, in essence ends not in existence. The beginning of what belongs in essence to the past is a pure formality. It is a false start remote in essence from change itself in essence; it transcends thought in essence. But this thinking now occurring in essence is itself the transcendent beginning terminating immediately in existence, repeating nothing in essence but the essence of history. Thought now is the perfectly transcendent repetition of existence itself in essence. It is this through its essential transformation from being itself transcendent to being the transcendent identity of another existence in essence. In this transformation the transcendent difference of past thought is incorporated as transcendent identity. The difference as such is not transcendent to an essentially transcendental thinking, which, indeed, is precisely the conception in essence of what is different in existence without being itself transcendent, that is, without being transcendentally identical with thought. Thought now comes face to face with existence itself in essence. The beginning of what belongs in essence to the past, not being thought in essence, is repeated in a merely *formally* tautologous manner, indefinitely without end, not world without end, but chaos without end. The beginning of what belongs in essence to the past is, as it were, an eternal stillbirth, no matter how endlessly repeated in time, no matter how forcefully it echoes in the still of its being the word spoken in essence, it remains in essence the embodiment of nothing. Even now that the world is able to be seen to have come to an end in essence, the absolute scepticism of the pure determination to exist *will not think* the beginning *in essence* of a completely new universe. Rather, it prefers to expend its energies on staving off the end of the old. It is incapable in essence, then, not only of creating a new universe (even that thinking now occurring for the first time in history is but the conception in essence of an essentially new universe: the formal, not yet material, repetition of the creation), but absolutely incapable of the conception itself of creation. It remains encapsulated within the horizonal essence of the past. It knows nothing at all of the repetition of existence itself except it be in the form of absolute self-denial. Excluding the fact that it will not think the beginning in essence, this absolute scepticism, out of the profoundest possible regard for truth in essence, will oppose nothing in the way of that thinking now occurring for the first time in history. This absolute scepticism is the very best of modernity, its purest expression, its defender in essence to the end. But that thinking beginning now in essence for the first time is answerable neither to God, nor to self, nor to the world. It responds, without choice, to the fact itself of existence now apparent in essence, namely, that the word is spoken in essence. This is the perception of truth itself.

The word is spoken in essence. The creation, formally, is thought for the first time in history. The essentially unconditioned actuality of the appearance of faith itself, essentially repeated in the essence of history, the irreducible element in the eucharist of existence, the divinity itself, is the fact in essence of the acceptance of being itself in the word, together with which, in the substantiality of faith, is, implicitly, the exaltation of being itself, together with being at the disposal of another. The fact that thinking is able to conceive in essence the eucharist of existence, in which it exists implicitly as the moment of absolute objectivity in thought, without contradiction, is intelligible in terms of the fact that difference no longer depends upon transcendence. The difference in essence is not *between* two or three things transcendentally identical. There is no transcendent interval, no interstice (however wide), no merely formal indifference to the apartness of things, no merely ideal energy of thought itself dwelling in the eternal recess of things; in essence, there is no time out. There is now nothing but the transcendent identity of thought's object, upon which thought now concentrates in essence, essentially without interruption. This identity is conceived to be the beginning, transcendentally differentiated, of a completely new universe. This is *the ready acceptance of being itself in thought for the first time in history; this is hearing the word spoken in essence in thought itself.* In this thinking now occurring the substantial function of the word spoken, *qua* spoken, the essence of history, is the repetition in essence of the creation. The word spoken in essence, then, history itself in essence, is conceived not to be in essence revelation, to be neither in essence communication, nor in essence remembrance. (Although, indeed, these three constitute essential functions of the spoken word, that is, functions formally conceived to belong of necessity to the word as spoken before now in that thinking in which the substance and function of what is thought do not constitute a transcendent identity differentiated without being other than itself in essence.) The substantial function of the word spoken in essence in this thinking now occurring is, formally, an invitation to the universe, to every-thing, to exist. However, *qua* substance, it is to create the form, to be the conception in essence of a world not ending in essence, but terminating in existence, to be not the essentially imperfect conception of an ideal world (something belonging in essence to the past), but *the essentially perfect conception of the point of departure in essence now existing in terms of which an essentially new universe actually coming into existence can be reckoned with hitherto inconceivable precision.* This is brought about through the synthesis of the fact itself of existence now occurring in thought for the first time in history in essence. Before now, Augustine apprehended the fact of existence in Truth. Thomas Aquinas made of the fact of existence the formal presupposition of thought. In Descartes thought transcendentally identifies itself with the formal presupposition of the fact of existence. This is the *material* synthesis of the fact itself as nothing in essence for thought itself, except it be purely formal. This is the negation of matter itself, except it be thought: the hypothesis that modernity sets out to prove positively. (It is an hypothesis unintelligible prior to the apprehension of the appearance of the essence of history, an apprehension in which the universe is shaken to its foundation in doubt.) In Leibniz the implication of Descartes' material synthesis is immediately *posited formally* as the substantial form of matter. The fact itself of existence is *something in essence* for thought. It is the ideal realization of the essential possibility of a universe of discrete entities. Leibniz realizes the universal implication of Descartes' *method* to be *position in existence.* Descartes' extension is integrated with mass or weight. In Kant the substantial form is *formally* synthesized, that is, *qua* form or appearance. The fact of existence as something in essence is a *possibility* actualized within the limits of a sensibly conditioned conception. This is the *position in fact* of matter taken up in thought as *specific* form or weight. In Hegel the formal synthesis of the fact is *posited in essence.* The fact of existence as something in essence for thought is the *necessity* of an absolutely unconditioned conception of matter itself as *position in essence.* In Hegel matter is essentially weight or the specific form of thought in essence: *the appearance in essence of thought.* In this thinking *now* occurring the apparent essence of thought is synthesized *essentially,* that is, *qua* essence or appearance itself, as the fact itself of existence comprising nothing in essence except thought, that is, as transcendental: being itself at the disposal of thought. In this thinking now occurring matter itself is time itself, the transcendent unity of indivisibles. What before now is mistakenly thought to be in fact the necessary position of existence in essence is *now* perceived to be the *disposition in essence* of thought itself to the fact of existence. It is seen to be the promise kept now essentially in thought itself for the first time in history. Indeed, *thought itself now comes into existence for the first time* through the realization of the *essential* difference separating being itself from transcendental thought effected by the appearance of the transcendental essence of existence itself. This is an essentially new thinking. Now nothing in thought is innate. There exist in essence no preconceptions; there is in essence no idea whatsoever. Neither is there pure transcendental ego. Now everything in thought is essentially historical—it is in fact being created. This is the *novitas mundi,* the essential state of the world's novelty, now for the first time coming to be perceived in a reflection itself essentially new, ending itself in the fact of its existence. In thought before now, the word, not spoken in essence in thought, remains substantially incomplete. Even, indeed, when, otherwise spoken, it says nothing in essence within thought's hearing. Now spoken in essence in thought itself, the word is the substantial form, the transcendent identity of every-thing complete in essence, that is, transcendentally differentiated. Before now thought heard tell of the word; now thought hears the word for itself.

Appendix γ

MISSA JUBILAEA: THE CELEBRATION OF THE INFINITE PASSOVER

Anyone who does eat my flesh and drink my blood has eternal life, and I shall raise him up on the last day. For my flesh is real food and my blood is real drink. He who eats my flesh and drinks my blood lives in me and I live in him. As I, who am sent by the living Father, myself draw life from the Father, so whoever eats me will draw life from me.

John 6:54–57

Here is the form of man: the infinite meekness of God beheld, through which everything exists in fact, in the blood of the lamb made forever new. With the bleeding of the lamb there exists for the first time the infinity of the fact of being there in the form of man, well disposed in essence to being the bread of life for others. This fact, before now inconceivable in essence, is now conceived for the first time in history in terms of the intelligibility of appearance itself as the unleavened bread of existence itself in the form of man. This bread of existence itself is unleavened in essence. Without depth or weight in essence it remains itself unchanged, change itself in essence, a perfect transparency in the form of man to existence itself now occurring to man in the form of thought itself. This is the knowledge of faith itself effected now for the first time in history in thought itself, the knowledge of the eucharistic essence of existence itself had by thought among those everywhere in whom existence has effected the sight of itself. This is the consciousness of the other in essence as being the absolutely, incomparably gentle friend who shares himself essentially with others in the form of man. This is the knowledge of the fact itself of existence. In the celebration of the infinite passover the appearance of the eucharist, *qua* appearance, is now seen for the first time in history to be itself the eucharist in essence: to transcend appearance, to be appearance itself. No longer are the elements on the table seen to be other than what they are in essence, namely, the flesh and blood of God in the form of man. In this thinking now occurring there is nothing beside the unleavened bread of existence itself, nothing beside the wine new in essence. *The appearance of the eucharist transcends its being simply for others not in being substantially different* (as it appeared before now in the natural objectivity of a formally transcendental reason), *nor in being contingently different* (as it appeared before now in the transcendental subjectivity of a purely natural faith), *nor, finally, in being transcendently different* (as it appeared before now in the formally transcendental thought of existence itself belonging in essence to reason). *But now, in being transcendentally different, it transcends the other in essence* (reducing, by the way, the Ockhamist's objection to silence in essence), *so that it exists for others essentially identical with itself, they, in turn, each with one another identical in the form of man with the eucharistic essence of existence itself.* This is the transcendent identity of an essentially transcendental thinking in which transcendence itself is essentially transitive, that is, thought's new object. This object is the essential unity of indivisibles, the form of an essentially new universe now existing for the first time in thought. The realization of this form is now but a matter of time itself in thought. This realization, in turn, is but the preparation in essence for the termination of history itself. In this thinking now occurring there are no elements in essence apart from the elements of Christ. The appearance itself of the eucharist is transformed in essence into the form of man, into Yahweh's flesh and blood, into his essential property, thereby obviating in essence the thought itself of appropriation, rendering it essentially unnecessary (appropriation itself) in every one of its forms, holy or unholy. Not now, in fact, is appropriation justifiable in the essential terms of the appearance itself in thought of an essentially intelligible existence, in the essential terms of the conception of the phenomenon of existence itself. This conception is the beginning, in essence, of the essential transcendence of existence itself in the form of a new world absolutely without self to begin with, absolutely different. In this light it is seen that appropriation itself is the final form of impenitence. Appropriation itself is the refusal to accept the change of thought itself in essence, now the absolutely evident fact of existence itself. It is the unwillingness to acknowledge the essentially transcendental existence of a thought-form transitive *in essence.* This form of final impenitence prefers nothing at all in the way of thought to the only alternative it recognizes, namely, the merely ideal conception of a new world by an essentially intransitive form of the thought of a transcendent existence, while transcendence itself remains a mere appearance, the essentially infinite reflection of thought upon itself. Appropriation itself, in transcending the intransitive form in essence, remains the recollection of Being upon itself, unable to think in essence beyond thinking in the past. Its transcendence to thought itself remains inconceivable in essence. Its existence is a poetic figment, the disappearance of thought in essence, unchanged in essence, suspended, as it is, in the form of its essential nonexistence. However, in this thinking now occurring for the first time in history the essentially intransitive form of thought is not transcended in essence, but it is transcended in fact in the substantial form of what now occurs to thought itself in essence, namely, faith itself in essence. In this thinking now occurring, in this critical occurrence to mankind, thought itself is now seen to be transitive in essence. It is essentially formal or transcendental. *What is essentially new about the form of thought now occurring is that it is form in essence not thought in essence: there is no essence beyond the formality of thought itself now seen to be transitive in essence, that is, the form itself of transcendence.* Now existing for the first time in history, in fact the termination of existence itself in thought, no longer is thought but the (essentially redundant) form of the form of transcendence. It is now the thought in essence of existence itself. As over against the essential finitude of past thought it may be said: *de facto tertium quid datur—exsistere ipsum.* What is thought *is* the unleavened bread of existence itself.

What happened before now in the Mass exclusively (*missa solemnis*) now happens in the Mass inclusively (*missa jubilaea*). Before now in faith the appearance of the transcendental essence of existence itself was predicated of the elements on the table *exclusively* in essence, that is, excluding in essence the appearance of the eucharist. Thus mystically understood, the substantial change of the elements remained essentially informal, or unintelligible in essence, save to the purely formal reason presupposing existence itself in essence in the light of faith. Such was the essential limitation upon the universality of the *missa solemnis* that the change in essence substantially transcended thought itself. At the same time, thought, under the impact of the reality of the *missa solemnis* (the *substantial* appearance of faith itself in essence, not, therefore, the *essential* appearance now occurring in thought), acquired, for the first time, its own transcendental form which, *qua* form of thought transcendent in essence, remained *essentially* intransitive to the *novitas mundi.* Thought at that time was the substantial appearance in the form of thought of the transcendent passion of faith itself in essence. It, therefore, essentially preceded in time modernity's self-conscious appropriation of the transcendental form of thought itself. Appropriation, in turn, was essentially followed in time, in the absence of faith itself, by the end of reflection in substance, or, the transcendent passion to exist. This passion clearly seen, in the essentially historical perception of thought itself now occurring, to have been substantially oblivious of the transcendent existence of the other in the form of man. At the end, *in extremis,* the irony of history was that the fate of the appearance of the eucharist, namely, the essential transcendence of thought itself to substantial change, became, in the course of time, the substantial experience of the world at large. The world itself in essence, in a variety of ways, experienced thought's essential incapacity to perceive the body in the form of man, to comprehend being beheld in essence, the infinite meekness of God in the form of man, the body itself. What now occurs in thought for the first time in history (transcending in fact the end of the world in essence) is *the perception itself of the body*—God in God in essence—the Temple of the New Jerusalem—effected now in essence inclusively in the *missa jubilaea,* the center of an essentially new consciousness in the conversion of the universe into an entirely new stuff. What is now seen to be essential in the appearance of the transcendental essence of existence itself is the appearance of appearance in thought: the intelligibility of appearance itself as the transcendent existence of the other in essence, at the same time, the transcendent identity of man in the form of man. In the *missa jubilaea* this essential intelligibility of appearance itself is predicated of the elements on the table *inclusively.* The elements themselves are transformed in essence to the resurrected Christ, the substantial repetition of the Son of Man. The essential distribution of the existence of man is essentially predicated of the world in the form of the elements on the table. The world itself in essence is the body, the identification in essence of man with God in essence. The world in essence itself is where God is with man. The world itself in essence is the body itself, is the living flesh of Jesus the Nazarene transformed into being here at the disposal of another in essence. The end of the world in essence in existence is the beginning of the world in the form of man. (The essential termination in existence is the beginning of transcendence; the converse is essentially unintelligible. As such, it belongs in essence to thinking in the past; whereas, in existence itself the end is *identically* the beginning. Neither before nor after, nor above nor below exist in essence apart from time itself, the form of what is essentially historical.) The beginning of transcendence is the predicament of existence itself in which the world discovers itself for the first time in the form of man, discovers that it is the body itself, that, as such, it is the bread of life upon which, in essence, others feed. To be essentially in the predicament of existence itself is to have no recourse whatsoever to being beyond itself. Transcending in essence the distinction body/soul, the body is substantially identical with the world itself in essence, from which it is transcendentally differentiated as that which the world itself in essence has become in the course of time. (Again, the converse is the essentially unintelligible assertion of the anti-Christ which says nothing in essence but speaks of a transcendent death, as if, in so doing, it were saying something. But nothing can be said without saying everything in essence.) To have become the body itself in the course of time, essentially without recourse to being itself beyond, is the world's being in essence the matter of a real transcendence, the food of an essentially transitive reality existing transcendentally. Having surrendered being itself beyond, being itself returns to it, as it were, from beyond in the form of its conception of a materially new essence. The demonstration that the world in fact exists depends, indeed, upon the world's discovering itself in essence to be in the predicament of existence unaccountably itself apart from being the living flesh of Jesus the Nazarene. The very conception of the world's existence requires the fact itself in essence, the existence of the essence of providence itself. Without this, faced with nothing at all (the essential infinity of things), there would be no conception whatsoever (on the part of the world) of the world's existence. The conception in essence of God's identification with man in the world in essence—the discovery in thought of the other in essence in the world, existing in formal transcendence to the world, in essence the other-transcendence of the world in the substantial form of man—the body itself—such, in fact, is the incomparable efficacy of the love of God in essence, the manna in the desert that this world is in essence. There is here no attempt to escape the difficulty in essence, the ineluctable constriction of possibility itself within the bounds of the historical actuality of thought itself. Neither is there any thought of positing a transcendent existence other than what appears in thought (for example, there is here no transcendent passion to exist), nor is there any refusal to posit an existence other than what appears transcendent in thought (for example, there is here no pure determination of the other to exist transcendently). There is nothing but the appearance itself of the transcendental essence of existence now in thought itself for the first time in history transcendent in fact, that is, essentially transitive, or transcending the formal limitations of the historical actuality of thought itself before now. The difficulty in essence disappears in fact in the Way that is the very existence of the world in essence. Now the Mass (*missa jubilaea*) proclaims what is in fact happening in essence. It becomes *essentially prophetic of the fact that now the appearance itself of faith itself in essence is effected in a transcendentally differentiated substance,* that is, without being other than itself in essence, in the appearance itself of the eucharist. What is in fact happening in essence is the transcendental repetition of the creation itself. This is the appearance itself in the world in essence of the repetition of existence itself in the form of man, of what, before now, even in its appearance in essence remained itself absolutely, or remained essentially intelligible, but which only now, in this thinking occurring for the first time in history in material consequence of the truth occurring in the Mass (*solemnis et jubilaea*), is actually intelligible, that is, absolutely nothing but itself in its appearance in essence without remainder. (In this thinking now occurring there is neither the embarrassment nor the encumbrance, neither the essential nor the formal limitations, of previous thought. The limitation of this essentially new form of thought is purely material: the proclamation in essence of the word, the *missa jubilaea,* the body itself.) In the *missa jubilaea* the appearance itself of the eucharist is effected in substance. The substantial transaction of the Mass is the proclamation of the word in essence. It is *the body itself,* the substantial reflection of the change itself of the universe now beginning for the first time in the form of a thought itself caught up in the essential predicament of existence itself. Now the body itself terminates in existence, that is, in the transcendental thought of being itself crucified with the crucified in essence, of sharing in essence in the pathos of the world, of being in the form of man essentially at the disposal of another. *What now occurs is not the elemental reconstruction of the world in the image of a mere man (as was the case with modern thought from its very inception), not the perception of the world as a mere appearance. What now occurs for the first time in history is the elemental reconstruction of the world essentially in the image of God. It is the perception of the world itself of its being the intelligible appear-*

ance itself of the essence of existence. What is celebrated in the *missa jubilaea* is not simply the human resurrection of the dead one (Jesus the Nazarene), but the divine resurrection of the dead messiah (Christ Jesus). The essential element in the reconstruction of the world now occurring is the Christ-element, the essential element of history. This thinking now occurring is the unreservedly pathetic form of an absolutely passionate essence. What is to be seen now is not simply existence completely changed in appearance, but the appearance itself of existence completely changed in itself. It is the perfect abstract understanding of the absolute essence of existence suffering itself, perceiving itself existing in the form of the other. The perfection of this abstraction is the perception of existence itself. Nothing in essence is lacking to this thought of existence, derived, as it is, in essence from the fact itself. Indeed, the transcendental repetition of the creation itself now occurring is nothing other than the mind of Christ in essence, the body itself, in its divine simplicity essentially the thought of God. This is the absolute reflection of the fact itself of a new universe shaped in essence (through the efficacy of the eucharistic element of existence itself, God's grace) to the form of thought itself. This universe as a matter of fact is not essentially transcendent to thought, but materially transcendental. That is, it is not yet conceived in essence, coming into existence for the first time in history in the form of thought itself—essentially transcendental. The conception in essence of the transcendental matter of the body itself is to be the fact itself in its utter finality. It is to be the end of an essentially transcendental thought in the body itself. That conception is to be the beginning of a substantially transcendent existence, now conceived to be thought's absolutely objective identity, to which, even now, it is drawn in the form of the infinite meekness of God, so, through no necessity whatsoever, but simply through the conception in essence of existence itself (which in this thinking now occurring involves as a matter of fact the beginning of substance). The transcendental substance of an essentially transcendental thinking is life itself. In the body itself the transcendental difference remaining, by which life itself is distinguished from the essence of thought, is to be itself absolutely overcome in essence in the utter finality of the fact itself, wherein an essentially transcendental thought is to be taken up in its essence perfectly so that there is no matter whatsoever (including sensible matter, or, the senses themselves) unclarified by thought itself, including the matter of the fact itself, including the matter of thought itself. Thus, reflection is to know nothing of the fact but what it itself is—to know the fact itself absolutely—full of knowledge beyond even transcendental difference, that is, beyond being itself without being other in essence. Indeed, reflection is to be beyond being the same as itself in essence while being transcendentally different (beyond even the implication of a transcendent difference, whether belonging essentially to the past or not, in the transcendent identity of that thinking now occurring). Reflection is to be absolutely different, to be in essence without reference. Reflection is to be itself materially other, to be the word itself appearing essentially in existence (neither, as before now, materially in time, nor, as now, formally in thought: absolutely neither, nor both, absolutely beyond the absolute [conceivable in terms of the *exsistere ipsum* in this thought now occurring *de facto*]): absolutely the thought of the other: the material repetition of the creation itself, *qua* material, uncreated . . . the Spirit of God, the Resurrection itself, the conception in essence of the transcendental matter of the body itself. In this thinking now occurring it is to be seen that it is not God, Father of Jesus Christ, Who 'in the beginning . . . created the heavens and the earth,' Who has passed away in the form of man, but, rather, that it is those very heavens and earth created in essence which passed away in the form of man before now with the appearance of the word. (Before now in time that appearance was the uncreated *essence* of the world. Now in thought it is the uncreated *form* of the world. Next it is to be the uncreated *matter* of the world.) Now the end of the world created in essence is confirmed with the appearance of the substantial form of the body itself. Now nothing remains to be effected but the absolute conception of the Spirit of Christ. The restriction on that thinking now occurring, while essentially not the self-restriction of that thinking belonging in essence to the past, is, nevertheless, the restriction of *the other* remaining in the form of matter, that is, in the form of the transcendental difference. The transformation of transcendental to essentially absolute difference is to be the appearance itself of the transcendental essence of existence *in existence itself*. Then there is to be no distinction whatsoever of matter from form, not simply in essence (by which the utter finality of the fact itself of this thinking now occurring might, for a moment, appear to have been, in the end, reclaimed by the absolutely self-relating ideal of thinking in the past), but in essence identically substance. In this thinking now occurring there exists in essence an identity of substance and function in the form of transcendence, in the utter finality of the fact itself the substantial identity of the transcendental essence. As a result, the essential indifference of this thinking now occurring to being other than itself remains a formal matter. It is not yet the substantial indifference of the completed thought in which, as opposed to the end of reflection, before now, in a substance thought to be nonexistent, reflection ends in a substance identically absolute thought itself, in which what previously was thought to be nonexistent in fact exists together with everything otherwise thought to be likewise nonexistent. This is what reflection is to be in the substantial identity of the transcendental essence: the resurrection of the dead. That universe now coming into existence for the first time in history in the form of thought itself is to be in its perfection essentially the conception of creation itself wherein nothing is to be known of an essence distinguished substantially from that of existence itself. This world, created in essence, was before now conceived in essence to exist substantially within the limitation of its own formality. It was conceived to be in essence the ideal substance, to be nothing at all in the event of the end of reflection in a substance thought to be nonexistent or in the event of the pure determination of the other to exist transcendently. This world, so conceived, is to have been forgotten in essence in the perpetual realization of the word, the universal essence (*to einai katholikon*) of existence itself. It is to have been remembered only in the substantial form of the body itself as eaten in essence by others existing substantially, that is, as the ingredient perception of an absolutely different life, or only insofar as it became in the course of time nourishment for life itself.

The prophetic transformation of the world into being real food, real drink, the substantial transaction of the *missa jubilaea,* the proclamation of the word in essence—the proclamation that is the word itself now in the form of thought—is the world's being made ready for the essential conception of the word in its priority to thought itself, for life itself, for the Kingdom of God the Father of Jesus the Messiah. The proclamation of the body itself is the final preparation for the messianic meal. The proclamation of the body itself is the world's being prepared to be admitted, through the essential receptivity of creation itself materially repeated (at the time appointed for the creation of Spiritual substance), to the essentially transcendental identity of substance constituent of the communion of the love of Father and Son, integral to the sharing of the Spirit of God in Christ. It is the preparation of the world's being for the absolute other-being of life itself, in which the essential identity of Father and Son is shared without interruption of thought or will, in which the Two are wholly rooted as in the very joy of their existence, in which they behold each the other as their absolute identity in being: absolutely different without being other than the One God. The absolute difference of Father and Son terminates in this Love, sprung from the essential identity of life itself, neither the one nor the other nor both, but a Third in which the exaltation of the One God is complete, root and flower: the Spirit of God in whom in fact this Trinity is reflected, the very joy of God's existence, indeed, the very essence of God's unicity, through whom the world before now has been essentially, now is formally, and is to be materially identified with life itself in the form of the essentially new identity it has become in the course of time, namely, the body itself or the Temple of the Spirit in which the historical structure of existence itself is reduplicated in the form of man. In the body itself, the temple that is God in God in essence (essentially, the risen Christ, formally, the perception of the promise of existence kept, and materially, the substance of history itself), is reflected in existence the Spirit itself, which is in existence the end of reflection.

In this existential reflection of the Spirit now in the world transformed in essence in the *missa jubilaea* is beheld the end in which the word is spoken in essence, in which creation itself exists in the form of man. What is now conceived is the very righteousness of God in which the Son stands before the Father, in which death in essence is completely defeated, bringing to nought the essentially infinite pathos of the negative, forgotten without hope of recall, obliterated absolutely from the Book of Life Itself. But this very righteousness of God, absolutely, is not yet conceived in essence. It remains essentially a reflection upon the perception of the body itself. It remains known to be there, or, the transcendental existence of the perceptible form of man, *qua* substance, a formally transcendental identity. It is *there is known,* it is not yet *there known,* or, the very identity of life itself, the identity in essence of the transcendental substance. In that thinking belonging in essence to the past God's transcendental existence, in one or another of an essentially infinite variety of forms (including, in essence, the forms of disappearance, essential as well as existential), was a necessity of thought itself in the latter's apprehension of the reality of the world. This, in turn, was everywhere self-related or related to being other than itself in essence, but in no event (including the event of the Nothing) was it other-related without being other than itself in essence. But in that thinking now occurring for the first time, reflecting in essence upon the substantial transaction of the Mass, God's transcendental existence is in no sense whatsoever a necessity of thought itself. Thought itself has been dispensed in essence from reflecting upon a world other than itself in essence, the world itself in essence having been transformed into the Temple of the Spirit, the identity of life itself, in essence other being. God's transcendental existence is simply a matter of this appearance itself in thought for the first time in history. Indeed, it is the substantial change itself of this world into being in essence at the disposal of another in the form of man, into being itself the bread of life here and now. It is not the purely formal embodiment of God himself (the embodiment of nothing in essence), but the embodiment of God in essence, that is, of Christ in fact resurrected. But, then, the embodiment of God in essence is in fact the embodiment of Christ in God, his being received of God in the form of man. But then, further, it is the embodiment of the Body Itself, the Temple of the Spirit, or the Spiritual Christ in God. Thus, finally, *the world in essence* is seen to be the *Christ embodied in God,* the *Christ specifically effected by the Spirit, that is, the Christ that the church essentially is, namely, the Lamb of God,* the scandal that the church essentially is in the world, the blood witness to the Love of God. Thus, the blood shed of the Lamb is seen for the first time in history in thought itself to recreate the transcendental world in essence, so that the world of our intelligible experience is made over into the substance in essence of Faith Itself. This perception itself is implicitly the complete exaltation of being itself, in which matter itself is to be itself exalted, is to cease to be in essence apart from being in essence identically the transcendental substance, appearance itself transformed in essence. In this thinking now occurring for the first time in history the universe itself in essence is transformed into being, *qua* substance, the essentially transcendental identity of the body itself of God, *ἡ ἐκκλησία καθολική*. The universe remains as a matter of fact for the time being formally transcendent to thought, the beloved object of God's love in essence, the material essence of this thinking not yet thought itself, but essentially anticipated in the form of this thought, that is, in time itself transcendentally differentiated. This thinking now occurring humbly acknowledges in essence the other-being, in fact, of the matter with which it is formally identified, that it is the Lamb of God, in fact not thought in essence, who takes away the sin of the world. Further, it is the Lamb of Christ (the Lamb of God in essence), the church, which, in the substantial transaction that is the *missa jubilaea,* takes away the death of the world in the proclamation of the word in essence. The efficaciousness of the word spoken in essence is witnessed to materially in this thinking now occurring in which is transcended the essential resolution of past thinking to reserve to itself in essence (in the form of being itself a position in essence) its final disposition (thereby, precluding in essence a conception of the fact of existence itself in essence). In this occurrence of truth itself the essential resolve of that thinking essentially in the past, to be in essence impregnable, indomitable, or, if subdued, self-subdued in essence, is, finally, overcome in essence in the form of the transcendental repetition of creation itself, through the appearance itself of the infinite meekness of God in essence (though the doors, as it were, remain still bolted shut in the citadel). Now itself, the body itself is in thought for the first time in history in the form of perception itself being there in essence: the conception in essence of existence itself transcendentally thought. This essential repetition of the essence of history now for the first time in thought is, in fact, the contradiction in essence of past thought's essential intention to be lord of its own death in essence, to be in essence immortal, absolutely not to die even in the event of the Nothing in which it disappears essentially in the recollection of Being upon itself. In the absolute clarification of the absolute now occurring in reality the absolute intention of that thinking essentially in the past is immediately comprehended (in the form of its truth [its immediate appearance], as well as in the disappearance of its substance [its self-clarification]) as formally divine, but essentially human. That is, that thinking is the essentially transcendental repetition of its own intention. Formally it is the transcendental thought of God, appropriated to all appearances without exception. But, in essence, it is the transcendental repetition of nothing but itself essentially unchanged. Essentially that thinking is the diversion to its own end of the transcendental repetition of the creation itself now occurring for the first time in history without being other than itself in essence. It is the temptation of faith itself in essence to be for another in essence rather than being simply at the disposal of another in essence. This final temptation to place one's self in the place of God is overcome in that thinking now occurring, absolutely unhindered, as it is, in essence by the Notion of Self. The absolute intention of this thinking is not merely formally divine, as was the case with the thinking belonging in essence to the past, but essentially divine, thereby transforming the form itself of the divine into the body itself in essence, essentially into the Lamb of God. In consequence of this essential origination the form of this thought is not merely not a repetition in essence of the sin of modern thought, but a nullification absolutely of the death of this world. What now occurs in thought is the transcendental perception of the church of God's very righteousness, the temple of the Spirit, as John described it in *Revelation* 21:22–27: "I saw that there was no temple in the city since the Lord God Almighty and the Lamb were themselves the temple, and the city did not need the sun or the moon for light, since it was lit by the radiant glory of God and the Lamb was a lighted torch for it. *The pagan nations will live by its light* and the kings of the earth will bring it their treasures. *The gates of it will never be shut by day*—and there will be no night there—and *the nations will come, bringing their treasure* and their wealth. Nothing unclean may come into it: no one who does what is loathsome or false, but only those who are listed in the Lamb's book of life."[396] Prior to this thinking now occurring for the first time in history there was no conception in essence of the spiritual reality of the church, that is, of the universe transformed in essence (*hē ecclesia katholikē*). Before now catholicity was understood to be an attribute of the church. Now in essence it is conceived to be the predicate itself of the church. The church itself, caught up, as it is, in the predicament of existence itself now occurring for the first time in thought, exists in essential continuity with the universe itself in its new actuality, in an identity transcendent substantially, but not essentially, to thought itself. Now the new world is conceived in essence. That is, it exists in fact in the form of thought itself essentially new, or in the form of the body itself. The new actuality of the world is its being in essence the temple of the Spirit thought. Before now the actuality of the world was either the substantial immediacy of its existence (possessing an identity of its own transcendent to thought) or the transcendental identity of thought itself (for which existence in essence was a mere appearance, in itself nothing substantial). So, in Kierkegaard, to take the substantial instance before now of the predicament of existence (not, therefore, the essential instance of the predicament of existence itself now occurring in thought),

[396] Ibid., 450.

we read (*Papirer* X^2 A 439): " 'Actuality' [*Virkeligheden*] cannot be conceptualized. Johannes Climacus has already shown this correctly and very simply. To conceptualize is to dissolve actuality into *possibility*—but then it is impossible to conceptualize it, because to conceptualize it is to transform it into possibility and therefore not to hold to it as actuality. As far as actuality is concerned, conceptualization is retrogression, a step backward, not a step forward. It is not as if 'actuality' were void of concepts, not at all; no the concept which is found by conceptually dissolving it into possibility is also in actuality, but there is still something more—that it is actuality. To go from possibility to actuality is a step forward (except in relation to evil); to go from actuality to possibility is a step backward. But in the modern period the baleful confusion is that 'actuality' has been included in logic, and then in distraction it is forgotten that 'actuality' in logic is, however, only a 'thought actuality,' i.e., is possibility. Art, science, poetry, etc., deal only with possibility, that is, possibility not in the sense of an idle hypothesis but possibility in the sense of ideal actuality. But is not the historical actuality? Certainly. But what history? Six thousand years of the world's history is certainly actuality, but a traversed actuality; it is and can exist [*vaere til*] for me only as thought actuality, i.e., as possibility. Whether or not the dead have actually realized [*realiseret*] existentially [*existentielt*] the tasks which were before them in actuality has now been decided, has been concluded; there is no more existential actuality for them except in what has been traversed, which for me, again, exists only as ideal actuality, as thought actuality, as possibility." [397] Before now actuality in essence was the absolute self-relation of the appearance conceived to be in itself nothing but a possibility, other than itself in essence. Indeed, in the moments of thought's absolute self-clarification what was denied in the first instance was that the essence of thought existed in substance. What was not denied (because in fact before now it could not have been denied) was that reflection existed *qua* essence, that it had an

[397] *Søren Kierkegaard's Journals and Papers* I (ed. Hong and Hong, Bloomington, Ind, 1967), 461.

actuality of its own, an ideal or possible existence, a purely formal existence, transcendent to being conceived in essence. Substantial immediacy conceded the actuality of thought since to do otherwise would have meant the absolutely impossible denial of the possibility of its own actuality, of its own dissolution in thought, since it self-evidently had no knowledge whatsoever of creation itself. It substituted for the fact itself of creation its transcendent passion to exist. Thereby it remained, although by way of the absurdity of an absolute self-contradiction, within the horizon of possibility itself, conforming in essence, therefore blindly, to the thought actuality of another's existence, transcendent in essence, a pure formality. Thus, darkly apprehending eternal existence, it existed before God. *But it is precisely this possibility itself that is annulled in the absolute clarification of the absolute now occurring in reality in the form of the transcendental repetition of creation itself* (the essential repetition of history in thought), *in substance the absolute nullification of everything including thought itself* (in which, therefore, the history of thought is for the first time actually thought in essence), *transcendentally distinguished as now occurring in essence.* (The difference between substance and essence is itself no longer transcendent, but merely a matter of time itself. In essence it is the conception itself of the Spirit implicit in the perception of the body itself.) *The annulment of the ideal actuality (actuality thought in essence) is the perception of the fact itself: the perception of a new transcendental,* exsistere ipsum. *The essence of the new transcendental involves the conception in essence of everything, the absolute nullification of the possible. Everything now comes to exist actually in the form of the body itself.* (It goes without saying that now absolutely nothing is possible but what actually exists in essence. But that what actually exists in essence [in the form of man] now is an other in whom every-thing possesses its transcendent identity.) Now nothing is a mere formality; everything is essential. There is no idle word, now that the word is spoken in essence. With the appearance of the transcendental essence of existence itself now in thought itself it is clearly seen that the actuality of substantial immediacy, that is, existential actuality as the contrary of the ideal actuality, in fact never existed. Rather, it was mere formality, a moment in the self-clarification of the absolute. It was the negative thought absolutely annulled in the transcendent passion to exist. Further, it is clear the traversed existential actuality of the dead never existed in essence, but remained the negative ideal of individual transcendental self-consciousness. Construed absolutely, this ideal was the prototype, Christ himself, nonexistence in essence, an eternal sign existing without being the church in essence, without being the Lamb of God, who takes away the death in essence of the world. This ideal existed without being the body itself, the resurrected Christ. Indeed, in this thinking now occurring it is clear that Christ himself has never existed in essential transcendence to the church. Before now he existed essentially in the form of the presupposition of the appearance of the transcendental essence of existence itself. He existed in the form of faith itself, understood to be, in the transcendental consciousness of the church, transcendent *qua* appearance. He was understood to be substantially immanent in the form of a complete change of the elements of this world, elements otherwise existing in time and space, in more or less specific locations. Before now Christ existed inessentially, that is, unchurched, in the form of the ideal of an intellectually desperate faith bereft of faith itself. This faith, just so, was precluded from understanding itself in essence. Now Christ himself exists in essence for the first time in history in thought itself in the form of the annulment of faith itself. He exists in essence in the conception of the body itself, the appearance itself of which is transcendentally distinguished from its substantial existence, *qua* matter, as *not yet conceived* in essence (so that the substantial existence of the body itself is materially intelligible in the form of this thought now fact-evidently existing for the first time in history, although not yet is it absolutely, *qua* matter, the transcendental essence, creation itself). *The annulment of faith itself absolutely coincides with the nullification of possibility itself.* This absolute coincidence is the thinking in which for the first time faith thinks for itself—in which thought hears the word for itself. This absolute coincidence is the perception of the fact itself in essence, transcendentally distinguished from being itself in its utter finality. It is the perception that now being itself is essentially historical, that neither sin nor death belong to the world in essence. What is now broken through in essence in thought itself is this world's dominion over its own fall. Even the world's absolute appropriation of sin itself is simply annihilated in this thinking in which the historical reality of Christ in essence is seen to be *the evidence of creation's being an act of absolute freedom in which everything comes into existence without exception, without precluding exception in essence*. Not that Christ himself in essence is that exception. Rather Mary, the virgin, is. In the virgin the appearance of the transcendental essence of existence itself came to be everything without exception for the first time in essence. She, in whose existence the essence of history is, *qua* essence, implicit, is the exception in essence. She, the virgin, is essentially implicated in the essential repetition of creation itself. For her nothing existed in essence but God himself. In her being, conceived before now in time without death in essence (extending even to the flesh), was perfectly prepared the way of the messiah. Christ himself was no exception to being conceived before now in essence in time as not being without death in essence. Indeed, he so suffered death in the flesh. He suffered the possibility of being conceived to be sin in essence or the annulment of absolute death in the form of appropriation (or, in the event of Nothing at all, the justification in essence of being disappointed with God). It was, then, in Mary, the bondmaid of God—in her person—that the faith of Israel was consummate; for her the messiah was everything. In her very being was planted the seed of faith itself in essence. In her perfect submission to the will of God this young woman came in the course of time, through the evidence of the Spirit, to a knowledge of the very righteousness of God, a knowledge which, since it encountered *nowhere* in Mary's exceptional being resistance in essence, bore fruit in the person of Jesus. In Jesus, in turn, the infinite meekness of God is beheld, now immediately in essence, without being an exception. Indeed, now in the form of the body itself is seen everywhere the form of man. The proof of the possibility of the transparency of the eucharistic essence of existence itself now occurring for the first time in history is the virgin birth. No proof of the actuality of what now occurs is possible other than the perception in essence of the fact itself. But, in any event, it is clear that the existence of sin in essence is not inherent in creation. What now occurs is the exception in essential form, an essentially transcendental exception, the exception that is thought itself now in essence in place of the absolute nullification of death itself. There is now in essence no obstacle whatsoever to the transcendental appearance of the essence of existence itself in thought, as, before now, there existed in Mary in essence no obstacle whatsoever to the identical appearance in time. Now it is clear that absolutely nothing whatsoever is necessary, including the necessity that God be compensated for creation by an imperfection in the form of the creature. God himself in essence is the incomparable evidence (implicitly the annulment of the absolute exception). God in essence is, entirely compatible with essential perfection in the form of the creature so disposed in absolute freedom, the *direct* evidence of the fact itself of creation, of the absolute freedom in which death in essence entered the world (in which sin itself is predicated of the world), there being in the Lamb of God explicitly the demand for perfection heard in essence. Nevertheless, in the exception in essence perfection is a purely formal phenomenon, since either, as before now, existence itself is a matter of the person or, as now, existence itself is a matter of being itself in essence, a matter absolutely of appearance in essence. In the exception in essence existence itself is now the matter of a new transcendental in the form of an essential predicate, the body itself, in terms of which the world is, for the first time in history, understood to exist in essence. The world is itself everywhere the body of God in the form of man, a conception of essential perfection not yet annulled in fact, that is, not yet absolutely unconditioned, but remaining for the time being the penultimate repetition of the end. In this thinking now occurring life itself is conceived to be, formally, the substantially transcendent identity of the transcendental (the essence, *exsistere ipsum*). In the transcendental repetition of creation itself now occurring the spiritual reality of the church in essence is conceived to be the absolute nullification of the Trinity now occurring for the first time in thought (not, therefore, the relative nullification as in that thinking belonging in essence to the past). This repetition is the absolute nullification of the historical structure of being itself, involving thought itself for the first time in history, the body itself of God, the actuality of an essentially new universe now in thought (essentially the Spirit dwelling in the temple). God is in fact (being there) in the absolute nullification of God—a conception in essence even now nullified in fact, but remaining still, formally, this absolute coincidence of faith itself with possibility itself; not yet the substantial transformation of time itself into motion itself in the absolute indifference of being itself annulled; not yet the absolute other-being of life itself. Now for the first time in history the communion of saints is seen to be the perception of the body itself in the form of man in thought itself so that the dead themselves, it is clear, do not escape the transformation of the world now occurring for the first time in history. But, indeed, themselves, the dead, are now, that is, formally, the substantial appearance of the transcendental essence of existence itself. That is, the dead are now transcendentally the resurrection, so that the dead are now for the first time able to understand themselves to be transcendentally identical with the body itself as the form of their essentially transcendent existence. At this point it can be clearly seen that modern thought, belonging, as it did, in essence to the past, was in its essential formality the abstract anticipation of the thought of the dead, never in essence the thought of the living, never the thought of God himself in essence which even the dead themselves can now be seen to share in as it occurs for the first time in the form of the body itself. The communion of saints is the indirect evidence of the fact itself of creation, of which communion of saints, precisely because of the severalty of its constituents, exceptional being in essence must be formally, not essentially, predicated. The communion of saints is the transcendental existence of the exception in essence, existing before now formally, now for the first time essentially, that is, now for the first time caught up in the essential predicament of existence itself. As such, it is the direct evidence of the essential repetition of history in which event it is now clear that it was not merely death (nothing in essence) that was overcome in the shedding of the Lamb's blood, but death in essence (sin, or, the essential nullification of nothing, or, the existence in fact of nothing in the form of man, or, the bread of life itself uneaten). As a result, the barrier separating the living from the dead is seen in fact to have been merely death (nothing in essence), a pure formality behind which in essence never simply existed the so-called traversed existential actuality of the dead. In fact the fruit of the transcendental repetition of creation itself implicit in the essence of history has been known before now in one or another form of faith itself. In fact it was, moreover, the sore temptation of faith itself to be thought to exist simply for another—a temptation historically conditioned in essence (of this fact faith itself before now would have been formally ignorant). This temptation materially obscured what is now the evident fact of thought itself for the first time in history, namely, that not only was death overcome, before now, in the blood shed of the Lamb, again, not only sin, but, as is immediately evident in the occurrence of the fact itself in essence for the first time in history in thought itself, faith itself in essence is now provided against its final temptation. It comes to an essential knowledge of itself as being in essence at the disposal of another. It comes to know itself as the body itself in essence in which form the blood shed of the Lamb is seen to overcome, in essence, sin in essence (both appropriation itself, and the absolute nullification of death, the bread of life eaten in the absence of the perception of the body itself), to absolutely annul sin, to bring it into existence for the first time in thought itself as that which (death in essence) belongs essentially to the universe now about to be left behind. This universe itself is now seen to be the perception of a thought itself belonging essentially to the past, a thought for which creation itself was a mere formality. But, it may be construed as the wrath of God that, while sin is now absolutely annulled (together with everything else, including thought itself, in substance in the transcendental repetition of creation itself now occurring), it is, just as everything else, including thought itself, not yet absolutely annulled in essence in substance: it *so* appears in existence not yet. Not yet is sin in essence under the necessity of appearing materially, although it was before now so essentially, and is now so in thought, where it is under the necessity of saying nothing in essence, under the necessity of bringing nothing into existence in essence . . . indeed, it dare not make a move lest it betray what appears to it to be its successful disappearance in essence. It remains absolutely still in essence . . . as it were, death itself playing dead. On the other hand, the motion now occurring for the first time in history occurs in essence in the transcendental substance, life itself, now the transcendent identity of thought itself. Motion itself is now seen to be in essence the material appearance of the identity that the Spirit of God is. While the glory of God itself is now for the first time in history essentially visible in thought, appropriation itself is yet identified formally with a mere appearance of being conceived in essence. Not yet, as is to be the case in the utter finality of the fact, is the absolute annulment of death the absolute annulment of absolutely nothing, visible *qua* essence, the pure disembodiment of light itself. Not yet is it not the body itself but the dark body, the spiritual holocaust, being nothing but death itself, unseeing and unseen, except in the form of the absolute absurdity of self, the negative absolute. Not yet is it in essence an absolutely superfluous world of necessity, an absolutely forgotten episode in the history of life itself—the mere occasion of a triumph in essence the reception of God: the absolutely eucharistic existence.

This perception itself is the essential constitution of the communion of saints (for the individual the motive power of its inclusion in fact). In light of this newly manifest intelligibility of the essence of faith it is clear that the specific locus of the body is the necessary superstition of a world whose thinking belongs essentially to the past. An expression of its will to survive death without faith in essence, without faith to survive sin, to be, after sin, immortal, in the aftermath of the appropriation of history itself to perpetuate the dead, to entomb them in one place, in one time, to equate death in its effect with the body, life with the soul (there being no communication between the two). But in this thinking which thinks the occurrence of time itself death is essentially equated with nothing. The locus of the body is not in the tomb; nor, in this perfect contradiction of the aftermath, is the soul, neither actually nor ideally, an alternative to the body. Rather, it is seen to be the place of the body's transcendental reconstitution in essence. For the first time in history the soul, now in essence other-conscious, is the plane of the resurrection upon which the body itself exists in essence. The soul is the place where the body itself is seen for the first time to exist in the form of man, to be in essence the resurrected Christ. The essential locus of the body is the soul, that is, the consciousness that the body is now essentially other. In essence the soul is the body itself perceived, in which perception it sees itself to be essentially historical, essentially the transcendent identity of substance now existing for the first time in the form of thought itself. Now, for the first time, the soul sees itself to be in essence the flesh and blood of Jesus the Nazarene thought, apart from which identity in fact it were absolutely impossible to conceive flesh and blood in essence, that is, *qua* creature, or in the form of man. In this new state of being there essentially intelligible, it is no longer a question of the merely formal faith of the specific individual. The substance of faith itself in essence is not the certitude of the individual (*a fortiori* not that of the congregation). But it is the transformation of the world in essence effected in the transubstantial action of the word being proclaimed in essence in the *missa jubilaea*. In this transformation the individual is essentially caught up beyond his own limitations, without being other than himself in essence, other-transcendent in essence, *qua* creature another individual in essence, the absolute annulment of another individual. Now (while sin in essence comprehends itself as not being perceived in essence, for the time being permitted its self-delusion) the individual is essentially transcendental; then, in the utter finality of the fact itself of existence, in the absolute annulment of another individual in essence, the individual is to be absolutely another, absolutely undifferentiated from life itself, without essential reference to another creature, the transcendental substance of existence itself *qua* creature, the exaltation of existence itself in the form of man in essence. There is, then, in the essential predicament of existence itself now occurring in the form of thought itself, no question of the merely formal faith existing before now wherein the dead were consigned to themselves or to God. But, in the form of the essential conversion of the world now occurring, the individual in essence is opened out, as a flower is exposed in its heart, into the community of saints. Essentially he is transformed into the perfect substitute itself, being in essence at the disposal of another, without being in essence for another. The individual is there in essence in the perception of the body itself, the very ground of which is the essentially perfect substitution that took place before now in time in the appearance of the transcendental essence of existence itself, the essential repetition of creation itself (the essence of history). *This transformation* qua *creature into the essentially perfect substitute* (the contradiction in fact of that disappointment in existence which to begin with substitutes its self for God in essence; therefore, as well, of the contrary form, namely, God's essentially identical self-substitution) *is the end of death's rationality, the demise of its necessity, its end in essence, that is, the end of sin in the form of an inessential thought of existence*. Thought is now essentially in the form of the end of sin in existence, that is, *is* the form of the absolute nullification of sin. Thought is now the absolute nullification of not yet in which the difference remains essentially transcendental. *No longer is the difference between the living and the dead. There is now nothing but the transcendent identity of the essentially perfect substitution of the living for the dead in which the dead themselves live in essence. That is, they share actively in the substance of life itself, share essentially in existence itself, share, that is, in the body itself.* In the body itself, in the essential predicament of existence itself now occurring for the first time in the form of thought itself, as God is Christ, as Christ is the other, so the one is the other in essence, nor is the one essentially substituted for the other in essence. Such is the clarity of the transcendental distinction that all confusion in essence is precluded. Such is the absolute condition of the fact itself that difference exists without being other in essence, identity with being other in essence. Apart from the life of the body itself there is no other in essence. There is nothing but the final loss of identity, nothing but the transcendent difference comprised in the thought of an inessential existence—in the utter finality of the fact itself existing in the form of absolute-self-recrimination—absolutely without the form of man. For the time being there exists not yet in essence the identity of self-perception, the pure formality of transcendental identity, essentially insubstantial, without love in the form of man in essence. Not yet does that exist in essence. The death of the individual remains substantially qualified by the fact of the body itself. In this thinking now occurring it is no longer possible to make sense of death itself. No longer has death a reason of its own for being of which we know nothing in essence. But, now for the first time in history, death is clearly perceived to be nothing in essence, a nothing which has its existence in sin (in the essential nullification of nothing). Death is no longer essentially transcendent (the transcendent is not nothing in existence, but what now for the first time in history comes into existence in thought, namely, the word made flesh in essence, the body itself, the substance of an essentially transcendental existence). So, the opposite of sin is now no longer as before now it appeared to be (necessarily, in the absence of a comprehension of the essential history of thought) substantially the faith of the individual, the transcendent passion to exist. Now the opposite of sin is seen to be for the first time in history (which is nothing other than to be such and so) the transcendental passion of existence itself in the form of the transcendental repetition of creation itself, sin's essential contradiction, the perception of salvation itself, the end of sin in the form of thought. Thought itself is nothing in essence except it be essentially suffering, the suffering in essence of history itself, infinite in being essentially transcendental existence, absolutely precluding, since we in fact exist transcendentally, the possibility of denying the same to others in essence. It is of the essence of the transcendental that it is shared without limitation. In the case of the absolute nullification of transcendental existence now occurring in thought, the substance of existence is seen to be infinite in essence—unconditionally the meekness of existence itself—the inception of a transcendent identity (there being in creation itself essentially no thought of death). What now occurs is the essentially distributive thought of existence itself: the thought in essence of the unleavened bread of existence itself, distributed before now by Jesus the Nazarene, in substance his very flesh and blood, the night before he died, the night he suffered death in essence, the night he suffered the essential nullification of nothing, sin, as it appeared before now in the person of Judas. Paul tells us (*1 Corinthians* 11:23–29): ". . . on the same night that he was betrayed, the Lord Jesus took some bread, and thanked God for it and broke it, and he said, 'This is my body, which is for you; do this as a memorial of me.' In the same way he took the cup after supper, and said, 'This cup is the new covenant in my blood. Whenever you drink it, do this as a memorial of me.' Until the Lord comes, therefore, every time you eat this bread and drink this cup, you are proclaiming his death, and so anyone who eats the bread or drinks the cup of the Lord unworthily will be behaving unworthily towards the body and blood of the Lord. Everyone is to recollect himself before eating this bread and drinking this cup; because a person who eats and drinks without recognizing the Body is eating and drinking his own condemnation." [398] In the suffering of sin the body itself appears, not in the suffering of death the next day; in death there is simply nothing. In the suffering of sin there is everything itself in its true identity, the love of Christ for others. The essentially transcendent identity of Jesus the Nazarene is the absolute *a priori* of Christ's death, that is, of God's death in essence, of God's suffering sin. The resurrected flesh and blood of Jesus the Nazarene is the fruition of the primordial intention of existence itself, existing in fact in essence prior to its realization in time. It is the fruition of the intention of other being in essence, the intention which exists in essence prior to its being perceived by others (and which, in its realization before now in time, now in thought itself, absolutely exists without reference in essence to the notion of self). Before now, the suffering of sin was the purity of heart in which God himself was seen. What now occurs is that purity of heart is the perception that God himself suffers appropriation itself. That is, Christ suffers sin, or the church itself now faces its death, in

[398] *The Jerusalem Bible: N.T.*, op. cit., 303.

anticipation of which essential death of Christ the universe itself is essentially transformed into the body itself in the *missa jubilaea* now occurring. In this occurrence, for the first time in history, the Lord's supper is essentially conceived. In this occurrence, it appears in existence. To share in this is the same as to exist. This is the final suffering of sin, the sin against the Spirit of God, in the midst of which essential provision is made by God in the form of the elemental Christ against the utter finality of the fact of existence itself. Before now, Judas, tempting God, sinned against the Son of Man. Then, in the death of Christ, God suffered the form of sin, died in the form of man, that is, *qua* creature. Now God suffers sin in essence, in the death of the church dies in the essential form of man to be resurrected in the form of an essentially new universe now existing for the first time in the form of the body itself through its being identified in essence, in this thinking now occurring, with a transcendent substance (there being in essence in this thinking no transcendent substance other than the flesh and blood of Jesus the Nazarene). Nor is the death of the church to be construed to be its end in essence. Neither church nor Christ end in essence (the simple essence: nonexistence), both, in essence, end in existence (the complex essence: existence itself). The church dies (not Christ) in *imitatio Christi* (Christ having died once and for all in the person of Jesus the Nazarene). Christ (not the church) now dies in essence, that is, the church does not now suffer sin except insofar as God himself suffers appropriation itself. Except insofar as the church itself is the absolute nullification of God himself, it does not suffer sin, that is, except it is essentially Christ. But then it suffers nothing in essence apart from Christ. The death of the church is its being Christ forsaken, the death in essence of Christ, the absolute annulment of the death of God, the death of God in the form of the church. It is God's being now in the essence of the church (God in God in essence) the transformation of the church in which God is seen to be in essence everything in everything. In this transformation God is seen to be in essence the perception of existence itself, in essence the middle term, the existence in essence of the predicament of existence itself, in essence the fact itself of existence, in essence the suffering of sin. Indeed, the possibility of the purity of the body itself is now, in the absolute nullification of everything, narrowed to this perception itself of the suffering of sin as being in essence God's. Apart from this absolutely historical perception there is in fact no salvation, in essence no exception to being itself, no transcendent form of the Word, no creation itself in terms of which being is made flesh in essence. Now the ban is lifted in the form of the conception in essence of the innocence of the world; now the ban is lifted in time itself in essence, that is, in the form of thought's construction of an essentially new universe. It occurs to the church to recognize that the nature of the situs in which it finds itself is now for the first time in history changed in essence through its having come to exist in the first place in that very place, that there is now no alternative to the predicament of existence itself in which it itself is caught up (there being before now an alternative in the conception of time itself as being in essence not yet). Now time itself is here and now in essence. There is nowhere in essence where the substantial transaction of the *missa jubilaea* is not effected in essence, where the word is not proclaimed in essence. No longer does there exist the necessity of faith's taking up a position outside of the world in order to move the world. What now exists is the motion itself of the world in which faith itself exists in essence. No longer is faith the spiritual device of the individual, the knight of faith, appealing from science to action, no longer the weird virtue of a fantastic realm. In the absolute clarification of the absolute now occurring, there is no action whatsoever (including the absolutely passionate action of faith itself in essence—nay, included by it in essence) transcendent to thought itself in essence (in a world in which the absolute itself exists in essence). There is now no refuge whatsoever from the essential predicament of existence itself, nowhere where thought itself has not now penetrated, where light itself does not now exist in essence, illuminating the utter universality of sin. So, it is now clear that there is no action, including that action which demonstrates its faith by being willing to be sacrificed, which is not essentially sin, which is opposed to sin in essence (appropriation itself), except that action which is being in essence at the disposal of another, that is, apparently, God's suffering Nothing himself, so that the Lamb, for all practical purposes, appears to be Christ forsaken. There is no immediacy, including the immediacy of faith, both before and after reflection, which is able to maintain itself in face of the absolute passion of God himself (indeed, the distinction of faith before and after reflection is now seen to belong in essence to the past, now that reflection itself is seen for the first time in history to end in existence itself in essence). Now that the word is seen to be spoken in essence, no longer is the word spoken out of necessity, but speech itself is grace. Now that the gate is opened in essence, it is no longer necessary to knock, knocking itself being encompassed by the fact of existence itself. To knock is to be seen to be within in essence, there being now no transcendence other than the substantial identity comprising admission to the fold, contradicting in fact the transcendence of nothing in essence. There is no transcendence of death. In fact there is nothing in essence but transcendental death, or Christ now suffering sin. This is the historical locus *par excellence* in which God himself suffers appropriation itself, in which the church suffers nothing in essence, that is to say, in which the church is the absolute nullification of God, that is, in which the church is what now occurs, or, what now occurs, *qua* absolute, is the church, God in God in essence, the temple of the New Jerusalem (while what now occurs, *qua* thought, is the appearance of the transcendental essence of existence itself, the reflection upon the church in the form of the body itself, the reflection that ends in existence in essence [the reflection that is in essence faith in the deliverance of God, the anticipation absolutely of the end of the world]). There now exists the transcendental annulment of the substance of death. Death is now thought to be life itself; the transcendent identity of transcendental death is the resurrection (the absolute contradiction of self-transcendent death). Transcendental death is the death of self which is the flesh and blood of Jesus the Nazarene. Death is nothing in essence except it be the eucharistic substance of the communion of saints, the absolute nullification of another individual in essence (the utter annihilation of the self in essence), the unleavened bread of existence itself. Death is now conceived to be the manna upon which the body itself lives, the sweet fruit which has come down from heaven (as we read in *John* 11:25–27: "Jesus said: 'I am the resurrection. If anyone believes in me, even though he dies he will live, and whoever lives and believes in me will never die. Do you believe this?' 'Yes, Lord,' she said 'I believe that you are the Christ, the Son of God, the one who was to come into this world.' " [399]) The blood of the martyrs, *semen ecclesiae,* is in essence the blood of Christ. That is to say, death, transcendentally comprehended, is the blood shed of the Lamb of God. The martyrs, Christ in essence, are now seen to be everywhere, although in the case of the individual the essential qualification for inclusion in the communion is the perception of God himself suffering appropriation itself, the purity of heart which suffers an essentially unnecessary death, a death of which existence itself is essentially predicated. The blood shed of the Lamb of God in which God himself remains, the death in which God himself remains, for the time being, without his Son, is the new covenant in the blood now conceived in essence. God is bereft of God in essence; the contradiction is overcome in appearance itself in essence, the world face to face with God in this magnification of his righteousness, in this perception of God's essential unicity, in which he is seen himself to suffer sin in essence. It may be said that God himself suffers the death of faith in essence, that he himself stands in the essential locus of the individual, that God absolutely is the essential individual, that God being Christ the individual in essence is God in essence, that God is there within the sound of every individual's voice, that Christ is the palpable reality of the world in essence. There now occurs the transcendental time of which before now Jesus spoke in essence, and of which he was the transcendent embodiment. That is, the time now occurs which before now existed in the flesh (the repetition of the creation in the form of man, or, the word made flesh), and which now exists in the flesh in essence (the transcendental repetition of creation itself): of

[399] Ibid., 171.

this time Jesus spoke in his reply to the Samaritan woman (*John* 4:20–26): " 'Our fathers worshipped on this mountain, while you say that Jerusalem is the place where one ought to worship.' Jesus said: 'Believe me, woman, the hour is coming when you will worship the Father neither on this mountain nor in Jerusalem. You worship what you do not know; we worship what we do know; for salvation comes from the Jews. But the hour will come—in fact it is here already—when true worshippers will worship the Father in spirit and truth: that is the kind of worshipper the Father wants. God is spirit, and those who worship must worship in spirit and truth.' The woman said to him, 'I know that Messiah—that is, Christ—is coming; and when he comes he will tell us everything.' 'I who am speaking to you,' said Jesus 'I am he.' "[400] In the essential perception of the divine unicity now occurring it is seen in essence that no movement is necessary, no journey need be undertaken. Not that no motion in fact takes place, but precisely because it exists in essence it is in essence unnecessary. The magnification of the divine righteousness in essence, the complete joy of the Father and Son in existence itself now perceived in the substantial reality of transcendental death, is the absolute acceptance of the world's being itself in essence, the sin in which God is God, in which God is spirit (in which it is manifest that in fact man in essence is not God, but that man in essence is Jesus the Nazarene, so that the individual absolutely is not simply man in essence nor is he God, but God in essence, or, *the essentially transcendental embodiment of God*). No longer is the individual confronted with the necessity in essence of positing his existence before God. Now, in fact, he is relieved of that necessity in essence by God's identifying himself with the existence itself of the individual (an identity in essence conceived in essence in that thinking now occurring for the first time in history), not, be it noted, with being the cause of the existence of the essence of the individual, nor with the sufficient reason for the existence of the individual essence, but with that existence itself which the individual in essence is. God identifies himself now with the

[400] Ibid., 153.

resurrected Christ, not simply, as before now, with Jesus the Nazarene. That is, not simply does he identify Jesus as his Son, as the Christ, as before now, but now identifies himself absolutely with the essence of history in the essentially transcendental identity of the transcendent individual with another in essence. God himself enters into the suffering of Christ in the transcendental repetition of creation itself now occurring. Before now, in the transcendental form of natural reason, God was thought to have *his* identity in *existence itself* (*suum esse ipsum*). Now God himself identifies himself with *existence itself in essence,* that is, *exsistere ipsum*. Before now Yahweh said to Moses (*Exodus* 3:14), " 'I Am who I Am.' "; before now God said within the hearing of some men, 'Jesus is who I Am'; now God says in essence within the hearing of every man, 'I Am Christ'; and no man in this world is now able to say in truth, 'I Am Christ,' for no man in this world now suffers sin but every such is essentially a sinner; this is the judgment of the word spoken in essence.

When God now says 'I Am Christ,' he says 'now the church is who I Am.' In this New Jerusalem, God together with the Lamb is the Temple, now the existence itself in essence of the individual. In the substantial transaction of the *missa jubilaea* now occurring for the first time in history the bread and wine on the table are transformed in essence into the body and blood in essence of Jesus the Nazarene, not, as before now, despite appearances to the contrary, nor as the matter, the form, nor, finally, as the essence of that purely natural faith that arose as the concomitant of modernity's radical subjectivity. Now they are transformed fact-evidently as comprehended in that thinking in which is seen in essence the fact that nothing in essence appears except what has been conceived in essence. But what now is conceived in essence is the transcendental substance of life itself, not yet appearing, therefore, in existence absolutely. So, breads and wines other than those on the table at mass are transcendentally differentiated from these latter material originals (for the time being) as being inessentially conceived to be the body and blood of Jesus the Nazarene (whether conceived to be so formally or materially), without being substantially different than these essential originals. That is, they are themselves in essence the identical transcendent substance. Now for the first time in history it is clearly perceived, that is, in essence, that all are saved in essence through the appearance of the transcendental essence of existence itself, that God is absolutely Christ. There is now absolutely no necessity to sacrifice those differences now seen to be in themselves inessential, to belong in essence to the past. Those differences are essentially a reflection of the ways in which before now God was necessarily conceived by a thought for which the time itself of its conception of the fact of existence itself in essence had not yet arrived in essence. Those differences are now set aside in essence in face of the fact of existence itself without either the affirmation or the denial of the self in essence. What is sacrificed in essence now is the very notion that those differences are other than themselves in essence. What now exists in place of that notion is the perception itself of the fact of being itself in essence at the disposal of another in essence without being absolutely for another in essence, *the fact of the perfect substitution of faith itself in essence for another in essence transcendent*. The perception in essence of the absolute passion of God himself now occurring, the transcendental repetition of creation itself in which everything is seen to be essentially new, the reduction of all transcendent differences to nothing in essence, was essentially foretold in the word with which Abraham was relieved of the burden of his faith (on the site of the then future temple of Jerusalem), as recounted in *Genesis* 22:11–18: ". . . the angel of Yahweh called to him from heaven. 'Abraham, Abraham' he said. 'I am here' he replied. 'Do not raise your hand against the boy' the angel said. 'Do not harm him, for now I know you fear God. You have not refused me your son, your only son.' Then looking up, Abraham saw a ram caught by its horns in a bush. Abraham took the ram and offered it as a burnt offering in place of his son. Abraham called this place 'Yahweh provides,' and hence the saying today: On the mountain of Yahweh he appears. The angel of Yahweh called Abraham a second time from heaven. 'I swear by my own self—it is Yahweh who speaks—because you have done this, because you have not refused me your son, your only son, I will shower blessings on you, I will make your descendents as many as the stars of heaven and the grains of sand on the seashore. Your descendents shall gain possession of the gates of their enemies. All the nations of the earth shall bless themselves by your descendents, as a reward for your obedience.' " [401] What occurs now is the formal repetition of the essence of history which occurred absolutely before now in time in essential fulfillment of God's promise. Before now the word was made flesh. Now the word is spoken in essence, that is, *Christ's imagination is now flesh* (an occurrence absolutely beyond the imagination of man, the existence itself of the essentially transcendental appearing now infinitely transcendent except for the synthesis of the fact itself in an essentially new thought now occurring for the first time in history in which man is essentially identified with a transcendent substance, that is, with the imagination of God in essence, with the imagination of Christ: Jesus the Nazarene). Now Christ suffers death in essence; now Christ is perceived to be embodied in God himself; now the world is seen to be the embodiment in essence of the transcendental passion of existence itself in essence. It is now the essentially transcendental perception of the body itself. The church in essence is the repetition in essence of the object of God's love. In the absolute objectivity of God's love now occurring, in the temple of the essentially new Jerusalem, he is no longer essentially beside himself in his passion, but is now conceived to be the absolute passion of existence itself in essence. What is now perceived in essence is Christ in the form of man in essence in the world. God in essence in the essential predicament of existence itself, *so that it is now absolutely evident that existence itself in essence is the existence itself of the finite in essence*, that existence itself in essence is unleavened bread changed in essence into the manna of the intelligible substance of a transcendent identity. Upon this eucharist of existence itself all men now feed in essence, although those who essentially recollect, rather than perceive in essence, another in essence feed essentially upon their own self-denial rather than upon the denial of the fact of existence in essence. This latter essentially can

[401] *The Jerusalem Bible: O.T.*, op. cit., 39.

neither be affirmed nor denied. (Indeed, it is clear that, faced with the fact of existence itself in essence in the conception in essence of the word, that is, in the mirror of the word in essence, or in the form of the transcendental imagination of Christ, the anti-Christ is essentially the recollection of another self in essence, essentially the double minded. The anti-Christ is the material silence in which Jesus the Nazarene is not conceived in essence, the darkness in which *God* is not the absolute thought in essence in existence but *the truth* is: truth but not the existence itself of the truth, the love of truth in essence. The anti-Christ is now transparently the substantial denial, that is, the renunciation in essence, of the absolute passion of existence itself.) What is now absolutely conceived in essence is the objectivity of existence itself from which there is in essence no appeal. Before now God was the existence itself of the concept in essence of the finite (as nonexistence), the infinite practical. But now God is the appearance itself of the finite in essence in existence, the essentially infinite conception of transcendence itself, the infinite passover conceived in essence in which God immediately involves himself, absolutely without reservation, God in God in essence, the absolute objectivity now occurring in thought in essence. Before now, in that thinking which belonged in essence in the past, in that thinking for which the repetition of existence itself remained a pure formality (a fact essentially substantiated in the transcendent passion to exist, in the phenomenon of the faith of the individual, *qua* individual—essentially a project), the name of God was essentially misused. That is, it was spoken inessentially to serve the purposes of thought. What before now *was* spoken in essence by the individual resolved in faith to believe in God (for whom his faith was an essentially unintelligible project) was not the word but an analogy to the word in the form of an analogy to the highest principle of an inessentially tautologous thinking (a thinking essentially recollective): so Kierkegaard writes (*Papirer* X^4 A 468): " '*I Am Who I Am*.' This is an analogy to the metaphysical point that the highest principles for all thought cannot be proved but only tautologically paraphrased: introverted infinity. As everywhere else, here also the highest and the lowest have a similarity, for tautology is the lowest kind of communication, is rubbish—and tautology again is infinitely the highest; in this case, then, anything other than tautology would be rubbish." [402] In the absence of the absolute clarification of the absolute now occurring for the first time in history, in terms of which an essentially tautologous thinking now exists in the word itself wherein is effected in essence the synthesis of the fact of existence itself in essence transcendent to thought not another in essence, the conception of the infinite transcendence essentially predicated by the name of God was formally inconceivable. Now that conception exists in essence in place of the infinite introversion of Kierkegaard, who was doubly removed, in fact and in intention, from what now occurs for the first time in history, the transcendental repetition of creation itself. This latter is the fact of existence itself in essence together with the word made flesh, *the synthesis in essence of history itself. It is the structure God himself is together with Jesus the Nazarene now conceived in essence. It is the body itself, the spiritual temple the world in essence is now perceived to be, the house which every-where, within the hearing of every man in essence,* hē energeia katholikē, *says of itself 'I Am Christ'—the unicity of every human individual—the foundation of an essentially new universe in the form of the transcendental imagination of God in essence (no longer, as before now, conceived to be indifferently the transcendental imagination of man, but of man essentially transformed into the word in essence)*. What now occurs is the transcendental imagination of the infinite transcendence of the transcendental essence. This is not the merely formal repetition of the transcendental imagination in that thinking belonging in essence to the past, but the repetition of the transcendental imagination in essence (nor is this perfect abstraction from the world of the past—appearance itself in existence—to be confused with the phenomenological abstraction derivative in essence from the repetition before now of the transcendental imagination). No longer is the infinite introversion of the individual the supposition of an existent world (except in that thinking essentially in the past). That supposition together with

[402] *Søren Kierkegaard's Journals and Papers* IV, op. cit. (1975), 511.

the expropriation of its substance by its finite counterpart, the supposition of existence itself together with the transcendent passion to exist, is now infinitely transcended in the discovery of the fact of the existence itself in essence of the absolute truth, that is, of the truth now occurring in essence in thought itself for the first time in history. Now the unicity of the individual man, his being the individual in essence, is the existence itself of the world in essence. This is the essential individual's infinite transcendence of the preexistent world, of the world in essence preconceived. The individual's being absolutely the word in essence is the existence of the world now conceived in essence for the first time in history. The middle term is the word in essence, the essential copula in which the two exist in one flesh in essence, in which everything is perceived in essence to be created. The burden of the word spoken in essence is light: the transcendent identity of everything now thought for the first time in history. (This is the transcendentally differentiated identity of substance and function, now existing in place of the transcendently differentiated unity of substance and function. This essentially perfect substitution of the passion of existence itself for its mere formality in and for thought its absolutely new essence is the elimination once and for all of the essential supposition of thought itself that there is such a thing as a profane creation, that there is something there to begin with, that there is an immediacy before thought. What is now eliminated is the Atlas of thought, that is, essentially, the self-denial in the face of the suffering of the world, the avoidance of that suffering in essence, the supposition that man in essence is not to be thought, ergo, is not there except he be God [or in the event that there is Nothing to begin with, that there is a immediacy after thought, that man in essence does not exist in fact except in the service of Being as the negative Atlas of finite thought, indispensable to Being in the coming to be of the world in essence in the event of Nothing, indispensable to Being in the construction of the temple of death, of the destruction that Nothing is together with Nothing in essence (the shrine of Nothing, death), indispensable to Being in the domination of the absolute relativity of everything, the identification of the world in fact with appropriation itself].) There is now no burden of existence in essence in thought, indeed, now nothing in essence exists in thought other than the finite in essence. In the transcendental repetition of creation itself now occurring death is nothing but the pure formality of being in the past, nothing but nothing in essence, absolutely the idol, death itself transcendent. That is, death posits its own transcendence in the world in essence, in the temple of death erected by Being for its own purposes, transparently sin suffered by God in essence (the disastrous abomination set up in the temple in essence). This transcendent death itself is essentially blind to its whereabouts, that is, to the body itself, within which for the time being it has its place, according to its own perception, in the body of death, or in the body of God himself. However, in fact, the body itself, now conceived in essence for the first time in history, is the love of God himself for others in essence, the object that is the absolute passion of existence itself in essence, the transcendent identity of thought in the form of the transcendental imagination of God in essence: the body of Christ, the end of theology in essence. It is not the idol in essence, but Christ himself, that is now embodied in God. There is here no attempt to transform the idol in essence into the bread of life itself. That bread comes from without the idol in essence absolutely. Indeed, that the world in essence exists absolutely without existence itself, that there is nothing but appropriation itself, this pure speculation in essence (reflection ending in nonexistence), is the essentially preposterous possibility now contradicted in fact by the appearance of the transcendental essence of existence itself in thought in terms of which there is essentially no transcendent death. In these terms the temple of death is absolutely nonexistent, there being nothing but the transcendental death suffered now in essence by Christ. *Before now Christ died in the flesh, now Christ dies in the flesh in essence. Now, for the first time in history, the death of God is in thought in essence, for thought a new essence. The overcoming of metaphysics in essence in and for thought is the word. There is now no thought of God in essence except for the body itself, the form of man in essence now in the world in essence, the existence itself of the world absolutely. The death of God is now no longer in thought in essence the thought of Being beyond thought in essence, of Being essentially beyond, indifferently, in its essential indifference to the distinction of essences, both man and God, as if the latter were essentially a being among others. Indeed, it is now clear that the body itself is everything's being in essence. The death of God in and for thought is the incontrovertible fact (nor is there possible contention of this fact) the enunciation of which provides the middle term in the transcendental demonstration that everything exists in fact. Now God himself suffers change itself in essence, undergoes (as thought itself now undergoes the time itself of its conception in essence), in his infinite transcendence, the time itself of the transcendental conception of creation itself,* begins in essence to exist absolutely in the form of *exsistere ipsum,* the body itself. Now God takes death into his own hands in essence. (Flesh, in essence, is God thought in the form of the body of God in essence; now flesh is the word in essence thought.) At the last it is now essentially intelligible how it is that God in fact does not take no for an answer. God help that world, together with those dwelling in it, whose final answer *is* no. There is in essence no alternative to the fact of existence itself, including its being conceived in essence now for the first time in history.

The tautology now existing in thought itself for the first time in history is the essential tautology of existence itself: not God in essence *or* God in essence (anti-Christ *or* Christ). This tautology, formulated in the words of Jesus the Nazarene, to wit, 'he who is not with me is against me' (*Matthew* 12:30) *or* 'he who is not against us is for us' (*Mark* 9:40), is resolved in essence ([(x = y) *or* (not y = z)] = (not x = z)): *He who is with me is for us*. The tautologous nature of this resolution of existence itself in essense is apparent immediately when rendered 'he who does not exist in himself is in fact included in the body itself,' and appears in essence in the rendering 'he who hears it said in essence "I Am Christ" is identified with the body whose essential constitution is God himself (the body, that is, constituted by Christ himself).' Or, quite simply, the essential tautology of existence itself is: He who is saved is the sinner (he who is with Christ Jesus, who hears it said in essence "I Am Christ," who weeps over his sin [who is purified in those tears], he it is whose consolation in fact is the love of God himself—the body of Christ *or* there is the weeping and gnashing of teeth of the sinner denying himself in essence [not to be confused with the formal self-denial of repentance, the being with him] which is the embodiment of the lie that the truth absolutely is without being loved when in fact it is the existence itself of the beloved in essence, which is the hatred in essence of the truth when in fact the truth identifies itself [before now in time, now in thought] with being loved in essence, with being at the disposal of another in essence, with absolute faith itself [so that it is now clearly perceived that the world itself in essence shapes itself to the expectation of absolute faith by the grace of existence itself *or* in the event of the final loss of innocence that the world collapses in essence under the burden that the word of God is taken to be in essence (a burden insupportable save in the form of the infinite self-division, that is, the infinite immanence, of existence itself, which different-transcendence replaces the infinite introversion of God with its own resignation to the impenetrable appearance of reality [as this latter is thoughtlessly taken to be constituted in essence], a movement from the affirmation of faith in essence to the essential negation of faith in which the transcendent other is a pure formality determined to exist in essence without death); either the essentially infinite world now conceived for the first time in history *or* the essential similarity of infinite chaos]; there is nothing in between in essence except the transcendent form of another with which he who exists in himself identifies himself [he who is not God in essence], appearing to disappear in appearance, i.e., a purely apparent reality, clearly perceived to be there in the supposition of its own imperceptibility, in the pure self-determination of another) *or* he is lost who hears nothing said in essence. That is, he is the enemy of the body itself for whom it is the embodiment of a transcendent death. *The word spoken in essence is the word heard in essence, "I Am Christ," the word made flesh of the body itself, the eucharistic essence of the communion of saints, the word of existence itself* (now for the first time in history heard everywhere in essence). *This word is the child, as it were, of Christ's body, the intelligible image of God existing in the world in essence through the essential power of the divine unicity* (the world transformed in essence in the conception of the infinite transcendence of God now occurring for the first time in history). This word is the unleavened bread of existence itself, the unconditioned itself appearing in essence bringing into existence in thought itself the essential condition of an essentially new world, the receptivity for the existence itself of another in essence, being itself in essence at the disposal of another in essence. The word spoken in essence sows the way itself of the word. *Silence is not of the essence of the word spoken in essence. That is, it is immaterial to the word spoken in essence, a pure formality to the word for which the gathering is immediately at hand. The word is no sooner spoken in essence than it is heard in essence—there is not a moment's delay, the receptivity for the word is not there to begin with. In the essential proclamation of the word the silence is unsaid, that is, the silence (which is not nothing) is itself spoken in essence,* or, comes into being together with everything else in the essentially creative action of the word. It is an essentially absolutely anachronistic conception derived from thinking belonging in essence to the past to suppose that the silence which is of the essence of the spoken word (which silence is nothing), that is, which before now was material to the word's being spoken, might, even now *in extremis,* in the form of a speech which says nothing in essence (in which nothing is unsaid), be confused with the conception of the body itself, to suppose that it is not clearly perceived in this thinking now occurring for the first time in history that such a conception is nothing in essence but the formal usurpation of God's body. Indeed, it is now clearly seen to be the final supposition contrary in essence to the fact of existence itself, opposing nothing in the way of the spiritual conception of the church now occurring for the first time in thought itself except an essentially inappropriate silence in the desperate hope of prolonging the discussion when in fact it is finished in essence now that the word is spoken in essence in thought itself. *When the word is spoken in essence the silence is spoken in essence, the two transcendentally differentiated in the unity of the word in essence as the advent of Christ Absolute in thought without the silence thought in essence, essentially without the necessity of being translated. What now occurs is the perfect substitution of the word spoken in essence for the word spoken* (essentially intelligible in an essentially tautologous thinking), *the perfect substitution of the word heard in essence for that silence which is of the essence of the spoken word.* For the translation that the word was essentially thought in the past to be in need of in essence, this is the perfect substitution: the transformation itself of the world in essence that the word now thought is conceived to be in essence in the form of the body itself. This is the absolutely unprecedented situation: The word itself in essence is in the category of existence itself. The world in essence is turned into Christ the absolute thought in essence (God in essence not God is man in essence; but God is God in essence; *ergo,* the word creation itself is the intelligible form of man). This is the concrete existence of an essentially new universe the very appearance of which evidences not only that Christ has never existed in essence in transcendence to the church, that is, that God has never existed in essence in self-transcendence (no more than God can now be conceived to be the infinite immanent except in the form of the essential negation of faith), but, further, that the antithesis Protestant/Catholic, whereby, in the words of Schleiermacher (*The Christian Faith* Prop. 24), "the former makes the individual's relation to the Church dependent on his relation to Christ, while the latter contrariwise makes the individual's relation to Christ dependent on his relation to the Church," [403] is now intelligible as belonging properly to the past. This is because the antithesis rests on the transcendent conceptions of, on the one hand, the essential immediacy of the word made flesh, on the other hand, the essential mediacy of the word made flesh. Both of these conceptions are transcended in essence in that thinking now occurring for the first time in history together with the essentially dichotomous relation of the other to the self which is the supposition of an essentially transcendent difference. Indeed, this now-occurring perfect substitution is the unicity of the individual man being absolutely the conception in essence of Christ Jesus, the absolutely transcendental conception of the word made flesh

[403] F. Schleiermacher, *The Christian Faith* (ed. Mackintosh and Stewart, Philadelphia 1928), 103.

wherein the immediacy is thought in essence, that is, the perception of the body itself. The transcendental existence of the church being Christ with which the individual is identified essentially as one of two in one flesh. This is a situation in which there is absolutely no difference between the existence itself of the individual and that of the church. Simply stated in the form of the essential resolution of the essential tautology of existence itself, *he who is with me is for us:* he who hears it said in essence "I Am Christ" comprehends in essence "now the church is who I Am." Now that it is thought in essence that Jesus is absolutely everywhere, that man in essence is existence itself everywhere in essence, that man absolutely is the historical essence (*to einai katholikon*), it is clear that the essential distinction of non-Christian/Christian transcendence is no longer intelligible. What is now essentially intelligible is the transcendental difference anti-Christ/Christ. The divine perfection now demanded of every individual man is the conception in essence of finite existence itself, the conception of the infinite transcendence of existence itself. What is now demanded is the conception of existence itself in essence absolutely the body itself and (now that God is all in all in essence) the conception in essence of Christ absolutely the way itself. (The transformation itself is the existence itself of the silence in which the word essentially sounds prior to being uttered: the speech which is absolutely the gift of tongues: the conception in essence of the immediate reception of the word.) This is the essential form of transcendental thought, the prophetic essence of transcendental speech, the utterly unselfconscious speech of the prophet in receipt of the eucharistic essence of existence itself. Now the word in essence is heard everywhere (*ho logos katholikos*). Indeed, the very form of an essentially tautologous thinking precludes in fact an intelligible alternative to the (essentially light) burden of individual human unicity. Now for the first time in history nothing else is thinkable in essence. Now for the first time in history thought is the individuation of the church, essentially the individuation of Christ, absolutely the individuation of God himself.

Before now, in that thinking belonging in essence to the past, individuation, or, the transcendental imagination, was thought to be absolutely the synthesis of life itself. Now for the first time in history the transcendental imagination is thought to be that in essence. What is now thought in essence is the absolute fact of existence itself (clearly, then, not the essentially derivative fact of the phenomenological reduction), the essence of history. What is now seen to belong in essence to the past is the necessity of being related in essence to existence itself in essence. There is now absolutely no necessity of being related to existence itself. Existence itself is immediately there: the unnecessary repetition of creation itself: creation itself absolutely unconditioned. That is, existence is there in an immediacy itself conceived in essence to be the way itself of the word, an immediacy not other in essence than thought itself, but the essential form of individuation. In the essential conception of the absolute synthesis of life itself now occurring everything exists in essence in the essential individuation of existence itself (the miracle of the loaves thought absolutely). Everything exists in essence in the embodiment of another existence itself, in being absolutely at the disposal of another (in being the servant of a master who reaps where he has not sown). Essentially faith is a superfluity. Existence is an absolute superfluity. Now everything exists in essence in the body itself, in Faith Absolute, the love for others in essence of existence itself. Now creation itself exists in essence in thought; the love of existence itself is absolutely intelligible: the absolute unicity of existence itself. Conceived in essence, existence absolutely spiritual is the body itself. Creation itself is now absolutely the temple of the Spirit. Thought is now not the Absolute Religion come to maturity in self-consciousness. But now for the first time in history thought is the Absolute Faith conceived as the essentially transcendental embodiment of existence itself, the perception itself of the body in which all things together with us 'live, and move, and have being.' The conception of existence itself is essentially transformative of the universe itself into a state of absolute perfection. In the Absolute Faith now conceived absolutely nothing absolutely exists including faith itself. Now faith itself exists in essence in thought contradicting in essence the appropriation of existence itself, an essentially irreversible conception, being the conception of existence itself, that is, essentially historical. Now the children of the father of faith, in the fulfillment of absolute existence itself, gain possession of the essential gates of their enemies, of the stronghold of the Absolute Religion. Now everything is an absolute phenomenon without being in fact nonexistent in essence. Here is the body of death, of a God-transcendent Christ (of a God who is Jesus not Christ), except for the fact of Christ now in thought for the first time in history. Now there is the beginning with absolutely nothing *or* there is the beginning absolutely with the body itself. That is, life itself is beginning to exist transcendently in the new thought of an absolutely transcendental universe. The dawn of the day of Yahweh is now occurring. Every-thing essentially is transformed in the unicity of existence itself into being in essence at the disposal of another. Before now, the thing in fact was not without Being beyond it, its nothingness a mere formality (its nothingness something in essence), the nothingness of the idea. Thus, in fact, the thing continued to exist long after it was supposed to have come to an end (in fact it never terminated in existence, uncreated as it was). Indeed, the thing in essence was what was remembered. Now, for the first time, the thing in essence is nothing in essence. Now, in fact, the thing is absolutely nonexistent. That is, it exists in fact without Being beyond it. Now it is clearly seen that nonexistence is the nonexistence of Being itself, not the nonexistence of the thing now appearing in existence itself substantially its transcendent identity. Now it is clear that nonexistence in essence is difference-transcendence. In fact nonexistence belongs to Being itself. The nothingness of Being itself conceived in essence is Being without history in essence, precluding absolutely thought itself seen now for the first time to be identical in essence with the essence of history, that is, substantially different (the substance of this thinking being the transcendental essence of existence itself). Now a thought exists in which there is absolutely no thought of apprehension, no thought of anything other than nothing except in essence the fact of existence itself, in which apprehensive thought is a thing of the past in essence, in which absolutely nothing is apprehended. But to apprehend sin (absolutely nothing) is impossible—there is nothing for it then but the fact of existence itself in essence fashioned in and for thought for the first time in history. Now it is comprehended that absolutely nothing terminates in the form of an essentially new thought, but that everything, including sin itself, terminates in existence itself, indeed, including thought itself. It is of the very essence of this thinking now occurring that everything exists perpetually. Perception absolutely ends not in essence, but essentially ends in existence. In the unicity of existence itself everything is changed into being absolutely at the disposal of another: in the intelligible love of existence itself, in the body itself, through the absolute displacement of space, of the potential for measuring motion by anything other than time itself, the new absolutely nonexistent thing everywhere perpetuated historically in the substantial identity of the infinite transcendence of existence itself, the memory of existence itself in essence in the form of the body itself. Faced with the impossibility of apprehension, thought absolutely comprehends for the first time in history that it is face to face with existence itself. Thought contemplates in itself in essence the absolute remembrance of history itself, the unicity of which experience is the perception of the body itself. In this conception the remembrance of the appearance of the transcendental essence of existence itself is essentially the transcendental imagination of existence itself in essence. That is, it is absolutely the individuation of existence itself, the essential form of which is the memory of existence itself in essence, time itself. Now for the first time this is the case. The substantial transaction of the mass is the absolute individuation of existence itself. Indeed, recollection itself is now seen for the first time to be essentially historical, to be the distribution of Being itself in fact. (Being itself now for the first time in history has been delivered from its nonexistence in fact absolutely by history itself. It is now an absolutely unexpected offspring, as it were, of the church, come into existence itself. Being itself is that absolute motion in which it is immediately perceived that God has committed himself in essence to history itself, and, without ceasing to be God, refrains absolutely from the determination of events and exists now absolutely through the body itself. This motion itself is materially existence itself, in essence the absolute displacement of space, the motion itself of the world's being, that is, of being absolutely placed, in which it is seen that the world is everything in essence, in which is beheld the absolute passion of existence itself.) Now thought in essence is the anticipation of the finality of the fact of existence itself. The distribution of Being itself in fact is the fact of existence itself dealing death wherever the blood of the lamb does not cover the door, wherever the perception of appearance itself in essence (transformed) does not exist, that is, wherever the infinite transcendence of existence itself is not conceived, wherever there is, instead, the conception of a different transcendence, an infinite self-division, a lifeless abstraction unalloyed with faith itself, the thought of the body of death, in which existence is thought to be absolute self-negation. But wherever the perception itself of appearance itself in essence exists, there it is seen the body and blood is eaten no longer merely in substance, but now for the first time substantially in essence. Thus, the body itself is transformed into something new in essence, into being itself absolutely loved. The memory of the body itself in essence exists beyond being itself without Being beyond itself: the essential form of the body changed into being time itself, into the transcendent form of thought's object, understood to be capable in essence of existing beyond the grave. It is only a matter of time until the absolute individuation of existence itself is thought in essence, until the imagination of existence itself in essence is thought absolutely transcendental, until existence itself shall be other than the transcendentally different without being at the same time transcendent. It is only a matter of time until there is no matter of time. (This is essentially intelligible as reflection in essence ending in existence itself. This matter of time is the object with which thought identifies itself, incipiently it is the end of time. This absolute displacement of space is now absolutely conceived as motion itself.) *It is only a matter of time until time itself ends in the absolute materialization of the objectivity that now is thought, until (though everything exists perpetually) memory itself ends in the absolute remem-*

brance of existence itself in essence. In the meantime we have come this much closer to the end, that its structure is now transparent to us. Time itself is the proportion of the thing to motion itself, the essentially transcendental conception of change itself as the object's identity, the memory in essence of existence itself, not of being before now in essence. In the meantime, for the first time in history, thought in essence is the absolute demonstration of existence itself (in terms of which the existence of everything including the world in essence is demonstrable). Thought in essence is the proof positive of existence itself (as distinct from modern thought's essentially negative proof of existence: a position taken up in essence in lieu of the fact itself) in the form of a thinking absolutely without sin. This is a thinking transcending absolutely in essence the essentially lifeless forms of (modern) thought's self-opposition, namely, in the first instance, the materially isolated thought of the individual in essence wherein the proof of existence is essentially affirmation in the form of immediacy (the absolute absurdity), in the second instance, the corporate rationality of the species wherein the proof of existence is essentially suspended in the form of a transcendental relation (the absolute science), and, in the third instance, the essentially nonexistent thought (the thing-thought) of the world wherein the proof of existence is essentially the negation of matter in the form of totality (the absolute speculation). What now occurs in the form of the perception itself in essence of the body itself, what now itself occurs in time, is that the so-called *prophetic doctrines* of Schleiermacher (*The Christian Faith* Prop. 159.2), the thought-forms of an inessentially transcendent matter, are displaced in essence by the appearance of existence itself in an absolutely transcendental thought. They are displaced absolutely by the materialization of the objectivity that now is thought. What to an essentially self-conscious, radically theological dogmatics would be nothing other than an essentially inappropriate, capricious fancy, the matter of which would no longer be subject in essence to interpretation, is now fact-evidently not the thought in essence of a new revelation. Now for the first time in history the word is thought absolutely (clearly, then, never before now subject essentially to exegesis). Absolutely, it is the mind of Christ without appropriation itself. The matter of this mind is not other than thought in essence but is the essential form of individuation, the memory of Christ, the substantial function of which is energy itself, creation itself. That is, it is the absolutely dynamic reality of the fact itself of history in essence in and for thought for the first time in history thought in essence. The transformation of the universe is now essentially in progress to a state of absolute perfection in satisfaction of the absolutely evident requirement of existence itself that there be that perfection. Thus this absolutely sinless form of thought (in essence the remembrance of Christ) transcends the essentially finite thought of modernity in taking to itself in essence the requirement for intelligibility in a way that modernity was unwilling in essence to follow. Namely, this form of thought submits itself to the conception in essence of being in time, thereby becoming for the first time in history (an essentially unprecedented thought) the essentially transcendental thought of existence itself appearing now. Modern thought, which demanded of everything that it demonstrate its existence, remained essentially indemonstrable with respect to existence itself. Indeed, it demanded what was essentially impossible so long as thought was not thought in essence but merely the form of thought in essence, that is, prior to the absolute difference of an essentially transcendental thought. Now in terms of this conception of difference modernity's inference of an essentially unconditioned thought from non-being (thought to begin with nothing, the result of nothing) is immediately converted into the acceptance of the fact of identity with the absolutely conditioned thought of existence itself. This state of thought before now was simply the result in fact of that thinking which began with the appearance of the transcendental essence of existence. Now with creation itself thought, thought absolutely exists, identically the transcendence in essence of past thought: the perfect inception in essence of existence itself. This, before now, would have been thought to be an intervention, but, now that it has occurred, this beginning of a perfect thought is clearly perceived to be everywhere in essence *ho logos katholikos*. It is the transcendental reality of the resurrected Christ. Now that thought is the perception of the body itself, now that an essentially eucharistic reality is the perfect object of thought itself, it is clear that reality is essentially contextual, that is, shaped *kata ton logon katholikon.* Reality is understood in essence in accordance with the word, that is, is essentially transcendental. Reality is existence itself. Essentially, it is not, therefore, a matter of the point of view, but in fact a different matter in essence which essentially does not exist at the expense of another. This is not yet another, albeit final, movement of the self in essence. *This reality, logically consistent in essence, is the perpetual weaning of everything in essence, everything absolutely transformed into independent being: into the memory of existence itself: the coming into being of the essential independence of the support of another in essence now thought:* the absolute individuation of the thought of existence itself now occurring in which there is transcendental thought absolutely differentiated, in which there is transcendental being absolutely differentiated, but in which, prior to everything else, there is absolute love for another in essence, identically the fact itself of an essentially transcendent existence (formally that of the thinking belonging in essence to the past). This is the infinite transcendence of existence itself: action itself, the absolute passion of existence itself conceived in essence (Christ Absolute suffering thought), the category of Christ himself, in which perfection of history itself everything exists absolutely, in which absolutely nothing escapes this essentially new species of being: the body itself (thought transcendentally, the creation of matter itself now [which thought is the essential constitution of the *novitas mentis*]; the substantial function of time itself now formally thought). Now that action itself is thought in essence it is clear that those are happy who suffer existence itself in this world, for of them is predicated the perfect joy of existence itself. Happy are those of whom a history is predicated, for theirs is a transcendent identity. Theirs is the materialization of a now essentially transcendental thinking in the essential unicity of which a new universe is founded for the first time in thought, the body itself absolutely inceived in thought: innocence itself (the loss of innocence the absolute oblivion of existence itself).

In the *missa jubilaea* now occurring the mind itself is touched in essence by the fact of God himself (such is the now occurring infinite passover of the creator's touch). Now the absolute illumination of existence itself is suffered *without the loss of self, but without the notion of self,* that is, without the necessity of being related. The liberty essentially predicated of salvation itself is now thought, the essential constitution of which is the absolute love of existence itself perceived. Now the world in essence is the conception of Christ Absolute. Now that the silence itself is thought for the first time in history, now, indeed, in essence, 'the stones cry out' (*Luke* 19:39–40), 'the sea thunders and all that it holds, and the world, with all who live in it.' In essence, 'all the rivers clap their hands and the mountains shout for joy, at the presence of Yahweh, for he comes to judge the earth, to judge the world with righteousness and the nations with strict justice' (*Psalms* 98:7–9). Now what is finite is not what is absolutely non-existent (that is the thing in essence). In the predicament of absolutely finite existence is discovered infinite identity itself. Existence itself be praised! What now occurs essentially in thought for the first time is the glorification of existence itself. Now for the first time in history the thought of the world is the Absolute Praise of God. In this absolute concretion of the thought of existence itself, there occurs in fact the elimination of the ontological difference materially. Now there is absolutely nothing but the ontological identity of God (the end of theology in essence, or the end of Christology, is now a reality), God's infinite passover, that is, Christ himself thought (clearly, then, not a metaphysics, real or imagined). This is the perception of the body itself. What is now for the first time thought in essence is the history of being. It is perceived that the history of being is the difference in essence of identity itself. The essential form of this history is transcendence, in turn, transcendentally differentiated from the finite transcendence of the past (which left to itself it would be) as the infinite transcendence of existence itself (which in fact it is in essence through no fault of its own). *The essentially circumspect statement of the thought now occurring is that in fact there is no word being beside itself, that in fact there is no paralogical being, that in fact there is no ontology other than the Word itself.* So be it now truth in the form of man beheld. The absolute thought now being itself. Existence itself in essence now the thought of an essentially transcendental imagination. Now it is, the eucharistic way of the world absolutely thought. Now thought is (an infinite ontology) the science of creation itself. Indeed, what now occurs is the absolute comprehension of the words of Jesus the Nazarene (*John* 15:12–16): "This is my commandment: love one another, as I have loved you. A man can have no greater love than to lay down his life for his friends. You are my friends, if you do what I command you. I shall not call you servants any more, because a servant does not know his master's business; I call you friends, because I have made known to you everything I have learnt from my Father. You did not choose me, no, I chose you; and I commissioned you to go out and to bear fruit, fruit that will last; and then the Father will give you anything you ask him in my name." [404] Absolutely comprehended now in thought, in essence, these words constitute the beginning of an essentially new universe: not, to be absolutely for another, but, to not be absolutely for another, precisely, to be absolutely at the disposal of another. And the fruit that will last in essence is now being absolutely thought; this now is the foundation stone; this is now the precious cornerstone; this now is the Stone of Witness. Now absolutely no one being at the disposal of another shall be disappointed. His being now is (the Living) God's Absolute Memorial, the body itself (resurrected). Now for the first time in history the material essence of thought is the complete joy of Christ, that is, now, *missa jubilaea.*

[404] *The Jerusalem Bible: N.T.,* op. cit., 180.

INDEX

INDEX

Given the complex nature of the text, this index, neither complete nor exhaustive, should be used only as a guide to certain important matters. It includes names and, by author, works quoted (the content *as such* of quotations is not indexed). Biblical references are grouped either under Old or New Testament. Further guidance to Section B can be had by using the *Précis* beginning on page 23.